Computational Methods in Applied Sciences

Volume 17

Series Editor

Eugenio Oñate, Universitat Politècnica de Catalunya, Barcelona, Spain

This series publishes monographs and carefully edited books inspired by the thematic conferences of ECCOMAS, the European Committee on Computational Methods in Applied Sciences. As a consequence, these volumes cover the fields of Mathematical and Computational Methods and Modelling and their applications to major areas such as Fluid Dynamics, Structural Mechanics, Semiconductor Modelling, Electromagnetics and CAD/CAM. Multidisciplinary applications of these fields to critical societal and technological problems encountered in sectors like Aerospace, Car and Ship Industry, Electronics, Energy, Finance, Chemistry, Medicine, Biosciences, Environmental sciences are of particular interest. The intent is to exchange information and to promote the transfer between the research community and industry consistent with the development and applications of computational methods in science and technology.

Book proposals are welcome at
Eugenio Oñate
International Center for Numerical Methods in Engineering (CIMNE)
Technical University of Catalunya (UPC)
Edificio C-1, Campus Norte UPC Gran Capitán
s/n08034 Barcelona, Spain
onate@cimne.upc.edu
www.cimne.com
or contact the publisher, Dr. Mayra Castro, mayra.castro@springer.com

Indexed in SCOPUS, Google Scholar and SpringerLink.

Tero Tuovinen · Pekka Neittaanmäki ·
Dietrich Knoerzer
Editors

Challenges in Design Methods, Numerical Tools and Technologies for Sustainable Aviation, Transport and Industry

Commemorative Publication Dedicated to the 80th Jubilee of Prof. Jacques Periaux

Springer

Editors
Tero Tuovinen
School of Technology
JAMK University of Applied Sciences
Jyväskylä, Finland

Pekka Neittaanmäki
Lappeenranta University of Technology
Jyväskylä, Finland

Dietrich Knoerzer
Aeronautics Consultant
Brussels, Belgium

ISSN 1871-3033 ISSN 2543-0203 (electronic)
Computational Methods in Applied Sciences
ISBN 978-3-031-98674-1 ISBN 978-3-031-98675-8 (eBook)
https://doi.org/10.1007/978-3-031-98675-8

This Springer imprint is published by the registered company Springer Nature Switzerland AG
The registered company address is: Gewerbestrasse 11, 6330 Cham, Switzerland

Contents

Chapter 1
Jacques Périaux—A Great Scientist, the Best Colleague and a Friend for Over 40 Years

(a personal message)

Eugenio Oñate

I write these lines on the occasion of the 80th birthday anniversary of Prof. Jacques Périaux, and I start by expressing my best wishes to him for this great occasion.

I have known Jacques for over 40 years. During this time, I have been able to appreciate his many personal talents. Among these I highlight his passion for research, his motivation to push the development of computational science and technology, mainly in the field of aeronautical engineering, beyond traditional academic and geographical frontiers, and his strive for promoting cooperation between applied mathematicians and engineers from all over the world.

He was one of the key architects in building the computational engineering community in Europe and beyond. He was one of the founders and first president of the European Community in Applied Sciences and Engineering (ECCOMAS). His work was essential for establishing cooperation between ECCOMAS and the main scientific organizations in the field of computational engineering at worldwide level, such as the International Association for Computational Mechanics (IACM), and also at national level, such as the Spanish Association for Computational Mechanics and Engineering (SEMNI).

E. Oñate (✉)
Professor at Universitat Politécnica de Catalunya, Barcelona, Spain
e-mail: onate@cimne.upc.edu

T. Tuovinen et al. (eds.), *Challenges in Design Methods, Numerical Tools and Technologies for Sustainable Aviation, Transport and Industry*, Computational Methods in Applied Sciences 17, https://doi.org/10.1007/978-3-031-98675-8_1

Eugenio Oñate, Jacques Périaux and Dietrich Knoerzer at CIMNE, Barcelona, 28th April 2017

I emphasize the role of Jacques in promoting research in the field of aeronautics at the International Center for Numerical Methods in Engineering (CIMNE) in Barcelona, Spain. After his initiative, several research projects were developed at CIMNE since 1987, many of them with financial support from the European Commission under the supervision of Dr. Dietrich Knörzer, and with participation of the main stakeholders at academia and industry in aeronautics. He has been very active in promoting scientific cooperation with universities and research centers in China in the aeronautics field.

Jacques is still very active as a UNESCO Professor on Numerical Methods in Engineering at CIMNE. I deeply thank him for his contributions, support and friendship during many years and wish him all the best at personal and professional level in the years to come.

Chapter 2
Personal Message to the 80th Anniversary of Jacques Périaux

Pierre Perrier

Dear Professors of the University of Jyvaskyla, Professors from many other universities as well as good friends of Professor Jacques Périaux, You are celebrating today the 80th birthday of our friend, aside from a scientific meeting on computations of climate prediction for years to come. First of all, let me apologize for not being here today. A toe broken last February on a dark and steep flight of stairs during the preparation for a scientific seminar in March in the south of France, which I havent completely recovered from, prevents me to be in your beautiful and efficient university with you all. It has become difficult over the years to bring together the six members of a famous team of academic and industrial workers. Their success story has gained some celebrity in the first great achievements in the field of Computational Fluid Dynamics …with the nickname of the "GP4B". It united 6 researchers involved in the advancements, 4 having a first letter P in their name and two others having a G and a B. Among these 4 Ps, two are in this room today, Pironneau and Périaux. But let me start with the beginning! In 1970 in France, the Direction for Research in the Military Field DRME (Directions des Recherches et Moyens d'Essais) of the Ministry of Defense wanted to set up a triangular program for support of better computational design in the Aeronautical industry, and first in fighters, with the cooperation of academic and industrial design office. And on a Saturday morning in 1970 took place the first meeting of a young candidate with a level in my office at Dassault Aviation for the selection of a mathematician (first employment!) as a link to a team of professors and researchers to be involved in the common program: A first challenge for Jacques Périaux. As an applied mathematician in Numerical Fluid Mechanics, he was invited to enhance the viability of the best ideas in method and algorithm with the Institute of Informatics INRIA and Paris Sorbonne University (the fluid being a continuous medium thus of infinite dimension and thus posing a problem of convergence for the nonlinear partial differential equation (PDE) models after discretization!) Jacques

P. Perrier (✉)
French Academy of Sciences, Paris Cedex, France
e-mail: pierreperrier78@gmail.com

T. Tuovinen et al. (eds.), *Challenges in Design Methods, Numerical Tools and Technologies for Sustainable Aviation, Transport and Industry*, Computational Methods in Applied Sciences 17, https://doi.org/10.1007/978-3-031-98675-8_2

Périaux being in DRME (with Ingénieur de l'Armement Vuillard), just graduated his DEA (diplôme d'études approfondies) in Numerical Analysis with Jacques-Louis Lions and de Possels during his military service at Balard (the location of the DRME south of Paris) and was providentially finding himself available just at this date. There was a lot of randomness in this situation, but some may also call it destiny, for this is how Jacques Périaux, a university student in the Department of Theoretical Fluid Dynamics, found himself in my office the unique place for what would be the Golden Age of CFD. Together with Olivier Pironneau and Roland Glowinski, you did very "useful" mathematics applied to aerodynamics in the collective context of a team that succeeded first in 1976 with the introduction of the finite element method in Fluid Mechanics applied in 3-D to a complete civil and military aircraft and later much more complex flows around re-entry vehicles returning from space!

That was quite a beginning, and you have kept on doing applied mathematics in the aeronautical field at first, and later with increasing involvement in teaching and researching in foreign universities spreading the French way of doing such mathematics to the benefit of many students.

Meilleurs voeux (best wishes),

Pierre Perrier Former Head of the Theoretical Aerodynamic Department of Dassault-Aviation and its Theoretical Aerodynamic research and development. Member of the French Academy of Technologies and of the US National Academy of engineering and of the European Air and Space Academy. Member of the French Academy of Sciences, France.

Chapter 3
An Introduction to Jacques Periaux Scientific Legacy

Olivier Pironneau

Abstract Our life long colleague Jacques Periaux has turned 80 recently and a scientific day is dedicated to him on the occasion of the CM3-conference in Jyvaskyla, early May 2023. It is an opportunity to recall some landmarks of his scientific career and his dedication to the scientific community.

3.1 Career

Jacques Periaux is born in France during WWII. He went to Université de Paris to study mathematics, and mechanics for his Master, completed in 1965. His first research position was at ONERA[1] for his military duties for 18 months. In parallel he decided to join the new Master program created by Prof. Jacques-Louis Lions in applied mathematics which he completed in 1969. In 1970 he joined the recently founded numerical simulation group at Dassault Aviation company headed by Dr Pierre Perrier.

Unlike many engineers, JP never left research. He obtained a PhD in applied mathematics in 1979 and a HdR (Hability to direct PhDs) in 1989. From 2007 to 2010 Jacques had a TEKES position as a Finnish Distinguished Professor at the University of Jyvaskyla. In 2010 he received a UNESCO chair position at CIMNE, the computational lab of the University of Catalonia UPC, in Barcelona–Spain. During the last three decades Jacques Periaux has initiated many scientific and technical cooperations in Europe (with the EU), Australia, China, Japan and the USA. He has written some 300 articles and published 20 Proceedings and one co-authored book.

---

[1] French NASA.

O. Pironneau (✉)
Applied Mathematics, Jacques-Louis Lions Lab, Sorbonne Université, Paris Cedex, France
e-mail: olivier.pironneau@sorbonne-universite.fr

T. Tuovinen et al. (eds.), *Challenges in Design Methods, Numerical Tools and Technologies for Sustainable Aviation, Transport and Industry*, Computational Methods in Applied Sciences 17, https://doi.org/10.1007/978-3-031-98675-8_3

3.2 Contribution to Science

Jacques attended a conference in Canada in 1971 where the first application to fluid flow of the finite element method (FEM) was discussed.[2] P. Perrier immediately saw the potential of the method for airplane design. Hence Jacques became perhaps the second scientist to use FEM for airplane design in 2D and 3D.[3]

Quickly JP & team headed towards nonlinear equations for transonic and compressible flows. A cooperation with the numerical team of J.-L. Lions & R. Glowinski at IRIA was initiated and in 1978 the first complete aircraft was simulated by the transonic equation. It was a world premiere for which the contribution of JP was essential.

To handle the nonlinearities a least square method in the Sobolev space H^{-1} was used because one could use distributed optimal control[4] and that space is adapted to second order equations like the transonic equation, the Navier-Stokes equations and the Maxwell equations for stealth airplanes. Many articles from that period are still up-to-date because the algorithms have not changed: GMRES, DDM (Domain Decomposition Methods), FDM (Fictitious Domain Methods), HUM (Hilbert Uniqueness Methods).

From 2000 onward JP concentrated on DDM and Evolutionary Algorithm for optimal design like Genetic Methods. These are summarized in his book *Evolutionary Optimization and Game Strategies for Advanced Multi-Disciplinary Design* published on 2015 in the Springer series *Intelligent Systems, Control and Automation.*

Consequently JP is a world expert in the following topics

1. Numerical Solution of Non Linear Partial Differential Equations;
2. Finite Element Methods for CFD and CEM;
3. Aerodynamic Design of manned/unmanned aircraft vehicles;
4. Domain Decomposition on parallel architectures;
5. Fictitious Domain Methods;
6. Mesh adaptation with a posteriori errors estimates;
7. Multidisciplinary Design Optimization;
8. Optimal Control Theory, Evolutionary Algorithms and Game Theory.

3.3 Contribution to the Organization of Science

Jacques has a gift for public relations which has allowed him to transport ideas and people from countries and continents. He has been instrumental in the creation or participation of France in a large number of projects, like:

[2] Norries D.H., G. de Vries: The finite element method. Academic pressNew-York (1973).

[3] J. Periaux, Int. J. Numer. Methods in Eng. 775–831 (1975).

[4] J.-L. Lions: Optimal Control of Distributed System, Springer 1988.

- HERMES (1986–92)
- ECARP (1994)
- INGENET (1998)
- FLOWNET (1999)
- GAMNI (1983)
- ERCOFTAC (1985)
- ECCOMAS (1992)
- MACSINET (2001)
- The DDM conference series (1998–2010).

As said earlier he has strong links with the UPC, the University of Jyvaskyla, the University of Houston-Texas and the University of Nanjin.[5]

3.4 Honors

Jacques' tremendous scientific energy and valor has been granted with the following recognitions:

- IMA Fellow, 1986;
- Doctor Honoris Causa, NUAA, Nanjing, P.R. of China, 1989;
- AIAA Associate Fellow, 1992;
- French Science et Defense Award, 1993;
- Chevalier de l'Ordre des Palmes Académiques, 1996
- Doctor Honoris Causa, BUAA, Beijing, P.R. of China, 1998;
- Friendship Certificate and Medal Award, Jiangsu Province, 2002;
- National "Friendship Award", SAFEA Beijing, 2010;
- Member of the International Academy of Astronautics (IAA): 2011.

We all wish him a very happy birthday and many returns.

[5] Open China EU aeronautical R&T cooperation.

Part I
Design and Optimization

Chapter 4
Prioritized Multi-objective Optimization of an Aircraft Flight Performance Based on Nash Games from Preponderant Pareto-Optimal Points

Jean-Antoine Désidéri, Julien Wintz, Mickaël Binois, Nathalie Bartoli, Christophe David, and Sébastien Defoort

Abstract A hierarchical multi-objective optimization process is applied to the performance of an Airbus-A320-type aircraft. Six design variables are used to calibrate wing geometry, mean aerodynamic chord and take-off thrust potential. Functional constraints are enforced (interval bound on static margin and upper bound on wing span). The system is subject to the classical Breguet laws of Flight Mechanics and to the numerical integration of the ascent phase. It is thus an Ordinary-Differential-Equation-constrained system and it is evaluated by the open-source software FAST-OAD. Three cost functionals are minimized: fuel mass at take-off, operational empty weight, and ascent duration. The first two, considered of preponderant importance are prioritized, and the third only secondary. The method permits, at moderate numerical cost, to identify quasi-Pareto-optimal by Nash games originating from Pareto-optimal solutions of the sole prioritized cost functions.

4.1 Introduction

In concurrent engineering, the numerical identification of the Pareto front by a robust method such as NSGA-II [4] is often a highly demanding numerical task, the cost of which may increase rapidly with the number of cost functions accounted for. However in many real design situations, a subset of the cost functions, also known as objective functions in the literature, has preponderant importance, and this incites to a hierarchical, or prioritized treatment. We have proposed to compute both the

J.-A. Désidéri (✉) · J. Wintz · M. Binois
Centre INRIA d'Université Côte d'Azur, Biot, France
e-mail: jean-antoine.desideri@inria.fr

N. Bartoli · C. David · S. Defoort
DTIS, ONERA, Université de Toulouse, Toulouse, France
e-mail: nathalie.bartoli@onera.fr

Fédération ENAC ISAE-SUPAERO ONERA, Université de Toulouse, Toulouse, France

T. Tuovinen et al. (eds.), *Challenges in Design Methods, Numerical Tools and Technologies for Sustainable Aviation, Transport and Industry*, Computational Methods in Applied Sciences 17, https://doi.org/10.1007/978-3-031-98675-8_4

Pareto set and Pareto front for these primary cost functions only, and to account subsequently for the secondary cost functions by Nash games [7, 9]. The efficacy and cost efficiency of the prioritized approach are presently assessed in a case of multi-objective optimization test-case aimed to be a representative preliminary design phase exercise.

The precise numerical construction is recalled in Sect. 4.2 where the convergence result is stated. The proposed algorithm is described in Sect. 4.3 from an implementation point of view. In Sect. 4.4, the methodology is applied to an aircraft performance optimization test-case involving three objective functions, The numerical results are commented in detail including a qualitative comparison of numerical cost efficiency depending on whether NSGA-II or Bayesian optimization is used to compute Pareto fronts. Finally, conclusions and perspectives are drawn in Sect. 4.5.

4.2 Theoretical Background

4.2.1 Mathematical Problem Definition and Objective

One considers a multi-objective differentiable optimization problem involving a set of M cost functions $\{f_j(\mathbf{x})\}$ $(j = 1, \dots, M;\ \mathbf{x} \in \Omega_a \subseteq \mathbb{R}^N)$, the first m of which $(1 \le m < M)$ are declared preponderant and named primary, and the remaining ones $(j = m+1, \dots, M)$, secondary; all are subject to K equality constraints, $c_k(\mathbf{x}) = 0$ $(k = 1, \dots, K)$. Objective and constraint functions are all smooth, say $C^2(\Omega_a)$.

A first phase of optimization (omitted here) has led to the knowledge of a point $\mathbf{x}_A^\star$ in the open interior of the admissible domain Ω_a. This point is assumed to be Pareto-optimal for the constrained minimization of the sole primary cost functions.

The second phase of optimization is aimed to construct a continuum of Nash equilibria $\{\overline{\mathbf{x}}_\varepsilon\}$ parameterized by ε $(\varepsilon \ge 0)$ in way such that: firstly, $\overline{\mathbf{x}}_0 = \mathbf{x}_A^\star$ (compatibility), and secondly, as ε increases but remains sufficiently small, the Pareto-optimality condition of the primary cost functions remains $O(\varepsilon^2)$, whereas the secondary cost functions decrease linearly, or faster.

4.2.2 Preliminary Calculations

The preliminary calculations have been defined in full in [7, 8]. They are made of three main steps:

- Definition of a convex "Primary Steering Function" $f_A^+(\mathbf{x})$
- Definition of a "Territory Splitting"
- Definition of a "Secondary Steering Function" $f_B(\mathbf{x})$.

The cost functions are all assumed to be strictly positive. This may require the preliminary application of an adequate exponential transform. This permits us to consider gradients in the logarithmic form, and examine the variations of these functions in terms of proportions.

The Pareto-stationarity condition satisfied at $\mathbf{x} = \mathbf{x}_A^\star$ is expressed as follows:

$$\sum_{j=1}^{m} \alpha_j^\star \frac{\mathbf{P}\nabla f_j^\star}{f_j^\star} = 0 \tag{4.1}$$

where the superscript * is used throughout to indicate an evaluation at $\mathbf{x} = \mathbf{x}_A^\star$; ∇ is the gradient operator w.r.t. $\mathbf{x}$, $\mathbf{P}$ is the projection operator onto the linear subspace orthogonal to all constraint gradients $\{\nabla c_k^\star\}$ $(k = 1, \ldots, K)$ (identified by application of the QR algorithm), and $\{\alpha_j^\star\}$ $(j = 1, \ldots, m)$ are the known coefficients of a convex combination. Then the primary steering function is defined as follows:

$$f_A^+(\mathbf{x}) = \sum_{j=1}^{m} \alpha_j^\star \frac{f_j(\mathbf{x})}{f_j^\star} + \frac{c}{2} \left\| \mathbf{x} - \mathbf{x}_A^\star \right\|^2 \tag{4.2}$$

where c is a nonnegative constant adjusted to enforce the local convexity of $f_A^+(\mathbf{x})$ and the associated Lagrangian. Then, by virtue of Eq. (4.1), the primary steering function is stationary and convex at $\mathbf{x}_A^\star$, and therefore achieves a local minimum.

The territory splitting is based on the orthogonal decomposition of the following $N \times N$ reduced Hessian matrix [8, 10]:

$$\mathbf{H}'_A = \mathbf{P}\mathbf{H}_A^{+,\star}\mathbf{P} = \boldsymbol{\Omega}\,\mathcal{H}\,\boldsymbol{\Omega}^t \tag{4.3}$$

where $\mathbf{H}_A^{+,\star} = (\nabla^2 f_A^+)^\star = \mathbf{H}_A^\star + c\mathbf{I}_N$ and $\mathbf{H}_A^\star = \nabla^2 f_A^\star$. Matrix $\mathbf{H}'_A$ is positive semi-definite, of rank $N - K$ exactly since matrix $\mathbf{H}_A^{+,\star}$ is strictly-positive definite by convexity fix. The eigenvalues in the diagonal matrix $\mathcal{H}$ are real and strictly positive. The eigenvectors, column-vectors of the orthogonal matrix $\boldsymbol{\Omega}$, are conventionally arranged in a special order: the first K are associated with the directions of the constraint gradients (null space of $\mathbf{P}$), and the remaining ones are ordered by decreasing eigenvalue. In this way, the tail eigen-modes are associated with the smaller sensitivities of the primary steering function. The splitting is associated with the following change of variables:

$$\mathbf{x} = \mathbf{x}_A^\star + \boldsymbol{\Omega} \begin{pmatrix} \mathbf{u} \\ \mathbf{v} \end{pmatrix} := \mathbf{X}(\mathbf{u}, \mathbf{v}) \tag{4.4}$$

where $\mathbf{u} \in \mathbb{R}^{N-p}$, $\mathbf{v} \in \mathbb{R}^p$, and p is an adjustable integer satisfying $1 \leq p < N - K$. A large p gives the secondary cost functions a great weight. In this setting, the following notations are used:

$$\boldsymbol{\Omega} = (\boldsymbol{\Omega}_{\mathbf{u}}\ \boldsymbol{\Omega}_{\mathbf{v}}), \quad \boldsymbol{\Omega}_{\mathbf{u}} = \begin{pmatrix} \boldsymbol{\Omega}_{\mathbf{uu}} \\ \boldsymbol{\Omega}_{\mathbf{vu}} \end{pmatrix}, \quad \boldsymbol{\Omega}_{\mathbf{v}} = \begin{pmatrix} \boldsymbol{\Omega}_{\mathbf{uv}} \\ \boldsymbol{\Omega}_{\mathbf{vv}} \end{pmatrix} \tag{4.5}$$

and scaled logarithmic gradients of the secondary cost functions are calculated:

$$\mathbf{g}_j = \mathbf{S}^{-\frac{1}{2}} \frac{\nabla_{\mathbf{v}} f_j^\star}{f_j^\star} = \mathbf{S}^{-\frac{1}{2}} \frac{\boldsymbol{\Omega}_{\mathbf{v}}^t \nabla f_j^\star}{f_j^\star} \qquad (j = m+1, \dots, M) \tag{4.6}$$

where the scaling matrix $\mathbf{S} = (\nabla\nabla_{\mathbf{vv}}^t f_A^+)^\star$ is the lower $p \times p$ diagonal block of matrix $\mathcal{H}$ of Eq. (4.3), hence it is guaranteed to be positive-definite by convexity fix. These gradients reflect the sensitivities of the secondary cost functions when $\mathbf{v}$ varies, and a scaling by elements related to the primary steering function as well. Following the strategy employed in the Multiple Gradient Descent Algorithm (MGDA) [7, 8], one identifies the minimum-norm element $\boldsymbol{\omega}_B^\star$ in the convex hull of the gradients $\mathbf{g}_j$:

$$\boldsymbol{\omega}_B^\star = \sum_{j=m+1}^{M} \alpha_j^\star \mathbf{g}_j \tag{4.7}$$

The coefficients $\{\alpha_j^\star\}$ of this convex combination are used to define the secondary steering function $f_B(\mathbf{x})$, thus completing the preliminaries:

$$f_B(\mathbf{x}) = \sum_{j=m+1}^{M} \alpha_j^\star \frac{f_j(\mathbf{x})}{f_j^\star} \tag{4.8}$$

4.2.3 Nash Game Formulation and Theoretical Results

One defines a continuation parameter ε $(0 \le \varepsilon \le 1)$ and the auxiliary cost function:

$$f_{AB}(\mathbf{x}) = f_A^+(\mathbf{x}) + \varepsilon \left(f_B(\mathbf{x}) - f_A^+(\mathbf{x}) \right) \tag{4.9}$$

For every fixed value of ε, two virtual players, Player A and Player B, respectively controlling the sub-vectors $\mathbf{u}$ and $\mathbf{v}$ in Eq. (4.4), are involved in a Nash game defined by the following concurrent strategies:

- Strategy A: *Player A* attempts to minimize $f_A^+\big(\mathbf{X}(\mathbf{u}, \mathbf{v})\big)$ by the strategy $\mathbf{u}$, subject to the equality constraint $\mathbf{c}\big(\mathbf{X}(\mathbf{u}, \mathbf{v})\big) = 0$, and by accounting for *Player B* 's fixed strategy $\mathbf{v}$,

concurrently with:

- Strategy B: *Player B* attempts to minimize $f_{AB}(\mathbf{X}(\mathbf{u}, \mathbf{v}))$ by the strategy $\mathbf{v}$, subject to no constraints, but by accounting for *Player A* 's fixed strategy $\mathbf{u}$.

The following results have been established [7, 8]:

- The Nash equilibrium point $\overline{\mathbf{x}}_\varepsilon$ exists for all sufficiently small ε, and

$$\overline{\mathbf{x}}_0 = \mathbf{x}_A^\star \tag{4.10}$$

  an essential property of the formulation referred to as compatibility or consistency.
- If $\phi_A(\varepsilon) = f_A^+(\overline{\mathbf{x}}_\varepsilon)$, $\phi'_A(0) = 0$ and

$$f_A^+ = 1 + O(\varepsilon^2) \tag{4.11}$$

  which reflects the fact that, as ε varies, the Pareto-optimality of the primary cost functions is degraded by $O(\varepsilon^2)$ only.
- If $\phi_B(\varepsilon) = f_B(\overline{\mathbf{x}}_\varepsilon)$, $\phi'_B(0) = -\sigma_B$, where $\sigma_B = \left\|\omega_B^\star\right\|^2 \geq 0$. The secondary steering function $f_B(\mathbf{x})$ decreases linearly as ε increases, and by virtue of properties of the MGDA construction, all the secondary cost functions do as well [7, 8].

Figure 4.1 illustrates this theoretical result, depicting in green the continuum of Nash equilibria originating from the point $\mathbf{x}_A^\star$ tangent at this point to the primary Pareto front (in red).

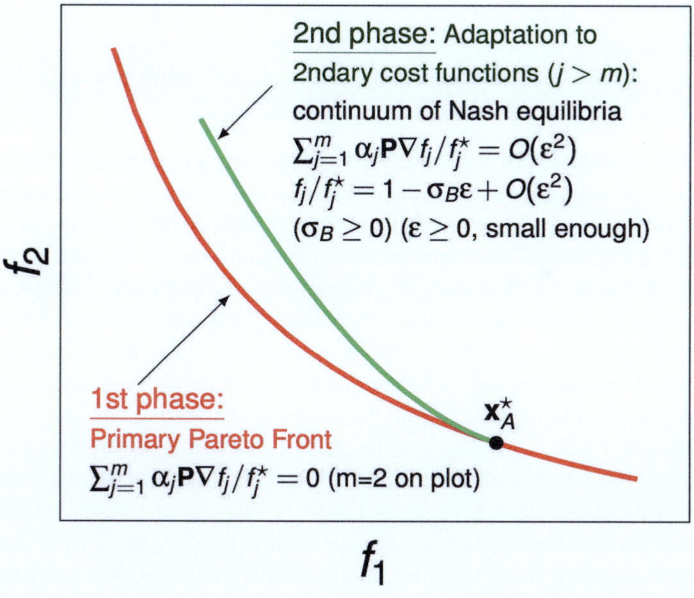

Fig. 4.1 Sketch of primary Pareto front (in red) and a representative continuum of Nash equilibria (in green) tangent to it

4.3 Numerical Implementation, Constraints and Coordination Loop

The present method can be applied directly to simple cases defined analytically. The software provided by the platform http://mgda.inria.fr can be used for this purpose according to procedures detailed in [6]. In the more general case in which the formal derivation is not entirely at hand, we proceed by applying the method to local metamodels some of which are upgraded in the course of the computational process. A customized variant of the software is then employed. We consider

- a split of the optimization variables $\mathbf{x} = (\mathbf{x}_\phi, \mathbf{x}_s) \in \mathbb{R}^N$ in which each component of $\mathbf{x}_\phi \in \mathbb{R}^n$ is in a one-to-one correspondence with one of the n natural physical variables of the problem, and those of $\mathbf{x}_s \in \mathbb{R}^s$ are slack variables ($N = n + s$);
- "physical functions" of $\mathbf{x}_\phi$ including cost functions and auxiliary functions in which the "physical constraints" of the problem are expressed; an evaluation procedure to compute these physical functions in terms of $\mathbf{x}_\phi$ is at disposal; the number of calls to this procedure is our measure of computational effort;
- "constraint functions" $\{c_k(\mathbf{x})\}$ $(k = 1, \ldots, K)$ devised to enforce the physical constraints by the somewhat artificial condition $c_k(\mathbf{x}) = 0, \ \forall k$; these functions are locally represented by metamodels. Once a candidate Nash equilibrium point $(\mathbf{x}_\phi, \mathbf{x}_s)$ is determined, the satisfaction of the physical constraints by $\mathbf{x}_\phi$ is tested; if the test is positive, in preparation of the determination of the next equilibrium point, the slack variables $\mathbf{x}_s$ are adjusted to reset the pointwise values of these metamodels to 0, upgrade the metamodels, and avoid accumulation of even minute errors on the $\{c_k(\mathbf{x})\}$; otherwise the process interrupts.

A Design of Experiment (DoE) is firstly conducted over a dataset of points in physical space ($\mathbb{R}^n$) combining a micro lattice, a medium-size lattice and a macro lattice. The micro lattice serves to approximate first derivatives of functions of interest at $\mathbf{x} = \mathbf{x}_A^\star$, and the diagonal elements of their Hessians by central differencing. It is made of $2n$ points at $\pm h$ (for a small h) about $\mathbf{x}_A^\star$ over each axis. The macro lattice and medium-size lattice have similar structure. They are made of 16 points per pair of coordinates $\{x_i, x_j\}$. The macro lattice is centered at the origin and extends over the entire admissible set. The medium-size lattice is centered at $\mathbf{x}_A^\star$ and its sides are adjusted to lie strictly within the domain. Altogether the macro lattice and the medium-size lattice provide a more global domain representation in view of approximating the off-diagonal elements of Hessians by least squares. In total, the evaluation of the entire dataset requires $16n(n-1) + 2n + 1$ additional calls to the simulation software. The dataset is then extended to dimension N by discretizing $\mathbf{x}_s$ only and by evaluating the $\{c_k(\mathbf{x})\}$ functions correspondingly; this evaluation is fast since the physical part has already been computed, and the remaining part is explicit. Quadratic metamodels are constructed once for all for the two steering functions. Similar quadratic metamodels are also constructed at first for the constraint functions, but these are upgraded throughout the computational process of Nash equilibria. We emphasize that this upgrade is essential to preserve the Pareto-stationarity condition.

The Hessian of f_A^+ and first derivatives of constraint functions are approximated by central differencing. This serves to define the orthogonal decomposition in Eq. (4.4).

At start ($\mathbf{x} = \mathbf{x}_A^\star$), the physical constraints are satisfied exactly since the assumed Pareto optimality of $\mathbf{x}_A^\star$ is subject to this condition, and the slack variables have been initialized accordingly. After identifying a number of Nash equilibrium points (outer loop), the metamodels for the constraint functions $\{c_k(\mathbf{x})\}$ are locally reset to 0; then, a situation similar to initial applies at $\mathbf{x} = \overline{\mathbf{x}}_\varepsilon$ for a certain $\varepsilon > 0$. The search for the next equilibrium point then proceeds with a coordination iteration for the pair $(\mathbf{u}, \mathbf{v})$. This inner iteration begins with the constrained minimization of the primary steering function f_A^+ by means of the subvector $\mathbf{u}$ (for fixed $\mathbf{v}$) in compliance with the strategy of virtual Player A (this minimization is carried out by a Newton-type algorithm); then follows the unconstrained minimization of the function f_{AB} by means of the subvector $\mathbf{v}$ (for fixed $\mathbf{u}$) in compliance with the strategy of virtual Player B; this is realized by inversion of the linear system of optimality. The coordination is judged converged when the variation in $\|\mathbf{v}\|$ has achieved a value less than a small tolerance. Then the pair $(\overline{\mathbf{u}}, \overline{\mathbf{v}})$ defines the candidate equilibrium point $(\mathbf{x}_\phi, \mathbf{x}_s)$. The test on physical constraints is made. If it is positive, the slack variables $\mathbf{x}_s$ are corrected, usually by minute variations, to enforce the rigorous satisfaction of the condition $c_k = 0,\ \forall k$. These small variations affect the slack subvector $\mathbf{x}_s$ only, and have no impact on the physical functions. This completes the definition of the new Nash equilibrium point. The constraint metamodels are then upgraded and the outer loop is continued. The process terminates when the constraints are violated, or when the determination of $\overline{\mathbf{u}}$ fails in the approach of the convexity limit, or because the coordination process has not converged after a prescribed number of iterations.

4.4 Application to an Aircraft Optimization Exercise

4.4.1 Test Case Definition

We consider the optimization of six design variables impacting wing and tail geometry, engine thrust and longitudinal stability of an Airbus A320-type aircraft to reduce two primary cost functions, and one secondary, subject to functional constraints. The test case is meant to be representative of a process taking place in a preliminary aircraft design phase.

Six design optimization variables in physical dimensions ($n = 6$):

- ϕ_1: distance from nose to 25% of wing Mean Aerodynamic Chord (MAC); to be in $[a_1, b_1] = [16, 18]$;
- ϕ_2: wing aspect ratio; to be in $[a_2, b_2] = [5, 11]$;
- ϕ_3: horizontal tail aspect ratio; to be in $[a_3, b_3] = [1.5, 6]$;
- ϕ_4: wing taper ratio; to be in $[a_4, b_4] = [0, 1]$;

- ϕ_5: wing sweep angle at 25% of mean aerodynamic chord; to be in $[a_5, b_5] = [20, 30]$;
- ϕ_6: maximum take-off thrust (denoted MTOT); to be in $[a_6, b_6] = [100000, 140000]$.

The above physical variables are the components of the vector $\mathbf{x}_\phi = \{\phi_i\}$ and are parameterized via dimensionless variables $\{x_i\}$ according to

$$\phi_i(\mathbf{x}) = \frac{1}{2}(a_i + b_i) + \frac{1}{2}(b_i - a_i)\sin\left(\frac{\pi x_i}{2}\right), \quad x_i \in [-1, 1] \quad (i = 1, \dots, 6) \tag{4.12}$$

Two primary cost functions ($m = 2$) are chosen:

- fuel mass, f_1 (by multiples of 10^4 kg);
- operating empty weight (OEW), f_2 (by multiples of 10^4 kg).

Physical constraints are considered:

- geometrical bound: wing span ≤ 36 (m);
- functional constraint for longitudinal stability: $0.05 \leq$ static margin ≤ 0.10.

One secondary cost function ($M = 3$) is added:

- duration of ascent to cruise altitude, f_3 (by multiples of 10^3 s).

A first slack variable x_7 to enforce the interval constraint on the static margin is used. Defining the following function observed to be locally convex

$$g_1 = g_1(\mathbf{x}) = \frac{0.10 - \text{static margin}}{0.05} \tag{4.13}$$

the constraint writes $0 \leq g_1 \leq 1$ and is enforced by letting

$$c_1 = g_1 - \sin\gamma_1 = 0, \quad \gamma_1 = (1 - x_7)\frac{\pi}{4}, \quad x_7 \in [-1, 1]. \tag{4.14}$$

A second slack variable x_8 ($s = 2$) is introduced to enforce the bound constraint on wing span. The function g_2 observed to be locally convex is defined:

$$g_2 = 1 - \frac{\text{wing span}}{36} \tag{4.15}$$

So the constraint writes $0 \leq g_2 \leq 1$ and is enforced by letting

$$c_2 = g_2 - \sin\gamma_2 = 0, \quad \gamma_2 = (1 - x_8)\frac{\pi}{4}, \quad x_8 \in [-1, 1]. \tag{4.16}$$

In this setting, the dimensionless admissible domain is $[-1, 1]^8$.

4.4.2 Evaluation and Optimization Software

The designs are evaluated by means of the FAST-OAD open-source software developed by ONERA and ISAE-SUPAERO [3] which combines the classical Breguet laws with a time integration of the ascent phase, and performs a full resizing (dimensions, masses) of the aircraft for each set of design variables through a multidisciplinary coupled approach. Hence, this evaluation is constrained by an ordinary-differential equation. The prioritized optimization is conducted by a customized version of the INRIA Nash-MGDA software [5]. From a point of Pareto-optimality of the sole pair (f_1, f_2), considered preponderant, the climb duration f_3 is to be reduced along a continuum of Nash equilibria.

4.4.3 Numerical Results Using NSGA-II

The first phase of optimization consisted in establishing the "Primary Pareto Front" (PPF) associated with the minimization of (f_1, f_2) alone under constraints. This was done firstly by the genetic algorithm NSGA-II [4] using a population of 50 individuals. A well-converged front was obtained using an indicative budget of 2500 calls to the FAST-OAD software, at a computation cost of about 22 h on a PC. Then, three sets of Nash equilibria were computed, each one originating from a point of the discrete PPF and appearing on Fig. 4.2 as a green path denoted L, C or R.

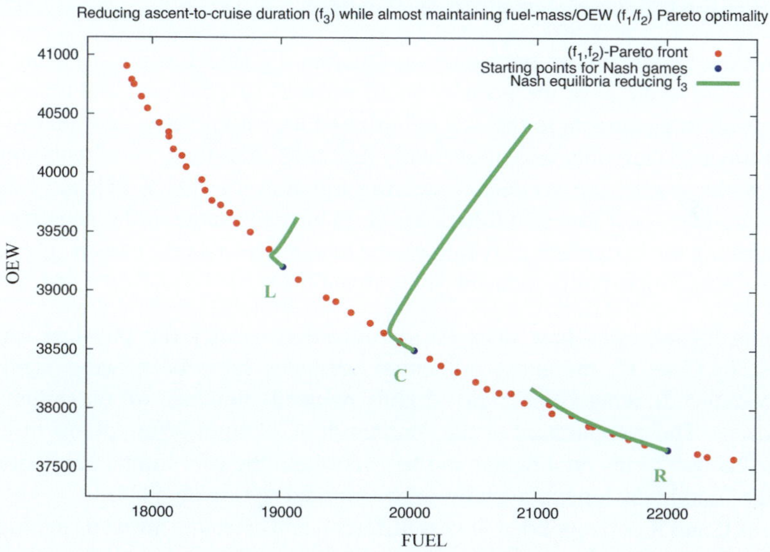

Fig. 4.2 Computed primary Pareto front and three continua of Nash equilibria denoted by L, C or R

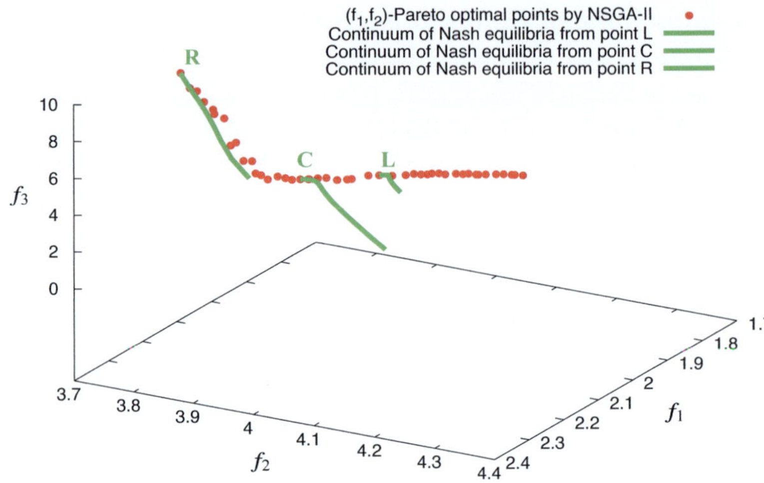

Fig. 4.3 Primary Pareto front and three continua of Nash equilibria represented in (f_1, f_2, f_3)

The three paths are tangent to the PPF and remain close to it initially over a variable length. On Fig. 4.3, the paths are represented in the (f_1, f_2, f_3) space.

These results have been described in details in [11], and we now focus our discussion mostly on the central case C. In this case, the DoE involved some 500 preliminary evaluations and this required about 4.5 h of computing time. Figure 4.4 illustrates the "funnel-shaped" convergence plot typical pattern:

- above the ε-axis, the primary steering function f_A^+ (black solid line), and below it, still above the axis, the primary cost functions (f_1, f_2) (red and orange solid lines), all three tangent to the axis by virtue of the initial Pareto-stationarity; both cost functions are only very moderately degraded (less than 5% at endpoint);
- below the ε-axis, the secondary steering function f_B (black hatched line) and the secondary cost function (blue solid line) have the same initial negative slope, confirming the linear decay; f_3 follows the metamodel f_B very closely; at endpoint f_3 has been importantly reduced (more than 60%).

Figure 4.5 indicates how does the optimization modify the physical variables with ε, in Case C: the most influential variables have been wing taper ratio (highly reduced), wing aspect ratio (slightly reduced), and take-off maximum thrust (increased). The computation of the continuum of 22 equilibrium points in Case C required about 7 mns on a laptop and has permitted the user interaction. Along the path, the functional constraints have been enforced rigorously [11].

Cases L and R corresponded to very different, and extreme, initial designs and the optimization resulted in very different modifications. Compared to the central case (C), the variations were briefly the following [11]:

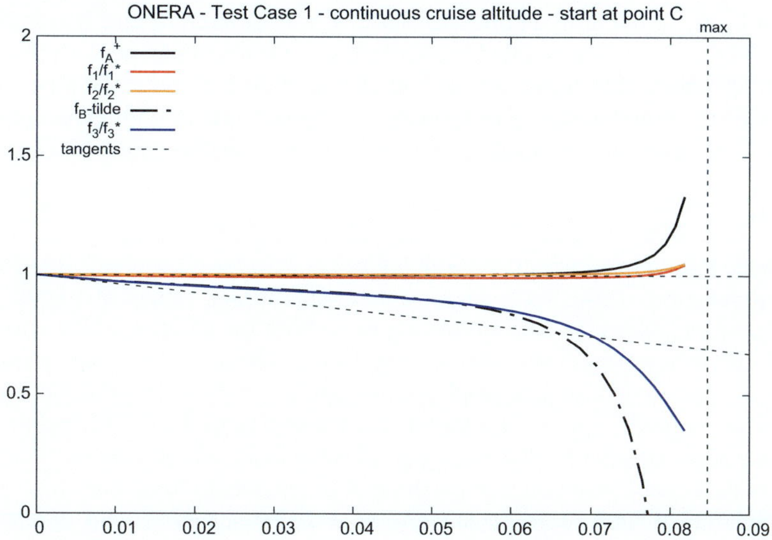

Fig. 4.4 Funnel-shaped variation plot of the cost functions along the continuum in Case C

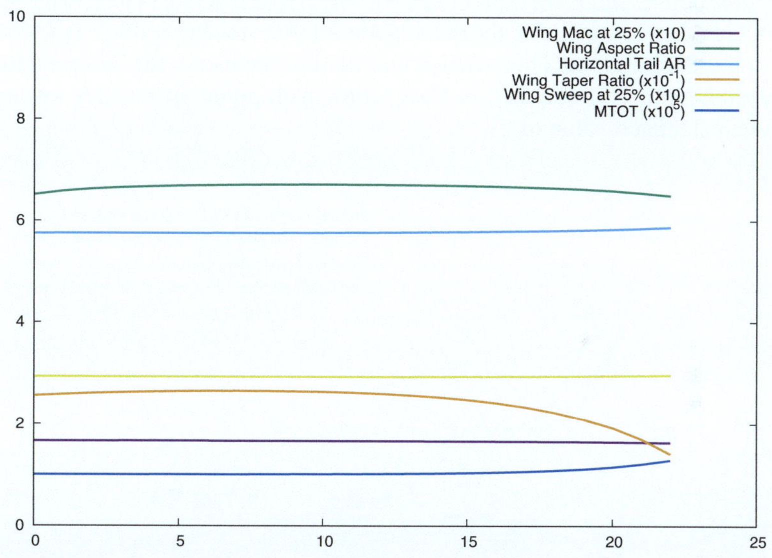

Fig. 4.5 Variation of the design variables in physical dimensions

– Case L: the initial design corresponded to a heavier aircraft with less embarked fuel. The optimization resulted in an increase of the maximum take-off thrust and small geometrical changes (mostly a reduction of the horizontal tail aspect ratio).
– Case R: the initial design corresponded to a lighter aircraft with more embarked fuel. The optimization resulted mostly in more sensible geometrical changes (increase of horizontal tail ratio and wing taper ratio, reduction of wing aspect ratio).

Lastly, the Pareto front associated with the triplet (f_1, f_2, f_3) was calculated with a larger population of 100, and a larger, somewhat arbitrary, budget of 10,000 evaluations, requiring some 77 h of computing time. With the addition of the cost function f_3, the Pareto front (blue dots on Fig. 4.6) is enhanced. The graphic software `gnuplot` permits to visualize an interpolated surface (light violet grid). This surface seems to contain the (f_1, f_2, f_3) triplets corresponding to the 3D-extension of the PPF over about two thirds of it, but a gap is visible in Case R and near it. This is the result of the better convergence of the three-objective Pareto front. Nevertheless, the three continua of Nash equilibria are tangent to the extended PPF, and all three tend to reduce f_3. The interpolated surface is clearly not convex, showing several local minima and a maximum inside the domain. This reveals that the cost functions are not all three convex, and this is particularly so near the peak of f_3 close to the starting point of Case R. Nevertheless, the green paths all three tend to reduce f_3 and remain close to the Pareto front. This verifies that the discrepancy with Pareto-optimality is very small ($O(\varepsilon^2)$ in theory). In Case C, the path points in a valley towards the minimum calculated value of f_3.

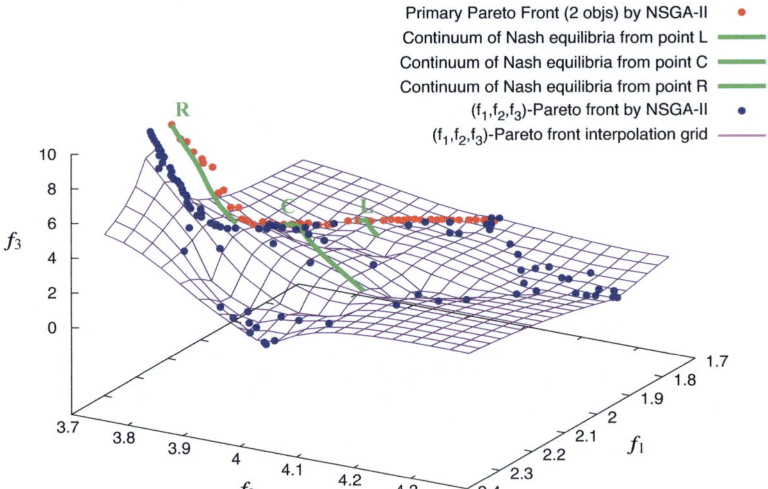

Fig. 4.6 Two-Objective Primary Pareto Front by NSGA-II, three continua of Nash equilibriums initiated from it, three-objective Pareto front and interpolated grid

4.4.4 Numerical Experiments Using Bayesian Optimization

We consider Bayesian optimization (BO) [13] as a replacement for NSGA-II to establish the PPF with less evaluations. BO relies on Gaussian process (GP) metamodels to predict the behavior of the objectives and guide the search towards promising regions of the parameter space. To handle multiple objectives, multi-output GPs are built in combination with sequential infill criteria aiming at finding Pareto optimal points, such as the hypervolume improvement [12] or the probability of improvement (PI [15]). The use of the SEGOMOE toolbox [1] with its recent extension to multi-objective problems [14] is illustrated in Fig. 4.7. An initial DoE of 20 points is used to build first metamodels and then 100 iterations are performed using PI based criterion, taking around 1h to complete. Compared to NSGA-II, BO required much less evaluations to obtain a three-objective Pareto front. The front is denser, but also less spread, especially at the extremities. This reflects the increasing difficulty of covering the entire Pareto front as more objectives are added, while searching for specific solutions instead is also advocated e.g., by [2]. With prioritized objectives, finding the PPF with BO is more cost-efficient than with NSGA-II. Further reductions of the number of evaluations in the prioritized approach could be obtained by considering GP metamodels rather than quadratic ones.

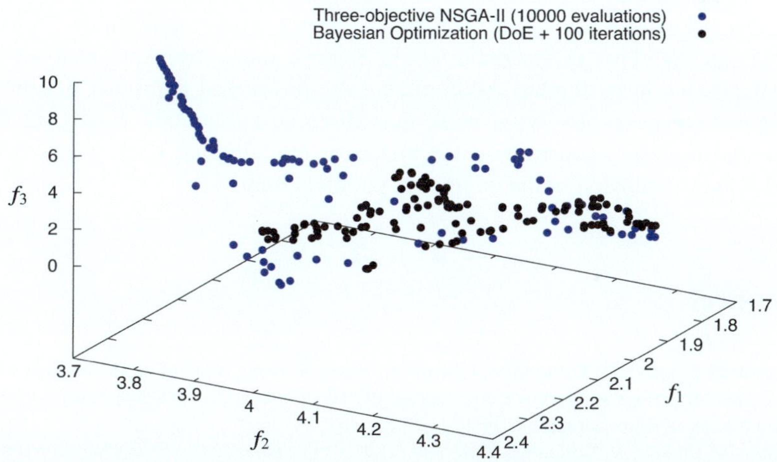

Fig. 4.7 Three-objective Pareto fronts obtained by NSGA-II (10000 evaluations) and Bayesian Optimization (DoE of 20 evaluations + 100 sequential iterations)

4.5 Conclusion and Perspectives

The efficacy and cost-efficiency of a prioritized approach have been explored in the context of concurrent engineering, in which several cost functions are to be minimized. The prioritized approach is justified particularly in the following situations:

1. when the cost functions are of different importance, which encourages to give greater importance to preponderant cost functions;
2. when the computational cost of using a specific software to identify Pareto fronts numerically increases rapidly with the number of cost functions accounted for.

Our prioritized approach is a two-step procedure consisting in firstly computing the Pareto front for the preponderant cost functions only, and to construct from it a continuum of Nash equilibria proved by asymptotic theory to be quasi-Pareto optimal for the whole set. The method has been applied to a test-case of aircraft flight performance optimization involving two primary cost functions and one secondary. Most of our numerical experiments were carried out using the classic NSGA-II for identifying Pareto fronts. Indeed, solutions very close to the Pareto front of the whole set of cost functions were identified by the prioritized algorithm (efficacy), far more economically than by computing the complete Pareto front (cost-efficiency). Our experiments with BO involved a relatively small initial database, and Pareto fronts could be computed much more economically, but resulting in a sparser Pareto front. The prioritized approach resulted in a much denser continuum focused in areas of interest. This comparison would deserve more extensive analysis using larger databases in particular. Additionally, the prioritized approach is promising for engineering cases involving more than three cost functions. Lastly, in Partial Differential Equation-constrained optimization, our ultimate goal, cost efficiency should also be evaluated in the context of parallel computing.

References

1. Bartoli N, Lefebvre T, Dubreuil S, Olivanti R, Priem R, Bons N, Martins JR, Morlier J (2019) Adaptive modeling strategy for constrained global optimization with application to aerodynamic wing design. Aerosp Sci Technol 90:85–102
2. Binois M, Picheny V, Taillandier P, Habbal A (2020) The kalai-smorodinsky solution for many-objective bayesian optimization. J Mach Learn Res 21(1):5978–6019
3. David C, Delbecq S, Defoort S, Schmollgruber P, Benard E, Pommier-Budinger V (2021) From FAST to FAST-OAD: An open source framework for rapid overall aircraft design. IOP Conf Ser Mater Sci Eng 1024(1):012062
4. Deb K, Pratap A, Agarwal S, Meyarivan T (2002) A fast and elitist multiobjective genetic algorithm: NSGA-II. IEEE Trans Evol Comput 6(2):182–197
5. Désidéri J-A, MGDA software platform for multi-objective differentiable optimization. http://mgda.inria.fr
6. Désidéri J-A (2019) Platform for prioritized multi-objective optimization by metamodel-assisted Nash games. Research Report 9290, Inria, September 2019. https://hal.inria.fr/hal-02285197

7. Désidéri J-A (2021) Adaptation by Nash games in gradient-based multi-objective/multi-disciplinary optimization. In: JANO13, mathematical control and numerical applications, Khouribga, Morocco, vol 372. Springer Proceedings in Mathematics and Statistics. https://hal.inria.fr/hal-03430972
8. Désidéri J-A, Duvigneau R (2019) Direct and adpative approaches to multi-objective optimization. Research Report 9291, Inria, September 2019. https://hal.inria.fr/hal-02285899
9. Désidéri J-A, Duvigneau R (2021) Prioritized optimization by Nash games: towards an adaptive multi-objective strategy. In: ESAIM: Proceedings and Surveys, vol 71, pp 54–63. EDP Sciences. https://hal.inria.fr/hal-03430912
10. Désidéri J-A, Duvigneau R, Habbal A (2014) Multi-objective design optimization using nash games. In: Becerra VM, Vassile M (ed)Computational intelligence in aerospace sciences. In: Lieuwen TC (ed.-in-Chief) Progress in astronautics and aeronautics, vol 244. American Institute for Aeronautics and Astronautics Inc., Reston, Virginia
11. Désidéri J-A, Wintz J, Bartoli N, David C, Defoort S (2022) Combining Pareto optimality with nash games in multi-objective prioritized optimization of an aircraft flight performance. Research Report RR-9490, Inria - Sophia Antipolis; Acumes
12. Emmerich MT, Giannakoglou KC, Naujoks B (2006) Single-and multiobjective evolutionary optimization assisted by gaussian random field metamodels. IEEE Trans Evol Comput 10(4):421–439
13. Garnett R (2023) Bayesian optimization. Cambridge University Press
14. Grapin R, Diouane Y, Morlier J, Bartoli N, Lefebvre T, Saves P, Bussemaker J (2022) Regularized infill criteria for multi-objective bayesian optimization with application to aircraft design. In: AIAA AVIATION 2022
15. Jones DR (2001) A taxonomy of global optimization methods based on response surfaces. J Global Optim 21:345–383

Chapter 5
Machine Learning-Based Biomimetic Optimization Models for Transport Networks

Shishir Barai, Joe Sgarrella, William Laplante, and Christian Peco

Abstract The slime mold *Physarum polycephalum* exhibits an innate ability to form networks that efficiently solve complex combinatorial problems and optimize transport pathways. This emergent behavior is largely attributed to the dynamic microstructural properties of its cytoplasm. In this study, we build upon a phase-field model of the slime mold and introduce a machine learning approach to optimize its microstructural responses, thereby capturing the essence of its transport network optimization. By conceptualizing transport networks as composite mesh structures–akin to biological organisms with specialized regions emulating functions such as nuclei and cellular matrices–we establish a connection between biological phenomena and computational modeling. Utilizing continuum equations, we simulate protoplasmic flux within these meshes, generating data that encapsulates the core dynamics of transport. A machine learning model is then trained to predict the resulting pressure and flow fields based on a given composite mesh structure. Once trained, this model rapidly approximates the physical behaviors, enabling iterative optimization of the mesh topology to enhance transport efficiency. Furthermore, encoding the mesh structures into a compact latent space accelerates the optimization process. This integrated framework combines composite mesh modeling, biophysical simulations, and machine learning to uncover novel bio-inspired strategies for designing optimized transport networks.

S. Barai · J. Sgarrella · W. Laplante · C. Peco (✉)
Department of Engineering Science and Mechanics, The Pennsylvania State University, University Park, PA, USA
e-mail: christian.peco@psu.edu

S. Barai
e-mail: svb6406@psu.edu

J. Sgarrella
e-mail: jfs5687@psu.edu

W. Laplante
e-mail: wxl5213@psu.edu

T. Tuovinen et al. (eds.), *Challenges in Design Methods, Numerical Tools and Technologies for Sustainable Aviation, Transport and Industry*, Computational Methods in Applied Sciences 17, https://doi.org/10.1007/978-3-031-98675-8_5

5.1 Introduction

Biological networks exhibit remarkable capabilities to dynamically self-organize and adapt to environmental stimuli. Organisms ranging from slime molds to vascular networks develop transport morphologies that maximize efficiency and resilience, solving highly complex optimization challenges with emergent decentralized intelligence [1, 2]. Mathematical models that encapsulate the underlying biophysical mechanisms have provided fundamental insights into natural network adaptations [3–5]. However, computational expense often limits iterative analysis and discovery over large parameter spaces. Recent advances in machine learning now offer pathways to develop surrogate models that rapidly approximate solutions to computationally intensive physics-based simulations [6, 7].

The plasmodium stage of the slime mold *Physarum polycephalum* (*P. polycephalum*) demonstrates notable abilities to adapt its tubular network morphology to optimize transport efficiency. When presented with spatially distributed nutrients, *P. polycephalum* will reconfigure its vein network to connect the nutrients through shortest possible pathways [2, 8]. Some examples of *P. polycephalum* forming optimized pathways can be observed in Fig. 5.1. This intelligent adaptation relies on simple decentralized rules enacted locally by the organism. Key mechanisms deduced from biological experiments include tube diameter expansion in response to higher fluxes of protoplasm and atrophy of tubes with lower fluxes [3, 9, 10]. Recent computational models integrating the dynamics of biological networks through systems of partial differential equations have shown good agreement with experimental observations [11–13]. These models combine a novel reference map-based phase-field approach [14] with a large-scale amenable description from fluid to solid phases [15, 16]. This approach eliminates the need for boundary tracking and greatly facili-

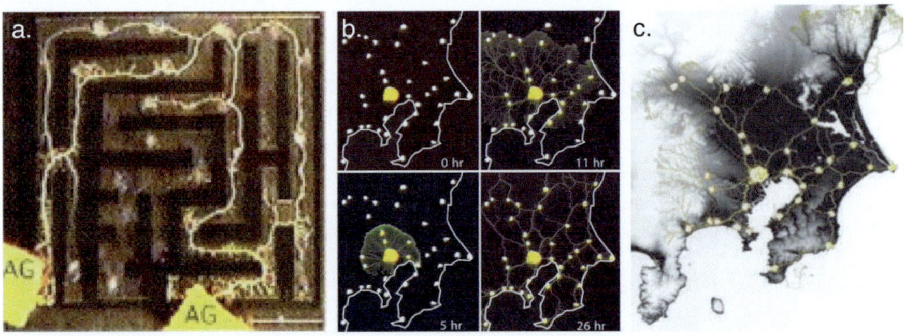

Fig. 5.1 **a** *P. polycephalum* solving a maze [3]. This is an intermediate state of the maze solving process. **b** Time-lapse images of *P. polycephalum* network formation of Tokyo railway system [26]. The network begins with a small plasmodium surrounded by food sources that mark major cities within the region. As time progresses, the plasmodium grows out from the initial food source to the additional food sources. An interconnected network of the food sources is left behind the growth front. **c** Reimagined Tokyo railway system

tates the representation of evolving geometries. Phase-field methodologies have been proven effective for tackling problems characterized by intricate geometries. They have been instrumental in advancing the modeling of various systems in soft matter, including biomembranes [17, 18] and particle raft fracture mechanics [19–23]. To numerically solve the associated partial differential equations, we adopt a variational approach and leverage MOOSE, a finite element method (FEM) implementation [24, 25]. However, challenges to fully characterize *P. polycephalum*'s adaptation space remain. These include capturing the complexity of the microstructural response and its associated computational cost, which makes direct simulation impractical.

In response to these challenges, recent works have begun leveraging machine learning to approximate solutions to the complex bio-physics models of slime mold network evolution for accelerated simulation and optimization [27, 28]. Transport networks are represented as weighted graphs, with edge weights encoding the conductivity of tubes. Continuum equations govern the complex mechanical behavior of the microstructure. Machine learning models are then trained on simulation data to predict network evolution dynamics orders of magnitude faster than direct numerical simulation, while being able to capture the response of very heterogeneous types of microstructure [29–31]. These learned models enable extensive iterative improvement of network parameters to optimize morphology for transport efficiency. In this work, we extend the principles observed in slime molds to a broader context of biological organisms, represented as composite mesh structures. These structures, analogous to biological organisms, consist of distinct regions mimicking functionalities like nuclei and cellular matrices. By simulating protoplasmic flux through these meshes using continuum equations, we generate data that mirrors the transport dynamics in such organisms. A machine learning model, trained on this data, is then capable of predicting how changes in the organism's microstructure might affect its internal transport pathways. The model's predictions, based on the mesh structure, can swiftly approximate complex biophysical behaviors, facilitating iterative optimization of mesh designs for enhanced transport efficiency. By encoding these mesh structures into a condensed latent space, we further expedite the optimization process, presenting a harmonious blend of composite mesh modeling, biophysics simulations, and machine learning. This integrated approach not only showcases the potential of machine learning in biophysical simulations but also highlights the remarkable adaptive capabilities of organisms like slime molds and their bio-inspired counterparts.

5.2 Slime Mold Modeling and Formulation

Continuum modeling provides a comprehensive framework to describe the behavior of systems with evolving geometries and complex interactions. In the context of slime mold transport networks, continuum modeling offers a mathematical representation of the mold's adaptive behavior in response to external stimuli. The growth dynamics of the slime mold are captured using a multispecies formulation [12]. This approach considers the availability and distribution of nutrients in the medium and the specific metabolism of each species. Within the domain of study, Ω, we consider the existence of $i = 1, 2, ..., N$ slime mold species. The free energy functional for the ith species is defined as :

$$\mathcal{G}_i(\phi, \nabla\phi, c_s, c_c) = \int_{\Omega} F_i(\phi, \nabla\phi, c_s, c_c)\, \mathrm{d}\Omega. \tag{5.1}$$

In this formulation, ϕ is a phase-field variable that represents the existence or absence of biomass. It ranges in values from $\phi = 1$ (existence of biomass) to $\phi = 0$ (absence of biomass). Additionally, the variables c_c and c_s represent the concentrated nutrient source and the substrate nutrient source respectively. The free energy density, F_i, that corresponds to the free energy functional is defined as:

$$F_i(\phi, \nabla\phi, c_s, c_c) = I_{\mathcal{G}}(\nabla\phi) + W_{\mathcal{G}}(\phi) + B_{\mathcal{G}i}(\phi, c_s, c_c). \tag{5.2}$$

The free energy density is composed of three terms that represent the interfacial energy ($I_{\mathcal{G}}$) and the contribution of the bulk phase energy ($W_{\mathcal{G}} + B_{\mathcal{G}i}$). The two terms that make up the bulk phase energy contribution are a symmetric double-well potential ($W_{\mathcal{G}}$) and the nutritional bias ($B_{\mathcal{G}i}$). These three terms are defined as:

$$I_{\mathcal{G}} = \frac{1}{2}\epsilon_{\mathcal{G}}^2 |\nabla\phi|^2, \tag{5.3}$$

$$W_{\mathcal{G}} = \phi^2(1-\phi)^2, \tag{5.4}$$

$$B_{\mathcal{G}i} = n_i(c_s, c_c)s(\phi). \tag{5.5}$$

In the interfacial energy equation, $\epsilon_{\mathcal{G}}$ is a parameter that is a length scale proportional to the width of the diffuse interface. In the nutritional bias equation, $n_i(c_s, c_c)$ is a tilting function that accounts for the growth or decay of the slime mold biomass due to the presence or absence of nutrients within the domain Ω. The function $s(\phi)$ in the nutritional bias equation is a monotonically increasing function with lower and upper bounds. The shape of the overall bulk free energy landscape ($W_{\mathcal{G}}^* = W_{\mathcal{G}} + B_{\mathcal{G}}$) double-well is determined by the presence or absence of the total nutrient.

The model also considers the metabolism and nutrient dynamics for the slime mold species. It is assumed that the concentrated nutrient source, c_c, is the primary nutrient source utilized by the slime mold for growth during foraging. The nutrient source present in the substrate, n_s, is responsible for the front morphologies and branching behavior exhibited by the slime mold's network. The evolution of the substrate nutrient source is dependent on the substrate nutrient diffusivity (D_{c_s}), the substrate nutrient (c_s), and a sink term that defines how the slime mold metabolizes the substrate nutrient ($\Xi(\phi, c_s)$). Similarly, the evolution of the concentrated nutrient source is dependent on the concentrated nutrient diffusivity (D_{c_c}), the concentrated nutrient (c_c), and a source term ($\varpi(\phi, c_c)$) that includes that replenishment and consumption of the concentrated nutrient source.

With the free energy potential and the metabolism and nutrient dynamics, we formulate the governing equations for the multispecies growth dynamics to be:

$$\frac{\partial \phi_i}{\partial t} = \lambda_{Gi}\left[\epsilon_G^2 \Delta\phi - \phi(4\phi^2 - 6\phi + 2) + 6(\phi - \phi^2) n_i(c_s, c_c)\right] \quad \text{in } \Omega \times [0, t_f], \tag{5.6}$$

$$\frac{\partial c_{si}}{\partial t} = \nabla \cdot (D_{c_s i} \nabla c_s) - \Xi_i(\phi, c_s) \quad \text{in } \Omega \times [0, t_f], \tag{5.7}$$

$$\frac{\partial c_{c_i}}{\partial t} = \nabla \cdot (D_{c_c i} \nabla c_c) + \varpi_i(\phi, c_c) \quad \text{in } \Omega \times [0, t_f]. \tag{5.8}$$

These governing equations are subjected to the following initial and boundary conditions:

$$\phi = \phi_D \quad \text{on } \partial\Omega_D \times [0, t_f], \tag{5.9}$$

$$c_s = c_{s_D} \quad \text{on } \partial\Omega_D \times [0, t_f], \tag{5.10}$$

$$c_c = c_{c_D} \quad \text{on } \partial\Omega_D \times [0, t_f], \tag{5.11}$$

$$\nabla c_s \cdot \mathbf{n} = q_{c_s} \quad \text{on } \partial\Omega_N \times [0, t_f], \tag{5.12}$$

$$\nabla c_c \cdot \mathbf{n} = q_{c_c} \quad \text{on } \partial\Omega_N \times [0, t_f], \tag{5.13}$$

$$\phi = \phi_0 \quad \text{in } \Omega(t = 0), \tag{5.14}$$

$$c_s = c_{s_0} \quad \text{in } \Omega(t = 0), \tag{5.15}$$

$$c_c = c_{c_0} \quad \text{in } \Omega(t = 0). \tag{5.16}$$

The Dirichlet and Neumann boundary conditions are denoted by $\partial\Omega_D$ and $\partial\Omega_N$, respectively. The boundaries representing the sidewalls of the Petri dish are considered to be impenetrable, therefore $\phi_D = 0$ for the simulations presented in this work. Additionally, to represent the slime mold growth being constrained to the Petri dish, $c_{s_D} = c_{c_D} = q_{c_s} = q_{c_s} = 0$ due to no-nutrient generating boundaries and no nutrient exchange with the surroundings. Table 5.1 summarizes the main parameters of the model and their corresponding values used in the simulations.

Within this framework, the slime mold's network dynamics are described using a Poisson equation that relates pressure differences across the network to the conductivities and lengths of the tubes [3]. This equation captures how the tubes' conductivities change over time based on the flux (flow rate) through them, affecting the pressure distribution across the network. For our project, the specific equations from the continuum modeling of slime mold dynamics will be adapted and utilized to simulate the behavior of the mold in our designed maze-like structure. To generate data for training and evaluating machine learning models, we represent the transport network constitutive morphology, inspired by the slime mold, as a 2D composite mesh. This

Table 5.1 Geometrical data and material properties

Property	Description	Value	Unit
ϕ	Slime mold marker	$0-1$	—
c_s	Substrate nutrient	$0-1$	none
c_c	Concentrated nutrient	$0-1$	—
c^*	Nutrient threshold	0.4	—
l	Domain size (diameter or length), 2D	100	mm
L	Domain size (diameter × height), 3D	100×20	mm
$\kappa_1^{c_s}$	Consumption rate, substrate nutrient	0.5, 1.0	h^{-1}
$\kappa_2^{c_s}$	Mass creation coefficient	0.7, 0.8, 2.0	—
κ^{c_c}	Consumption coefficient, concentrated nutrient	0.05	None
$\lambda_{\mathcal{G}}$	Growth mobility	$3.33 \cdot 10^3$	mm·h
D_{c_s}	Substrate nutrient diffusivity	1.0	$mm^2 \cdot h^{-1}$
$\epsilon_{\mathcal{G}}$	Growth regularization parameter	0.01	mm

representation is achieved using the open-source Gmsh package [32]. Within this mesh, distinct regions are demarcated to exhibit heterogeneous mechanical properties (elastic properties, conductivities), symbolizing the plasmodium microstructure, as shown in Fig. 5.2. The mechanical response of these composite meshes is then simulated using the MOOSE Frameworks, an open-source, parallel finite element framework [33]. By discretizing the poroelastic flow equations over the mesh geometry, we can solve for the resulting pressure and velocity fields in a realistic, complex constitutive behavior. The data generated from these simulations serves as the foundation for training a deep neural network model using TensorFlow in Python [34]. This data becomes the input to the neural network, while the outputs are the predicted displacements, which induce pressure and velocity fields. As the network undergoes training on a plethora of simulations, it becomes adept at swiftly predicting the physics of protoplasmic flux based on the given mesh topology.

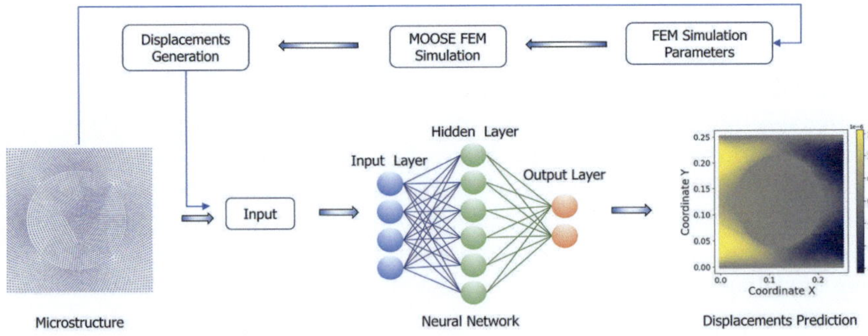

Fig. 5.2 Schematic of feedforward neural network and MOOSE FEM simulation

5.3 Numerical Examples: Model Testing and Validation

In this work, we generated composite mesh structures representing simplified transport network morphologies inspired by the slime mold. An example structure was shown in Fig. 5.1, containing an inclusion region with distinct material properties mimicking the vein-like tubes. The remarkable adaptation abilities of slime molds like *P. polycephalum* emerge from the intricate microstructural changes in their cytoplasm and tubular networks.

Capturing this complex mechanical response is key to elucidating the biophysical mechanisms behind their emergent intelligence. However, directly simulating the dynamics at the microstructural level over large iterative spaces is computationally prohibitive. To address this challenge, we leverage machine learning to develop surrogate models that accurately approximate the material's complex behavior under various conditions. This machine learning framework bypasses expensive simulations to enable extensive topology optimization driven by data-science approximations of the intricate multiscale phenomena. The integration of biophysics, composite material modeling, and machine learning provides both novel biological insights and tools to engineer smart, resilient systems.

We represent the microstructure as a composite mesh containing distinct domains, as inspired by biological cells. Physics-based simulations on these meshes generate datasets reflecting the rich nonlinear dynamics of protoplasmic flux and network remodeling. Protoplasmic flux through these composite meshes was simulated using the finite element module in the MOOSE framework. The poroelastic flow equations were solved to obtain pressure and velocity fields reflecting the composite transport physics. The inclusion region with higher conductivity exhibited greater pressure gradients and flux compared to the matrix, as expected. The simulation data was used to train a fully connected deep neural network model in TensorFlow to predict the constitutive response of the slime mold's microstructure. The matrix and inclusion regions were assigned Young's moduli of 210 GPa and 9 GPa and Poisson's ratios of 0.22 and 0.33, respectively. This composite mesh was discretized using 4,449 nodes for simulation. Uniaxial tensile displacement of 10^{-6} was applied in the y-direction.

As depicted earlier in the schemartic of Fig. 5.2, the neural network consists of one input layer, one hidden layer, and an output layer. The model takes the mesh structure nodes as input data for training and displacements as target data in order to predict the displacements of each node. After training on numerous mesh simulations, the model achieved high accuracy in rapidly inferring the composite transport physics. Figure 5.3 illustrates the strong agreement between the simulated and predicted pressure fields for a sample test case. To validate its predictive capability, the trained model was evaluated on a distinct composite mesh containing an inclusion of diameter 0.1. As shown in Fig. 5.3, the neural network accurately predicted the nodal displacements compared to those obtained from direct finite element simulation. The loss tends to decrease as the model goes under training which can be shown from the Loss versus Epoch plot in Fig. 5.3a. Also, our model achieved average mean

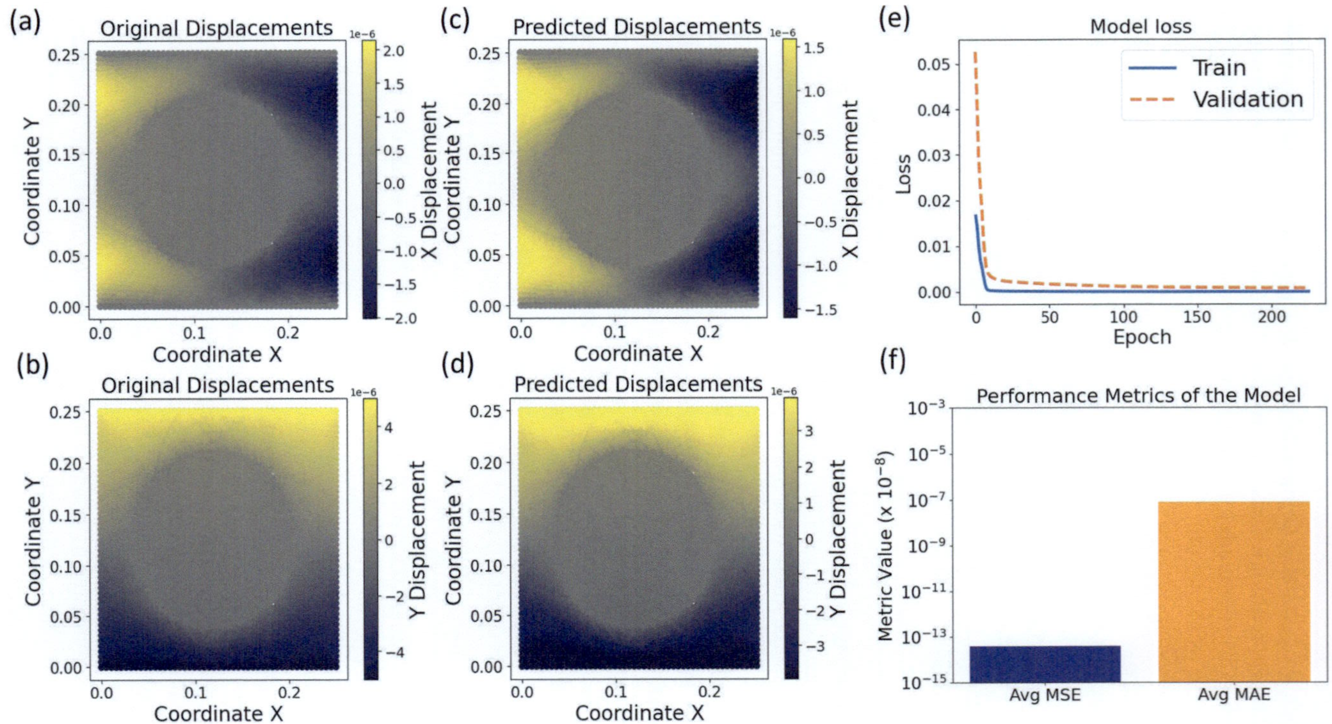

Fig. 5.3 Results of machine learning prediction model. **a–d** Original and predicted displacements of nodes. **e** Train versus validation model loss with epochs. **f** Average mean squared errors and average mean absolute errors of the prediction model

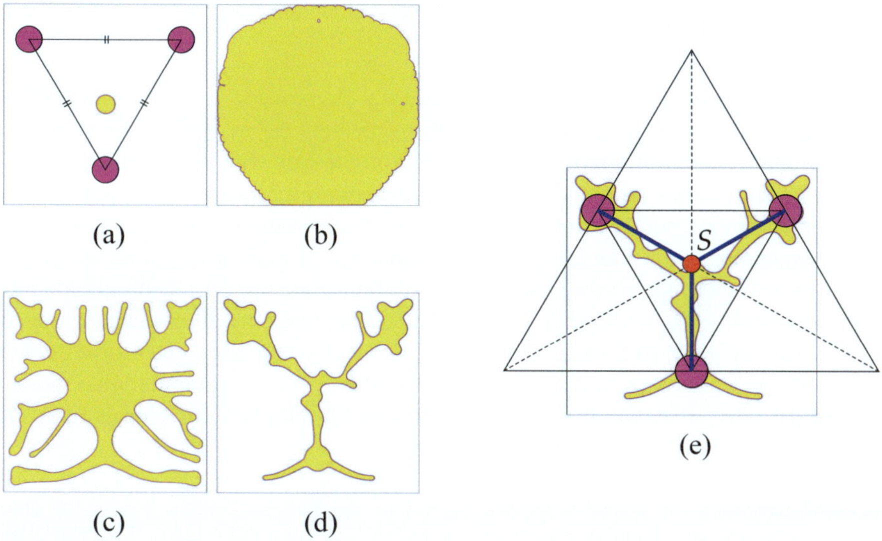

Fig. 5.4 Numerical results for a three-point optimization problem. **a** The nutrient sources, indicated by pink circles, are placed on the vertices of an equilateral triangle. The initial slime mold culture is placed at the center of the equilateral triangle. **b–d** Display the growth of the slime mold and the process of optimizing the branching. **e** Numerical results overlapped with the geometry of the Steiner point, marked by *S* and solid blue lines [12]

absolute and mean squared errors of 7.78 $\times 10^{-8}$ and 3.95 $\times 10^{-14}$, respectively, demonstrating reliable generalization, Fig. 5.3b.

This machine learning framework establishes a foundation to leverage data-driven approximations of complex biomimetic transport physics for accelerated discovery of optimal network designs. Ongoing work is focused on encoding the meshes into a compact latent space to enable differentiable topology optimization through the trained neural network model. As an example and benchmark of the results that can be achieved we present here the triangle of Steiner, a three-point network optimization problem. In this example, three nutrient sources are placed at the corners of an equilateral triangle. A slime mold culture is then placed at the center of the triangle. The results of this simulation can be seen in Fig. 5.4. In the simulation, the slime mold grows to the nutrient sources during the foraging phase but begins to reconfigure to optimize the transport network between the three nutrient sources. The final results of this simulation agree nicely with the solution to the Steiner Tree problem.

5.4 Conclusions

This work demonstrates a machine learning framework to rapidly optimize biomimetic transport network designs. By representing networks as composite meshes and generating simulation data of protoplasmic flux, we train models to approximate complex biophysics. The trained networks then enable iterative topology improvements purely through fast inference. Encoding the meshes accelerates this optimization loop. The integrated methodology synergistically leverages composite mesh modeling, physics simulations, and machine learning to illuminate bio-inspired pathways for intelligent network adaptation. Ongoing efforts involve applying the framework to additional transport problems and exploring generative models to directly propose optimized designs. Overall, this fusion of computational modeling and data-driven techniques provides both novel biophysical insights and tools to engineer smart, resilient systems.

Acknowledgements We acknowledge the support of the National Science Foundation grant 2339373. Any opinions, findings, conclusions, or recommendations expressed in this material are those of the authors and do not necessarily reflect the views of the National Science Foundation. Computations for this research were performed on the Pennsylvania State University's Institute for Computational and Data Sciences' Roar supercomputer.

References

1. Clauset A, Moore C, Newman MEJ (2008) Hierarchical structure and the prediction of missing links in networks. Nature 453(7191):98–101
2. Nakagaki T, Yamada H, Tóth Á (2000) Maze-solving by an amoeboid organism. Nature 407(6803):470–470
3. Tero A, Kobayashi R, Nakagaki T (2007) A mathematical model for adaptive transport network in path finding by true slime mold. J Theor Biol 244(4):553–564
4. Adamatzky A (2010) Physarum machines: computers from slime mould, vol 74. World Scientific
5. Jones J (2010) Characteristics of pattern formation and evolution in approximations of physarum transport networks. Artif Life 16(2):127–153
6. Raissi M, Perdikaris P, Karniadakis GE (2019) Physics-informed neural networks: a deep learning framework for solving forward and inverse problems involving nonlinear partial differential equations. J Comput Phys 378:686–707
7. George EK, Kevrekidis IG, Lu L, Perdikaris P, Wang S, Yang L (2021) Phys-Inf Mach Learn Nat Rev Phys 3(6):422–440
8. Adamatzky A (2012) Slime mold solves maze in one pass, assisted by gradient of chemo-attractants. IEEE Trans Nanobiosci 11(2):131–134
9. Kamiya N (1950) The rate of the protoplasmic flow in the myxomycete plasmodium. i. Cytologia 15(3-4):183–193
10. Alim K, Andrew N, Pringle A, Brenner MP (2017) Mechanism of signal propagation in physarum polycephalum. Proc Natl Acad Sci 114(20):5136–5141
11. Ghanbari F, Costanzo F, Hughes DP, Peco C (2020) Phase-field modeling of constrained interactive fungal networks. J Mech Phys Solids 145:104160

12. Ghanbari F, Sgarrella J, Peco C (2023) Emergent dynamics in slime mold networks. J Mech Phys Solids 179:105387
13. Ghanbari F, Sgarrella J, Peco C (2023) Coding soft matter with bionetworks-inspired emergent principles. Bioinspiration Biomimetics and Bioreplication XIII 12481:5–14
14. Ghanbari F, Peco C (2023) Eulerian finite-strain elasticity with phase-field and the reference map technique. Trans Jpn Soc Comput Eng Sci 27:8–13
15. Shimada T, Nishiguchi K, Peco C, Okazawa S, Tsubokura M (2021) Eulerian formulation using lagrangian marker particles with reference map technique for fluid-structure interaction problem. In: 9th edition of the international conference on computational methods for coupled problems in science and engineering, pp 1–7
16. Shimada T, Nishiguchi K, Peco C, Okazawa S, Tsubokura M (2022) Eulerian unified formulation for fluid-structure interaction problems using marker particles with Reference map
17. Rosolen A, Peco C, Arroyo M (2013) An adaptive meshfree method for phase-field models of biomembranes: Part I: Approximation with maximum-entropy basis functions. J Comput Phys 249:303–319
18. Peco C, Rosolen A, Arroyo M (2013) An adaptive meshfree method for phase-field models of biomembranes. Part II: A Lagrangian approach for membranes in viscous fluids. J Comput Phys 249:320–336
19. Peco C, Chen W, Liu Y, Bandi MM, Dolbow JE, Fried E (2017) Influence of surface tension in the surfactant-driven fracture of closely-packed particulate monolayers. Soft Matter 13(35):5832–5841
20. Peco C, Liu Y, Rhea C, Dolbow JE (2019) Models and simulations of surfactant-driven fracture in particle rafts. Int J Solids Struct 156–157:194–209
21. Liu Y, Peco C, Dolbow J (2019) A fully coupled mixed finite element method for surfactants spreading on thin liquid films. Comput Methods Appl Mech Eng 345:429–453
22. Spencer BW, Jiang W, Dolbow JE, Peco C (2016) Pellet cladding mechanical interaction modeling using the extended finite element method. In: Top Fuel 2016: LWR fuels with enhanced safety and performance, pp 929–938
23. Guével A, Meng Y, Peco C, Juanes R, Dolbow JE (2023) A Darcy–Cahn–Hilliard model of multiphase fluid-driven fracture. J Mech Phys Solids 181:105427
24. Gaston D, Newman C, Hansen G, Lebrun-Grandié D (2009) MOOSE: a parallel computational framework for coupled systems of nonlinear equations. Nucl Eng Des 239(10):1768–1778
25. Sgarrella J, Ghanbari F, Peco C (2023) I-STL2MOOSE: From STL data to integrated volumetrical meshes for MOOSE. SoftwareX 21:101273
26. Tero A, Takagi S, Saigusa T, Ito K, Bebber DP, Fricker MD, Yumiki K, Kobayashi R, Nakagaki T (2010) Rules for biologically inspired adaptive network design. Science 327(5964):439–442
27. Whiting JGH, Jones J, Bull L, Levin M, Adamatzky A (2016) Towards a physarum learning chip. Sci Rep 6(1):19948
28. Jones J (2015) The emergence and dynamical evolution of complex transport networks from simple low-level behaviours. arXiv:1503.06579
29. Ghanbari F, Rodriguez EG, Millán D, Simonetti F, Argüelles AP, Peco C (2023) Modeling of wave propagation in polycrystalline ice with hierarchical density gradients. Finite Elem Anal Des 217:103916
30. Rodriguez EG, Millán D, Peco C (2023) Representative Volume Element (RVE) analysis for mechanical characterization of ice with metallic's inclusion of micro/nano particles. In: XXXIX Congreso Argentino de Mecánica Computacional y I Congreso Argentino Uruguayo de Mecánica Computacional, vol 40, p 506
31. Rodriguez EG, Peco C, Millán D (2023) The influence of material properties distribution of waves in 1d: Application to cryoultrasonics. Mecánica Computacional 39(7):217–217
32. Geuzaine C, Remacle J-F (2009) Gmsh: A 3-d finite element mesh generator with built-in pre-and post-processing facilities. Int J Numer Meth Eng 79(11):1309–1331
33. ...Lindsay AD, Gaston DR, Permann CJ, Miller JM, Andrš D, Slaughter AE, Kong F, Hansel J, Carlsen RW, Icenhour C, Harbour L, Giudicelli GL, Stogner RH, German P, Badger J, Biswas S, Chapuis L, Green C, Hales J, Hu T, Jiang W, Jung YS, Matthews C, Miao Y, Novak A,

Peterson JW, Prince ZM, Rovinelli A, Schunert S, Schwen D, Spencer BW, Veeraraghavan S, Recuero A, Yushu D, Wang Y, Wilkins A, Wong C (2022) 2.0 - MOOSE: Enabling massively parallel multiphysics simulation. SoftwareX 20:101202

34. Abadi M, Barham P, Chen J, Chen Z, Davis A, Dean J, Devin M, Ghemawat S, Irving G, Isard M, et al (2016) {TensorFlow}: a system for {Large-Scale} machine learning. In: 12th USENIX symposium on operating systems design and implementation (OSDI 16), pp 265–283

Chapter 6
Multibundle Method for Constrained Nonsmooth Multiobjective DC Optimization

Kaisa Joki and Outi Montonen

Abstract In this paper, we propose a multibundle method. The new method is of the descent type and it is designed for constrained nonsmooth multiobjective optimization problems whose objectives and constraints can be represented as a difference of convex (DC) functions. The method combines pieces from the multiple subgradient descent bundle method [49], the double bundle method [26], and the multiobjective double bundle method [47]. The idea is to find descent directions for every individual objective by using a single-objective bundle method designed for DC functions, and then, form a common descent direction. The novelty in our approach is that the decision maker has an option to steer the solution process by indicating whether an individual descent direction should be used to improve one objective or should the method test the optimality. Furthermore, we present two alternative constraint handling strategies. The multibundle method is proven to have a finite convergence to an approximate weakly Pareto stationary solution under mild assumptions. Finally, the new method is compared with a multiobjective DC method as well as a multiobjective nonconvex method to demonstrate the numerical capability of the proposed method. In addition, the constraint handling strategies are compared and some interactive examples are given.

Keywords Multiobjective DC optimization · Bundle methods · Descent direction algorithms · Interactive multiobjective optimization · Constraint handling via improvement functions

6.1 Introduction

In many practical applications, the optimization problem under consideration involves several objectives to take into account. These problems are called multiobjective optimization problems and some examples of areas where these problems arise are

K. Joki (✉) · O. Montonen
Department of Mathematics and Statistics, University of Turku, Turku, Finland
e-mail: kjjoki@utu.fi

T. Tuovinen et al. (eds.), *Challenges in Design Methods, Numerical Tools and Technologies for Sustainable Aviation, Transport and Industry*, Computational Methods in Applied Sciences 17, https://doi.org/10.1007/978-3-031-98675-8_6

machine learning [24], economics [53], chemical engineering [58], cancer treatment planning [13, 21], and final disposal of the spent nuclear fuel [46, 50]. In addition, many of these problems involve nonsmoothness causing a possible failure of classical gradient-based methods since functions do not need to be continuously differentiable (or smooth).

A wide subclass of nonsmooth nonconvex functions is formed by DC functions which are functions that can be represented as a difference of two convex functions. The benefit of this class is that the convex analysis can be applied to some extent. However, the DC decomposition of a DC function is not unique and it might be hard to find even one decomposition. Luckily, many practical applications are naturally in the explicit DC form such as clustering [5, 27, 30], classification [2, 17], energy management [1], wireless sensor network planning [3], and data visualization [10]. Thus, we are motivated to consider multiobjective DC optimization.

Some general strategies to solve a DC problem in the single-objective case are DCA utilizing duality [33, 34, 54], proximal point based methods [60], bundle methods with a special cutting plane model [16, 25, 26, 28, 52], and branch-and-bound and outer approximation algorithms [19]. Many of these methods find only critical solutions which are obtained by investigating whether the zero vector belong to the difference of the DC components' subdifferentials. However, in general, nonconvex optimization methods generate stationary solutions with the zero vector in the subdifferential of the objective being a stronger stopping condition. The relation between critical and stationary solutions is exemplified in [26] in the single-objective case and in [47, 48] in the multiobjective case. To overcome this theoretical drawback, the double bundle method (DBDC) guarantees stationarity by employing a special escape procedure proposed in [26, 28]. Considering multiobjective DC optimization, the variation of methods is more narrow. Some recent methods include the proximal point methods in [11, 22, 23] and the multiobjective double bundle method (MDBDC) [47, 48]. Like its single-objective counterpart DBDC, MDBDC finds stationary solutions. Furthermore, in [11] together with the proximal point method, an application to the group dynamic problem is demonstrated.

The focus of this paper is on descent methods improving all the objectives at every iteration. Recently, different descent methods have been presented for nonsmooth multiobjective problems in the convex [6, 7, 31, 49, 55] and nonconvex cases [20, 39, 40, 44, 57, 62]. Usually, descent methods do not involve an interaction with a decision maker (DM) and this makes them relatively fast solvers yielding one solution amongst (weakly) Pareto optimal solutions. A common feature of descent methods is that they are sensitive for the selection of the starting point. Due to the descent property, these methods project the starting point to the Pareto set such that a solution lies in the negative orthant from the starting point in the objective space. Thus, we may obtain an approximation of the Pareto front by performing the method repeatedly from different starting points. Descent methods can also be used as a part of some interactive method like in [44, 50, 51] or projecting the nonoptimal solution to the Pareto front [8].

Here we propose a new multibundle method (MB-DC) for constrained nonsmooth multiobjective DC optimization problems combining the descent idea and interactivity. The method employs the similar idea to detect descent directions than the multiple subgradient descent bundle method (MSGDB) for unconstrained convex multiobjective optimization problems from [49]. MSGDB again bases on the multiple gradient descent algorithm (MGDA) presented in [14, 15]. The idea in brief is to form a common descent direction improving all the objectives simultaneously from the descent directions of the individual objectives. This idea has been applied also in the nonsmooth stochastic [55] and nonconvex inexact [20] cases. Since in MSGDB only convex optimization problems are considered we carry on the idea to DC optimization by utilizing DBDC to detect individual search directions. In order to ensure a Pareto stationary solution, the escape procedure [26, 28] is applied as in [47].

Descent methods as no-preference methods are usually relatively fast. However, if the DM has any expectations towards the final solution, they probably will not be met if a no-preference method is used. The interactive methods are presented to face this challenge (see e.g. [9, 42–44, 56, 61, 63]). In order to increase the contentment of the DM, in this paper we introduce an interactive approach to our descent method. In this approach, the DM can express if the negative orthant given by the current iteration point is not satisfactory and select which objective should be improved. Thus, the DM can react to the unpleasant starting point before the solution process is ended up to the Pareto front as well as to the Pareto optimal solution. The communication is carried over by offering individual descent directions obtained during the direction finding procedure as an alternative for the escape procedure if a common descent direction is not found. Even if the DM can take a part in the solution process, MB-DC works perfectly fine as a no-preference method.

Another new feature compared with MSGDB is that MB-DC is also capable to handle constraints. In particular, two alternative constraint handling strategies are introduced and both of them use some aid from the improvement function. The improvement function is known as a tool to handle constraints both in single-objective [29, 32, 35, 41, 45, 59] and multiobjective optimization [31, 37, 39, 40, 44, 47, 62]. In the multiobjective case, it can be used as a technique to handle not only constraints but also several objectives simultaneously. A nice feature of the improvement function is that the descent direction of the improvement function is descent for all individual objectives. Both constraint handling strategies proposed in this paper check the feasibility during the line search. If the direction is infeasible, we check whether the current solution is Pareto stationary by using the escape procedure with the improvement function handling the constraints. In addition to this, the second strategy uses modified objectives which are formed as an improvement function containing an individual objective and constraints of the problem. Thus, the individual descent directions are feasible which is not guaranteed in the first strategy.

The outline of this paper is following. In Sect. 6.2, the main concepts of nonsmooth multiobjective optimization are given. The core ingredients of MB-DC are discussed, the algorithm is described, and the convergence analysis is provided in Sect. 6.3. The numerical experiments are considered in Sect. 6.4. There the efficiency of the method is investigated by comparing it with other methods, the alternative strategies for the constraint handling are contrasted, and examples of the interactive case are given. Finally, the concluding remarks are given in Sect. 6.5.

6.2 Preliminaries

We consider a *constrained nonsmooth multiobjective DC optimization problem* of the form

$$\min_{\boldsymbol{x}\in X} \quad f_1(\boldsymbol{x}), \ldots, f_k(\boldsymbol{x}) \tag{6.1}$$

where $X = \{\boldsymbol{x} \in \mathbb{R}^n \mid g_c(\boldsymbol{x}) \leq 0,\ c \in C\}$ and C is a set of constraints. In addition, the objectives $f_i : \mathbb{R}^n \to \mathbb{R}$, for all $i \in I = \{1, \ldots, k\}$ and the constraints $g_c : \mathbb{R}^n \to \mathbb{R}$ for all $c \in C$ are assumed to be *DC functions*. This means that they can be decomposed as a difference of two convex functions. The *DC decomposition* of f is $f = f^1 - f^2$, where $f^1 : \mathbb{R}^n \to \mathbb{R}$ and $f^2 : \mathbb{R}^n \to \mathbb{R}$. The functions f^1 and f^2 are convex and they are called *DC components*. DC functions are known to be *locally Lipschitz continuous* at $\boldsymbol{x} \in \mathbb{R}^n$ [18]. In other words, for a DC function f there exists a *Lipschitz constant* $K > 0$ and $\varepsilon > 0$ such that $|f(\boldsymbol{y}) - f(\boldsymbol{z})| \leq K\|\boldsymbol{y} - \boldsymbol{z}\|$ for all $\boldsymbol{y}, \boldsymbol{z} \in B(\boldsymbol{x}; \varepsilon)$, where $B(\boldsymbol{x}; \varepsilon)$ is an open ball with a center $\boldsymbol{x}$ and a radius ε.

A DC function $f : \mathbb{R}^n \to \mathbb{R}$ is *directionally differentiable* meaning that at a point $\boldsymbol{x} \in \mathbb{R}^n$ in every direction $\boldsymbol{d} \in \mathbb{R}^n$ there exists the *directional derivative* [18]

$$f'(\boldsymbol{x}; \boldsymbol{d}) = \lim_{t\downarrow 0} \frac{f(\boldsymbol{x} + t\boldsymbol{d}) - f(\boldsymbol{x})}{t}.$$

For a convex function, this can be written as well in the form (see e.g. [4])

$$f'(\boldsymbol{x}; \boldsymbol{d}) = \max\{\boldsymbol{\xi}^T\boldsymbol{d} \mid \boldsymbol{\xi} \in \partial_c f(\boldsymbol{x})\},$$

where $\boldsymbol{\xi}$ is a *subgradient* of a convex function f at the point $\boldsymbol{x}$. The subgradient is an element of the set called *subdifferential*

$$\partial_c f(\boldsymbol{x}) = \{\boldsymbol{\xi} \in \mathbb{R}^n \mid f(\boldsymbol{y}) \geq f(\boldsymbol{x}) + \boldsymbol{\xi}^T(\boldsymbol{y} - \boldsymbol{x}) \text{ for all } \boldsymbol{y} \in \mathbb{R}^n\}$$

generalizing the gradient in the nonsmooth convex case. The subdifferential is a nonempty, convex, and compact set. In the nonconvex case, the *generalized subdifferential* of a locally Lipschitz continuous function f at a point $\boldsymbol{x} \in \mathbb{R}^n$ [12] is

$$\partial f(\boldsymbol{x}) = \operatorname{conv}\left\{\lim_{i\to\infty} \nabla f(\boldsymbol{x}_i) \mid \boldsymbol{x}_i \to \boldsymbol{x} \text{ and } \nabla f(\boldsymbol{x}_i) \text{ exists}\right\},$$

where conv denotes the convex hull. If a function f is convex, then $\partial_c f(\boldsymbol{x})$ coincides with $\partial f(\boldsymbol{x})$, and thus, the subscript c is omitted. In some cases, it is beneficial to approximate the generalized subdifferential with the *Goldstein ε-subdifferential* [41]

$$\partial_\varepsilon^G f(\boldsymbol{x}) = \text{cl conv}\,\{\partial f(\boldsymbol{y}) \mid \boldsymbol{y} \in B(\boldsymbol{x}; \varepsilon)\},$$

where cl denotes the closure of the set and $\varepsilon \geq 0$. The smaller the parameter ε is, the better is the approximation for the generalized subdifferential.

The solution $\boldsymbol{x}^* \in X$ of the multiobjective problem (6.1) is *globally Pareto optimal* if there does not exist any $\boldsymbol{x} \in X$ such that $f_i(\boldsymbol{x}) \leq f_i(\boldsymbol{x}^*)$ for all $i \in I$ and $f_j(\boldsymbol{x}) < f_j(\boldsymbol{x}^*)$ for at least one $j \in I$. If there does not exist a solution $\boldsymbol{x} \in X$ such that $f_i(\boldsymbol{x}) < f_i(\boldsymbol{x}^*)$ for all $i \in I$, the solution $\boldsymbol{x}^*$ is *globally weakly Pareto optimal*. The solution $\boldsymbol{x}^*$ is *a locally* (weakly) *Pareto optimal solution* if $\boldsymbol{x}^*$ is a globally (weakly) Pareto optimal on $B(\boldsymbol{x}^*; \delta) \cap X$, where $\delta > 0$.

Let $K_S(\boldsymbol{x})$ be a contingent cone at the point $\boldsymbol{x} \in S$ and $S^{\leq}$ be a polar cone of the set S (see e.g. [4]). Next, we can state the following necessary optimality condition for local weak Pareto optimality.

Theorem 1 ([38]) *If $\boldsymbol{x}^* \in X$ is a local weakly Pareto optimal solution for the problem* (6.1) *and the constraint qualification*

$$\Big(\bigcup_{c \in C(\boldsymbol{x}^*)} \partial g_c(\boldsymbol{x}^*)\Big)^{\leq} \subseteq K_X(\boldsymbol{x}^*)$$

is valid, where $C(\boldsymbol{x}^*) = \{c \in C \mid g_c(\boldsymbol{x}^*) = 0\}$, then

$$\boldsymbol{0} \in \text{conv} \bigcup_{i \in I} \partial f_i(\boldsymbol{x}^*) + \text{cl cone} \bigcup_{c \in C(\boldsymbol{x}^*)} \partial g_c(\boldsymbol{x}^*). \tag{6.2}$$

The point $\boldsymbol{x}^*$ satisfying the condition (6.2) is *weakly Pareto stationary*. Similarly, for a locally Lipschitz continuous function f a point $\boldsymbol{x}^*$ satisfying the well-known necessary local optimality condition $\boldsymbol{0} \in \partial f(\boldsymbol{x}^*)$ is called *Clarke stationary*. Furthermore, a point $\boldsymbol{x}^*$ is an *approximate Clarke stationary solution* if $\|\boldsymbol{\xi}^*\| \leq \delta$, where $\boldsymbol{\xi}^* \in \partial_\varepsilon^G f(\boldsymbol{x}^*)$ and $\varepsilon, \delta > 0$.

A useful tool for multiobjective and constrained optimization is an *improvement function*. Given the problem (6.1), the improvement function is defined by

$$H(\boldsymbol{x}, \boldsymbol{y}) = \max\{f_i(\boldsymbol{x}) - f_i(\boldsymbol{y}),\ g_c(\boldsymbol{x}) \mid i \in I,\ c \in C\}. \tag{6.3}$$

Since the objectives f_i and constraints g_c for all $i \in I$ and $c \in C$ are DC functions, the improvement function is a DC function [47]. The usefulness of the improvement function is validated with the following theorem giving a link between the minimum of the improvement function and a global weakly Pareto optimal solution.

Theorem 2 ([39, 47, 62]) *If the solution $\boldsymbol{x}^* \in X$ is a globally weakly Pareto optimal solution of the problem* (6.1)*, then*

$$\boldsymbol{x}^* = \text{argmin}_{\boldsymbol{x} \in X}\ H(\boldsymbol{x}, \boldsymbol{x}^*).$$

Furthermore, if $\boldsymbol{0} \in \partial H(\boldsymbol{x}^*, \boldsymbol{x}^*)$, then the solution $\boldsymbol{x}^*$ is weakly Pareto stationary.

6.3 Multibundle Method

In this section, we propose the new multibundle method (MB-DC) for constrained multiobjective DC optimization. The basic idea is to combine the outline of the multiple subgradient descent bundle method (MSGDB) for unconstrained convex multiobjective optimization [49] with the single-objective double bundle method for DC optimization (DBDC) [26, 28]. This way we can benefit from a DC cutting plane model [25] used in DBDC capturing the convex and concave behavior of a DC function. In order to ensure the weak Pareto stationarity of the solution, the escape procedure presented in [26, 28] is utilized for the improvement function as in [47]. Compared with MSGDB, the new method can address DC functions, handle constraints and take advantage of the input gained from the DM by way of the individual descent directions. In Sect. 6.3.1, the core ingredients of the method are discussed following with Sect. 6.3.2 presenting the algorithm of MB-DC. The convergence is proved in Sect. 6.3.3.

6.3.1 *Core Ingredients of the Method*

As said, MB-DC uses a similar direction finding procedure as MSGDB which in its turn generalizes the ideas of the multiple gradient descent algorithm (MGDA) [14, 15] for nonsmooth convex unconstrained multiobjective problems. The fundamental idea in this procedure is to define the minimum norm element of descent directions of individual objectives. A direction $\boldsymbol{d} \in \mathbb{R}^n$ is a *descent direction* for a function f_i if at a point $\boldsymbol{x} \in \mathbb{R}^n$ there exists $\varepsilon > 0$ such that $f_i(\boldsymbol{x} + t\boldsymbol{d}) < f_i(\boldsymbol{x})$ for all $t \in (0, \varepsilon]$. If a direction $\boldsymbol{d}$ is descent for f_i at the point $\boldsymbol{x}$, then

$$\boldsymbol{\xi}_i^T \boldsymbol{d} < 0 \quad \text{for all } \boldsymbol{\xi}_i \in \partial f_i(\boldsymbol{x}).$$

By defining the minimum norm element of the individual descent directions, we aim to obtain a *common descent direction* meaning a direction which is descent for all the objectives simultaneously. In the smooth case, a negative gradient gives a descent direction of an individual objective. Unfortunately, in the nonsmooth case the similar result does not hold for an arbitrary subgradient. Due to this, in MSGDB and in MB-DC, a descent direction of an individual objective is defined by utilizing a bundle approach. The same phenomenon occurs when the common descent direction is under the scope: in the smooth case, the common descent direction can be ensured, while in the nonsmooth case, we have to be content with a candidate for the common descent direction. However, similar to Theorem 3.1 in [49] for convex functions, we can prove the following theorem for locally Lipschitz continuous functions. This motivates to use the negative direction of the minimum norm element as a candidate for the common descent direction. Here we adopt the motivation from an unconstrained multiobjective problem. Indeed, in MB-DC the direction finding is done by using unconstrained problems and the feasibility of a direction is not considered until the line search.

Theorem 3 *Consider the problem* $\min_{\boldsymbol{x}\in\mathbb{R}^n}\{f_i(\boldsymbol{x}) \mid i \in I\}$. *Let* $f_i : \mathbb{R}^n \to \mathbb{R}$ *be a locally Lipschitz continuous function for all* $i \in I$ *and let* $\boldsymbol{p}^* = \operatorname{argmin} \|\boldsymbol{p}\|$, *where* $\boldsymbol{p} \in conv \bigcup_{i\in I} \partial f_i(\boldsymbol{x})$ *and* $\boldsymbol{d}^* = -\boldsymbol{p}^*$. *Either we have*

(i) $\boldsymbol{d}^* = \boldsymbol{0}$ *and the point* $\boldsymbol{x}$ *is weakly Pareto stationary, or*
(ii) $\boldsymbol{d}^* \neq \boldsymbol{0}$ *and the vector* $\boldsymbol{d}^*$ *is a common descent direction for every objective function at the point* $\boldsymbol{x}$.

Proof We divide the proof of the theorem in two cases:

(i) From the fact that $\boldsymbol{d}^* = \boldsymbol{0}$ it follows that $\boldsymbol{0} \in \operatorname{conv} \bigcup_{i\in I} \partial f_i(\boldsymbol{x})$. Thus, by Theorem 1 the point $\boldsymbol{x}$ is a weakly Pareto stationary solution.
(ii) Now $\boldsymbol{d}^* \neq \boldsymbol{0}$ and $\boldsymbol{0} \notin \operatorname{conv} \bigcup_{i\in I} \partial f_i(\boldsymbol{x})$. Then $\boldsymbol{0} \notin \partial f_i(\boldsymbol{x})$ for any $i \in I$. Recall that the subdifferential is a convex and compact set. Then by Lemma 2 in [38] the set $\operatorname{conv} \bigcup_{i\in I} \partial f_i(\boldsymbol{x})$ is compact as well. Therefore $\boldsymbol{p}^T\boldsymbol{p}^* \geq \|\boldsymbol{p}^*\|^2$ for all $\boldsymbol{p} \in \operatorname{conv} \bigcup_{i\in I} \partial f_i(\boldsymbol{x})$ by Lemma 5.2.6 in [41]. This implies that by taking an arbitrary subgradient $\boldsymbol{\xi}_i \in \partial f_i(\boldsymbol{x}) \subset \operatorname{conv} \bigcup_{i\in I} \partial f_i(\boldsymbol{x})$ we obtain

$$\boldsymbol{\xi}_i^T\boldsymbol{d}^* = -\boldsymbol{\xi}_i^T\boldsymbol{p}^* \leq -\|\boldsymbol{p}^*\|^2 < 0$$

and $\boldsymbol{d}^*$ is a descent direction for f_i for all $i \in I$. Thus, $\boldsymbol{d}^*$ is a common descent direction.

Thus, we either obtain a weakly Pareto stationary point or a common descent direction. □

The new method has major differences compared with MSGDB. First, MB-DC solves problems with DC objectives instead of convex objectives. Due to this, DBDC designed for DC problems is utilized. Second, MSGDB is designed for unconstrained problems whereas MB-DC can handle DC constraints. We present two different strategies for this: default one in the general outline of the method and the sophisticated one in Sect. 6.3.1.5. Another major difference is the inclusion of the interaction with the DM in Sect. 6.3.1.4. These new features together with the main properties of MB-DC are discussed in details in the following.

6.3.1.1 Individual Descent Directions

We start by discussing how the individual descent directions $\boldsymbol{d}_i \in \mathbb{R}^n$ for each objective function $f_i, i \in I$ are obtained. Here the procedure is presented for one objective $i \in I$, but in the algorithm, this is done for all $i \in I$ separately. Note that in the default constraint handling strategy, we omit constraints when individual descent directions are calculated (see Sect. 6.3.1.5 for the sophisticated strategy). In order to find an individual descent direction $\boldsymbol{d}_i \in \mathbb{R}^n$ for one objective f_i, $i \in I$, we solve the problem

$$\min_{\boldsymbol{d}_i \in \mathbb{R}^n} f_i(\boldsymbol{x}_h + \boldsymbol{d}_i), \tag{6.4}$$

where $\boldsymbol{x}_h$ is the current iteration point. Furthermore, we assume the following assumption throughout the paper:

A0: At every point $\boldsymbol{y} \in \mathbb{R}^n$, we can evaluate only the function values $f_i^l(\boldsymbol{y})$ and $g_c^l(\boldsymbol{y})$ as well as arbitrary subgradients $\boldsymbol{\xi}_i^l \in \partial f_i^l(\boldsymbol{y})$ and $\boldsymbol{\xi}_c^l \in \partial g_c^l(\boldsymbol{y})$ for all $i \in I$, $c \in C$, and $l = 1, 2$.

During the execution of the algorithm, we obtain information for the objective f_i and store it separately for both DC components in the *bundles* of the form

$$\mathcal{B}_{i,h}^l = \{(\boldsymbol{y}_{i,j}, f_i^l(\boldsymbol{y}_{i,j}), \boldsymbol{\xi}_{i,j}^l) \mid j \in J_{i,h}^l\} \quad l = 1, 2,$$

where $\boldsymbol{y}_{i,j}$ is an auxiliary point, $\boldsymbol{\xi}_{i,j}^l \in \partial f_i^l(\boldsymbol{y}_{i,j})$ is an arbitrary subgradient, and $J_{i,h}^l$ is an index set. The bundles are defined similarly to other objectives as well, but the auxiliary points may vary. However, at each iteration, all the bundles must contain the triplet associated to the current iteration point $\boldsymbol{x}_h$.

In order to solve the problem (6.4), we form a *nonconvex DC cutting plane model* for the objective function f_i according to the cutting plane model proposed in [25, 26]. This cutting plane model is of the form

$$\hat{f}_{i,h}(\boldsymbol{x}) = \hat{f}_{i,h}^1(\boldsymbol{x}) - \hat{f}_{i,h}^2(\boldsymbol{x}), \tag{6.5}$$

where

$$\hat{f}_{i,h}^l(\boldsymbol{x}) = \max_{j \in J_{i,h}^l} \{f_i^l(\boldsymbol{x}_h) + (\boldsymbol{\xi}_{i,j}^l)^T(\boldsymbol{x} - \boldsymbol{x}_h) - \alpha_i^l(\boldsymbol{x}_h, \boldsymbol{y}_{i,j})\} \quad \text{for } l = 1, 2. \tag{6.6}$$

Here $\alpha_i^l(\boldsymbol{x}_h, \boldsymbol{y}_{i,j})$ is a *linearization error* defined as

$$\alpha_i^l(\boldsymbol{x}_h, \boldsymbol{y}_{i,j}) = f_i^l(\boldsymbol{x}_h) - f_i^l(\boldsymbol{y}_{i,j}) - (\boldsymbol{\xi}_{i,j}^l)^T(\boldsymbol{x}_h - \boldsymbol{y}_{i,j})$$

for all $j \in J_{i,h}^l$ and $l = 1, 2$. Since the DC components are convex, all the linearization errors are non-negative.

The problem (6.4) can be approximated with the following nonsmooth problem:

$$\min_{\boldsymbol{d} \in \mathbb{R}} \quad P_i^h(\boldsymbol{d}) = \hat{f}_{i,h}^1(\boldsymbol{x}_h + \boldsymbol{d}) - \hat{f}_{i,h}^2(\boldsymbol{x}_h + \boldsymbol{d}) + \frac{1}{2t}\|\boldsymbol{d}\|^2, \tag{6.7}$$

where $t > 0$ is a *proximity parameter*. The term $\frac{1}{2t}\|\boldsymbol{d}\|^2$ is a *stabilizing term* and its role is to keep the approximation local enough and to guarantee the existence of the solution. As we see, P_i^h is a nonconvex DC function with the DC components $\hat{f}_{i,h}^1(\boldsymbol{x}_h + \boldsymbol{d}) + \frac{1}{2t}\|\boldsymbol{d}\|^2$ and $\hat{f}_{i,h}^2(\boldsymbol{x}_h + \boldsymbol{d})$. Since the second DC component is polyhedral convex (i.e. it is of the form $\max\{\boldsymbol{u}_k^T\boldsymbol{x} + v_k \mid k \in \mathcal{K}\}$, where $\boldsymbol{u}_k \in \mathbb{R}^n$, $v_k \in \mathbb{R}$, and $\mathcal{K}$ is finite), we can use the solution approach given in [33, 34, 54] and applied to the similar problem in [25, 26]. In the approach, we solve $|J_{i,h}^2|$ convex yet nonsmooth subproblems to ensure the global optimum of (6.7). Thus, by adjusting the size of the bundle $\mathcal{B}_{i,h}^2$, the number of subproblems solved can be controlled. The only restriction is that the element corresponding to the current iteration point must be included. As a result, each subproblem gives a solution and the solution yielding the smallest value for the problem (6.7) is selected as the final solution $\boldsymbol{d}_i^*$.

Next, we calculate a new auxiliary point $\boldsymbol{y}_i = \boldsymbol{x}_h + \boldsymbol{d}_i^*$. The direction $\boldsymbol{d}_i^*$ is selected as an individual descent direction $\boldsymbol{d}_i$ if $\boldsymbol{d}_i^*$ satisfies the condition

$$f_i(\boldsymbol{y}_i) - f_i(\boldsymbol{x}_h) \leq m_1(\hat{f}_{i,h}(\boldsymbol{y}_i) - f_i(\boldsymbol{x}_h)), \tag{6.8}$$

where $m_1 \in (0, 1)$. If the condition (6.8) is not satisfied, we perform a null step for the individual objective f_i by adding a new triplet related to the new auxiliary point into both bundles $\mathcal{B}_{i,h}^l$, $l = 1, 2$. Furthermore, if at some point we obtain $\|\boldsymbol{d}_i^*\| = 0$, the escape procedure will be performed for the objective f_i to obtain an individual search direction $\boldsymbol{d}_i$.

6.3.1.2 Common Descent Direction and Serious Step

After all individual descent directions are obtained, we test whether some of them are common descent directions as such. Based on this, we calculate a *decrease ratio*

$$\rho_i = \begin{cases} \sum_{j \in I} \frac{f_j(\boldsymbol{x}_h) - f_j(\boldsymbol{x}_h + \boldsymbol{d}_i)}{|f_j(\boldsymbol{x}_h)| + 1}, & \text{if } \boldsymbol{d}_i \text{ is a common descent direction} \\ -1, & \text{otherwise} \end{cases} \tag{6.9}$$

for each $i \in I$. If $\max\{\rho_i \mid i \in I\} > 0$, we use the direction giving the maximum decrease ratio as a common descent direction $\boldsymbol{d}_h$. In the numerical experiments in Sect. 6.4, this is usually utilized in the beginning of the solution process. If none of the individual descent directions is suitable, a candidate for a common descent direction is formed. This is done by calculating the minimum norm element of the convex hull of the individual descent directions by solving the problem

$$\begin{aligned} \min \quad & \left\| \sum_{i \in I} \lambda_i \boldsymbol{d}_i \right\|^2 \\ \text{s.t.} \quad & \sum_{i \in I} \lambda_i = 1 \\ & \lambda_i \geq 0 \text{ for all } i \in I. \end{aligned} \tag{6.10}$$

Let $(\lambda_1^*, \ldots, \lambda_k^*) \in \mathbb{R}^k$ be a solution of the problem (6.10). Then the candidate for the common descent direction is $\boldsymbol{d}_h = -\sum_{i \in I} \lambda_i^* \boldsymbol{d}_i$. Since $\boldsymbol{d}_h$ is only a candidate, we test whether this direction indeed is descent for all the objectives or not. If the descent requirement is met, we set $\boldsymbol{d}_h$ as the new common descent direction. Otherwise, we select one of two options. The first option is that we perform either a null step or the escape procedure discussed in Sect. 6.3.1.3. The second alternative is to consult the DM as described in Sect. 6.3.1.4.

Once the new common descent direction $\boldsymbol{d}_h$ is found, a serious step is performed if a suitable stepsize is found. A new iteration point is calculated by $\boldsymbol{x}_{h+1} = \boldsymbol{x}_h + \tau \boldsymbol{d}_h$, where $\tau \geq \tau_{tol}$ is a stepsize and $\tau_{tol} > 0$ is a stepsize tolerance. The stepsize is determined by finding $\tau_i \geq \tau_{tol}$ for each individual objective f_i, $i \in I$ with some line search strategy such that $\boldsymbol{x}_{h+1}$ is feasible and improves the objective. Then, we set $\tau = \min\{\tau_i \mid i \in I\}$ guaranteeing that $\boldsymbol{x}_{h+1}$ is feasible for the problem (6.1) and improves all the objectives. If any suitable $\tau \geq \tau_{tol}$ is not found, we do not make a serious step but we continue as in the case where the common descent direction is not obtained.

6.3.1.3 Null Step and Ensuring Weak Pareto Stationarity

In the case when the candidate for the common descent direction is not descent for all the objectives or the point obtained would be infeasible, our first aim is to perform a null step similar to bundle methods in general. However, in this case we have several objectives, and thus, we calculate two new auxiliary points $\boldsymbol{x}_h + \boldsymbol{d}_h$ and $\boldsymbol{x}_h + \boldsymbol{d}_i$ for every objective $i \in I$. The triplets associated to these points are added to the corresponding bundles and new individual descent directions are calculated as well as the candidate for the common descent direction. We count the number of consecutive null steps with n_{null}. If $n_{null} > n_{null}^{\max}$, where $n_{null}^{\max}$ is the maximum number of consecutive null steps, we have a hint that we might have reached a weakly Pareto stationary solution. Thus, we test whether the current iteration point is weakly Pareto stationary or not.

Based on Theorem 2, in order to say that a solution $\boldsymbol{x}^*$ is weakly Pareto stationary for the problem (6.1), it should be a Clarke stationary solution for the improvement function (6.3). Thus, we move on to consider the improvement function (6.3) and its subdifferential. First, the improvement function is DC and its DC components are

$$H^1(\boldsymbol{x}, \boldsymbol{y}) = \max \Big\{ f_i^1(\boldsymbol{x}) + \sum_{\substack{j \in I, \\ j \neq i}} f_j^2(\boldsymbol{x}) + \sum_{t \in C} g_t^2(\boldsymbol{x}) - f_i(\boldsymbol{y}),$$

$$g_c^1(\boldsymbol{x}) + \sum_{\substack{t \in C, \\ t \neq c}} g_t^2(\boldsymbol{x}) + \sum_{j \in I} f_j^2(\boldsymbol{x}) \mid i \in I,\, c \in C \Big\}$$

$$H^2(\boldsymbol{x}) = \sum_{j \in I} f_j^2(\boldsymbol{x}) + \sum_{t \in C} g_t^2(\boldsymbol{x}),$$

where $\boldsymbol{y}$ is treated as a fixed point. For the further details about this decomposition, like how it is formed and how to calculate its subgradients, we refer to [47]. Second, due to the assumption **A0** we can only obtain subgradients of DC components $H_1(\cdot, \boldsymbol{y})$ and H_2. Unfortunately, there is no guarantee that $\boldsymbol{\xi}^1 - \boldsymbol{\xi}^2 \in \partial H(\boldsymbol{x}, \boldsymbol{y})$ for arbitrary $\boldsymbol{\xi}^1 \in \partial H^1(\boldsymbol{x}, \boldsymbol{y})$ and $\boldsymbol{\xi}^2 \in \partial H^2(\boldsymbol{x})$. Therefore, we utilize the escape procedure proposed in [26, 28] for the improvement function as is done in [47].

Next, we describe the general idea of the escape procedure and the full details are presented in [26, 28]. In order to use the escape procedure, we make the following mild assumption:

A1: The subdifferentials $\partial f_i^l(\boldsymbol{x})$ and $\partial g_c^l(\boldsymbol{x})$ for all $i \in I$, $c \in C$, and $l = 1, 2$ are polytopes at each $\boldsymbol{x} \in X$.

Note that since $\partial f_i^l(\boldsymbol{x})$ and $\partial g_c^l(\boldsymbol{x})$ for all $i \in I$, $c \in C$, and $l = 1, 2$ are polytopes, then $\partial H^1(\boldsymbol{x}, \boldsymbol{y})$ and $\partial H^2(\boldsymbol{x})$ are also polytopes.

Since the escape procedure might be used for some individual objective as well as for the improvement function, we describe the idea for a general DC function $f = f^1 - f^2$. We define the following set

$$G^l(\boldsymbol{x}; \boldsymbol{d}) = \big\{ \boldsymbol{\xi} \in \partial f^l(\boldsymbol{x}) \mid \boldsymbol{\xi}^T \boldsymbol{d} = \max\{\boldsymbol{w}^T \boldsymbol{d} \mid \boldsymbol{w} \in \partial f^l(\boldsymbol{x})\} \big\} \quad l = 1, 2$$

for any $\boldsymbol{d} \in \mathbb{R}^n$, $\boldsymbol{d} \neq \boldsymbol{0}$. This gives subgradients yielding the directional derivative. Furthermore, let T_{DC} be a set of full measure at the point $\boldsymbol{x} \in \mathbb{R}^n$ such that $G^1(\boldsymbol{x}; \boldsymbol{d})$ and $G^2(\boldsymbol{x}; \boldsymbol{d})$ are singletons for any $\boldsymbol{d} \in T_{DC}$. Based on Theorem 4.1 in [26] by taking $\boldsymbol{\xi}^1 \in G^1(\boldsymbol{x}; \boldsymbol{d})$ and $\boldsymbol{\xi}^2 \in G^2(\boldsymbol{x}; \boldsymbol{d})$ their difference $\boldsymbol{\xi}^1 - \boldsymbol{\xi}^2 \in \partial f(\boldsymbol{x})$ for $\boldsymbol{d} \in T_{DC}$. Due to Theorem 4.8 in [26], at the point $\boldsymbol{x} \in \mathbb{R}^n$ we can always make a small change into the direction $\boldsymbol{d}$ such that we are able to construct a direction $\bar{\boldsymbol{d}} \in T_{DC}$ close to the direction $\boldsymbol{d}$. In addition, the subgradients of the DC components obtained by using $\bar{\boldsymbol{d}}$ define also the directional derivative of the DC function in the direction $\boldsymbol{d}$.

In the escape procedure, the subdifferential of f at the point $\boldsymbol{x}$ is approximated with the Goldstein ε-subdifferential. In practice, this approximation is denoted by U_j, where the index j refers to the corresponding iteration of the escape procedure and the elements of this set are composed of the subgradients formed as described above. By solving the problem

$$\min_{\boldsymbol{u}\in U_j} \quad \frac{1}{2}\|\boldsymbol{u}\|^2 \tag{6.11}$$

and denoting its solution with $\bar{\boldsymbol{u}}_j$, we get a search direction $\boldsymbol{d}_{j+1} = -\bar{\boldsymbol{u}}_j/\|\bar{\boldsymbol{u}}_j\|$. If Clarke stationarity is not achieved nor we have obtained a descent direction, the approximation U_j is improved by calculating a new subgradient. If Clarke stationarity is achieved, the escape procedure stops. However, if the descent direction for the improvement function is found, we have found a common descent direction for all the objectives [31]. Note that this direction is also feasible.

6.3.1.4 Interactivity

While MB-DC works well without a decision maker (DM), (s)he may have expectations which are not met without an additional input. Thus, we introduce a strategy to allow the DM to steer the solution process. This strategy bases on the fact that descent methods are sensitive for the selection of the starting point since the solution will be found in the negative orthant from the starting point in the objective space. Hence, the current solution gives upper bounds for the objective values in the final solution and the aim of the DM is to analyse whether a solution in this range is a good enough. Once the DM's input is needed, the DM tells whether (s)he is satisfied with the current solution (or a solution lying in the negative orthant from the current solution in the objective space) or which objective needs to be improved. The improvement in the objective selected is obtained by using the corresponding individual descent direction as the search direction.

The input from the DM is asked before entering to the escape procedure. If the current course of the method is suitable for the DM, we perform the escape procedure. Otherwise, the DM selects which objective will be improved and we set $\boldsymbol{x}_{h+1} = \boldsymbol{x}_h + \boldsymbol{d}_{\hat{\imath}}$, where $\hat{\imath} \in I$ indicates the objective selected. Here the DM has to consider the trade-offs since $f_i(\boldsymbol{x}_{h+1}) < f_i(\boldsymbol{x}_h)$ will not be valid for every $i \in I$. However, if for some objective there does not exist an individual descent direction suggesting that this objective is in its optimum, the corresponding objective cannot be selected to be improved. Another case when an individual descent direction cannot be selected is when the direction is not feasible. As the entrance to the escape procedure may be seen as a hint of the weakly Pareto stationary solution, instead of just guiding the iteration path towards the more desirable region, we may change from one weakly Pareto stationary solution to another, more satisfactory, one.

In practice, the number of the interactions n_{dm} is limited with $n_{dm}^{\max} \geq 0$ enabling an option to perform the method as a pure no-preference method. However, even if the maximum number $n_{dm}^{\max}$ is reached it is important to ensure that the DM is content with the results. Furthermore, if the DM selects the escape procedure before the limit $n_{dm}^{\max}$ is met, the DM is assumed to be happy with any solution from the given orthant and the input from the DM will not be asked any more.

6.3.1.5 Constraint Handling

Until now, the constraints are handled by checking the feasibility during the line search and if the feasibility cannot be guaranteed, the null step or the escape procedure is applied as is described in Sect. 6.3.1.3. We call this strategy as a *default strategy*. In this section, we present an alternative strategy for constraint handling called a *sophisticated strategy*.

This sophisticated strategy utilizes the improvement function (6.3) while the individual search directions are calculated. Instead of the objective f_i, we consider the modified objective function of the form

$$\tilde{f}_i(\boldsymbol{x}, \boldsymbol{y}) = \max\{f_i(\boldsymbol{x}) - f_i(\boldsymbol{y}),\ g_c(\boldsymbol{x}) \mid c \in C\} \quad \text{for all } i \in I, \tag{6.12}$$

where the improvement function is used to combine the objective f_i and the constraints g_c for all $c \in C$. The nice feature of the improvement function is that if we have found a descent direction for the modified objective function, it is descent for the original objective as well. In practice, the use of the modified objective function has an impact, for example, to the cutting plane model (6.5), the DC decomposition, and the construction of the bundles. Indeed, the DC components are of the form

$$\begin{aligned} \tilde{f}_i^1(\boldsymbol{x}, \boldsymbol{y}) = \max \Big\{ & f_i^1(\boldsymbol{x}) + \sum_{t \in C} g_t^2(\boldsymbol{x}) - f_i(\boldsymbol{y}), \\ & g_c^1(\boldsymbol{x}) + \sum_{\substack{t \in C, \\ t \neq c}} g_t^2(\boldsymbol{x}) + f_i^2(\boldsymbol{x}) \mid c \in C \Big\} \qquad (6.13) \\ \tilde{f}_i^2(\boldsymbol{x}) = f_i^2(\boldsymbol{x}) + & \sum_{t \in C} g_t^2(\boldsymbol{x}). \end{aligned}$$

For the second component $\tilde{f}_i^2$, the cutting plane model is constructed as previously for f_i^2. However, in order to form the cutting plane model for the first component $\tilde{f}_i^1(\cdot, \boldsymbol{x}_h)$ at the current iteration point $\boldsymbol{x}_h$, we build a separate convex cutting plane model for each component inside the maximum in (6.13) similarly than in (6.6). The final cutting plane model for $\tilde{f}_i^1(\cdot, \boldsymbol{x}_h)$ is obtained by taking a maximum from these separate models. For the further details we refer to [47].

After finding individual descent directions by using the modified objectives, the procedure continues as before by determining a candidate for the common descent direction and whenever the line search is performed we still need to take the feasibility into account. If we cannot find a suitable stepsize, we have a hint that we might have reached the optimum and continue as would be done when we did not find a common descent direction. However, if the constraints are convex, we can show that the candidate for the common descent direction $\boldsymbol{d}_h$ obtained from the problem (6.10) is feasible. Thus, we can omit the feasibility check in the line search if the stepsize is at most 1. This is considered in the following theorem.

Theorem 4 *Consider the problem* (6.1). *Let the objectives f_i for all $i \in I$ be DC functions, the constraints g_c for all $c \in C$ be convex, and the current iteration point $\boldsymbol{x}_h$ be feasible. Assume that each individual descent direction $\boldsymbol{d}_i$ for $i \in I$ is obtained by utilizing the modified objective $\tilde{f}_i(\cdot, \boldsymbol{x}_h)$* (6.12) *in the problem* (6.4). *Then by solving the problem* (6.10), *we obtain a candidate for the common descent direction $\boldsymbol{d}_h$ such that a point $\boldsymbol{x}_h + \tau \boldsymbol{d}_h$, $\tau \in [0, 1]$ is feasible.*

Proof Due to the procedure described in Sect. 6.3.1.1, by using the modified objective $\tilde{f}_i(\cdot, \boldsymbol{x}_h)$ as an objective of the problem (6.4) for each $i \in I$, we obtain a direction $\boldsymbol{d}_i$ such that $\tilde{f}_i(\boldsymbol{x}_h + \boldsymbol{d}_i, \boldsymbol{x}_h) < \tilde{f}_i(\boldsymbol{x}_h, \boldsymbol{x}_h)$ for each $i \in I$. By (6.12) and the fact that $\boldsymbol{x}_h$ is assumed to be feasible, we see that $\tilde{f}_i(\boldsymbol{x}_h, \boldsymbol{x}_h) = 0$ for each $i \in I$. Thus $\tilde{f}_i(\boldsymbol{x}_h + \boldsymbol{d}_i, \boldsymbol{x}_h) < 0$ for all $i \in I$, and especially, $g_c(\boldsymbol{x}_h + \boldsymbol{d}_i) < 0$ for all $c \in C$ meaning that the point $\boldsymbol{x}_h + \boldsymbol{d}_i$ is feasible.

Now, we consider the individual descent directions $\boldsymbol{d}_i$ starting from the current iteration point $\boldsymbol{x}_h$ and form a convex hull of these directions. Since at the point $\boldsymbol{x}_h$ all the directions $\boldsymbol{d}_i$, $i \in I$ are feasible with stepsizes $\tau_i \in [0, 1]$ and the constraints are assumed to be convex, the convex hull of the directions $\boldsymbol{d}_i$, $i \in I$ belongs to the feasible set at the point $\boldsymbol{x}_h$. Thus, especially the minimum norm element of the convex hull $\boldsymbol{d}_h$ is feasible. Then, the point $\boldsymbol{x}_h + \tau \boldsymbol{d}_h$, $\tau \in [0, 1]$ is feasible. □

6.3.2 Algorithm

The presentation of MB-DC is divided into four different algorithms. The main part of the method is described in Algorithm 1. Algorithm 2 is similar to the main iteration of DBDC [26] and it is used to find the individual search directions. The escape procedure for checking whether the Clarke stationarity is obtained is given in Algorithm 3. Finally, the scaling procedure is given in Algorithm 4.

Before presenting the algorithms, we make another assumption in addition to the assumption **A1** which is related to the level set of the starting point $\boldsymbol{x}_0 \in X$:

A2: For the enlargement parameter $L \geq 0$ the level set $\mathcal{F}_{i,0} = \{\boldsymbol{x} \in \mathbb{R}^n \mid f_i(\boldsymbol{x}) \leq f_i(\boldsymbol{x}_0) + L\}$ for each $i \in I$ is compact.

This assumption is mild and with $L = 0$ it is commonly used in bundle methods. If MB-DC is executed without interactivity, the selection $L = 0$ is sufficient. However, if interactivity is included, the options of the DM can be broadened by giving a positive L since the starting point may be close to some non-preferred solution. As an example, see Problem 1 in Sect. 6.4.3.

Comments related to Algorithm 1 are in order. In practice, the numerical efforts can be slightly decreased. Indeed, if in Step 1 some $\boldsymbol{d}_i = \boldsymbol{0}$, we could enter straight Step 5 since by solving the problem (6.10) the solution will be $\boldsymbol{d}_h = \boldsymbol{0}$. In addition, the parallel computing could be utilized in Step 1 to compute all individual search directions simultaneously as well as in Step 2 in Algorithm 2 to solve the problem

Algorithm 1 MB-DC

Data: The descent parameters $m_1, m_2 \in (0, 1)$, the decrease parameters $r_1, r_2 \in (0, 1)$, the increase parameter $R > 1$, the maximum number of consecutive null steps $n_{null}^{\max} \geq 0$, the stopping tolerances $\delta_1, \delta_2 \in (0, 1)$, the enlargement parameters $\theta > 0$ and $L \geq 0$, the escape tolerance $\gamma > 0$, the proximity measures $\varepsilon_1, \varepsilon_2 > 0$, the stepsize tolerance $\tau_{tol} > 0$, and the maximum number of interactions $n_{dm}^{\max} \geq 0$

Step 0. *(Initialization)* Select a starting point $\boldsymbol{x}_0 \in X$ and execute Algorithm 4 if $n \geq 250$. Calculate $\boldsymbol{\xi}_i^l(\boldsymbol{x}_0) \in \partial f_i^l(\boldsymbol{x}_0)$ for all $i \in I$ and $l = 1, 2$. Initialize the bundles $\mathcal{B}_{i,0}^l = \{(\boldsymbol{\xi}_i^l(\boldsymbol{x}_0), 0)\}$ for all $i \in I$ and $l = 1, 2$. Set $h = 0$, $n_{null} = 0$, and $n_{dm} = 0$.

Step 1. *(Individual directions)* For each $i \in I$, find an individual search direction $\boldsymbol{d}_i$ by executing Algorithm 2 and calculate the decrease ratio ρ_i in (6.9).

Step 2. *(Search direction candidate)* If $\max_{i \in I}\{\rho_i\} > 0$, then set $\boldsymbol{d}_h = \boldsymbol{d}_{i^*}$, where $i^* = \operatorname{argmax}_{i \in I}\{\rho_i\}$. Otherwise, solve the problem (6.10) to obtain $\boldsymbol{d}_h$. If $\|\boldsymbol{d}_h\| < \gamma$, then set $\tau = 0$ and go to Step 5. Otherwise, go to Step 3.

Step 3. *(Stepsize)* Calculate stepsize τ. If $f_i(\boldsymbol{x}_h + \tau \boldsymbol{d}_h) - f_i(\boldsymbol{x}_h) > -\gamma$ for some $i \in I$, then set $\tau = 0$.

Step 4. *(Null step)* If $\tau < \tau_{tol}$ and $n_{null} < n_{null}^{\max}$ then calculate $\boldsymbol{\xi}_i^l(\boldsymbol{x}_h + \boldsymbol{d}_h) \in \partial f_i^l(\boldsymbol{x}_h + \boldsymbol{d}_h)$ and $\boldsymbol{\xi}_i^l(\boldsymbol{x}_h + \boldsymbol{d}_i) \in \partial f_i^l(\boldsymbol{x}_h + \boldsymbol{d}_i)$ for all $i \in I$ and $l = 1, 2$. Insert the elements $(\boldsymbol{\xi}_i^l(\boldsymbol{x}_h + \boldsymbol{d}_h), \alpha_i^l(\boldsymbol{x}_h, \boldsymbol{x}_h + \boldsymbol{d}_h))$ and $(\boldsymbol{\xi}_i^l(\boldsymbol{x}_h + \boldsymbol{d}_i), \alpha_i^l(\boldsymbol{x}_h, \boldsymbol{x}_h + \boldsymbol{d}_i))$ into the bundle $\mathcal{B}_{i,h}^l$ for each $i \in I$ and $l = 1, 2$. Set $n_{null} = n_{null} + 1$. Go to Step 1.

Step 5. *(Escape procedure and interactivity)* If $\tau < \tau_{tol}$, then execute the following procedure:

(a) If $n_{dm} < n_{dm}^{\max}$, set $n_{dm} = n_{dm} + 1$ and ask the DM either to select whether to improve an individual objective and go to (c) or to check if the weak Pareto stationarity is achieved and go to (d).
(b) If $n_{dm} \geq n_{dm}^{\max}$, the go to (e).
(c) Set $\boldsymbol{x}_{h+1} = \boldsymbol{x}_h + \boldsymbol{d}_{\hat{\imath}}$, where $\hat{\imath} \in I$ is the objective to be improved. Go to Step 7.
(d) Set $n_{dm} = n_{dm}^{\max}$ and go to (e).
(e) Execute Algorithm 3 for the point $\boldsymbol{x}_h$ with the improvement function $H(\cdot, \boldsymbol{x}_h)$ and the proximity measure ε_2 to obtain $\boldsymbol{x}^+$. If $\boldsymbol{x}^+ = \boldsymbol{x}_h$, then STOP with $\boldsymbol{x}^* = \boldsymbol{x}_h$ as the final solution. Otherwise, set $\boldsymbol{x}_{h+1} = \boldsymbol{x}^+$ and go to Step 7.

Step 6. *(Serious step)* If $\tau \geq \tau_{tol}$, then $\boldsymbol{x}_{h+1} = \boldsymbol{x}_h + \tau \boldsymbol{d}_h$ and $n_{null} = 0$.

Step 7. *(Bundle update)* Update linearization errors by (6.14), calculate $\boldsymbol{\xi}_i^l(\boldsymbol{x}_{h+1}) \in \partial f_i^l(\boldsymbol{x}_{h+1})$ for all $i \in I$ and $l = 1, 2$. Select $\mathcal{B}_{i,h+1}^l \subseteq \mathcal{B}_{i,h}^l$ and insert the element $(\boldsymbol{\xi}_i^l(\boldsymbol{x}_{h+1}), 0)$ into the bundle $\mathcal{B}_{i,h+1}^l$ for each $i \in I$ and $l = 1, 2$. Set $h = h + 1$. Go to Step 1.

(6.7) globally. In Step 3, a feasible stepsize τ is defined. In our implementation, we simply try $\tau = 1$ to test whether $\boldsymbol{x}_h + \boldsymbol{d}_h$ gives a better value for each objective than $\boldsymbol{x}_h$ and is feasible. If that serves, we increase τ as long as we have a feasible point having better value for each objective than the previously tested one. Otherwise, τ is decreased until the direction is descent and $\boldsymbol{x}_h + \tau \boldsymbol{d}_h$ is feasible. If τ would be smaller than τ_{tol}, we set $\tau = 0$. In Step 7, the bundles are updated. In practice, our implementation resets the bundles $\mathcal{B}_{i,h}^2$ for all $i \in I$ but that is allowed since the new element related to the new iteration point $\boldsymbol{x}_{h+1}$ is added to each bundle before starting a new iteration. Additionally, since the linearization error depends on $\boldsymbol{x}_h$, it needs to be updated whenever $\boldsymbol{x}_{h+1}$ is found with the formula

$$\alpha_i^l(\boldsymbol{x}_{h+1}, \boldsymbol{y}_{i,j}) = \alpha_i^l(\boldsymbol{x}_h, \boldsymbol{y}_{i,j}) + f_i^l(\boldsymbol{x}_{h+1}) - f_i^l(\boldsymbol{x}_h) - \boldsymbol{\xi}_i^l(\boldsymbol{y}_{i,j})^T(\boldsymbol{x}_{h+1} - \boldsymbol{x}_h), \tag{6.14}$$

where $i \in I$, $l = 1, 2$, and $j \in J_{i,h}^l$. Thus, it is enough to store only subgradients and linearization errors in the bundles while auxiliary points and function values can be omitted.

Remark 1 In Step 5a of Algorithm 1, the DM has an option to steer the solution process. We enter Step 5 either when the maximum number of consecutive null steps is performed and the stepsize is close to zero or when the common descent direction is close to a zero vector. If the DM decides to improve some individual objective, then the objectives with $\|\boldsymbol{d}_i\| > 0$ are available for the selection. However, if for some $i \in I$ the selection would lead us on an infeasible solution or away from any compact level set assumed in **A2**, the objective f_i cannot be selected.

If the DM selects the escape procedure, (s)he is considering the current solution rather good. Thus, the algorithm stops if Clarke stationarity is achieved for the improvement function or continues by improving all the objectives. In addition, the DM is not able to steer the solution process after selecting the escape procedure.

During the execution of Algorithm 2, the iteration point $\boldsymbol{x}_h$ stays untouched and only a descent direction for the individual objective is achieved. However, if the direction $\boldsymbol{d}_{\hat{\imath}} = \boldsymbol{0}$ for some $\hat{\imath} \in I$, then it is gained from the escape procedure in Algorithm 3 during the Step 3 of Algorithm 2. This implies that the current iteration point is Clarke stationary solution for the objective $f_{\hat{\imath}}$. Additionally, we reset all the bundles $\mathcal{B}_{i,h}^2$ for $i \in I$ in Step 7 of Algorithm 1.

Algorithm 3 can be executed in two places: while determining an individual search direction and if a common descent direction is not found. In the first case, the objective in the escape procedure is an individual f_i, $i \in I$ and in the latter case the improvement function. Furthermore, different proximity measures $\varepsilon_1 > 0$ and $\varepsilon_2 > 0$ are used. Similarly to Algorithms 2, the escape procedure is executed inside an iteration implying that the current iteration point will not change during its execution. As an output, we receive a point $\boldsymbol{x}^+$. This either implies Clarke stationarity or includes the information about the descent direction of the objective used in the escape procedure. Depending on the objective, we either obtain a descent direction for the individual objective or a common descent direction.

In [47], it is noticed that one DC function can dominate the others and hide their effects if the magnitudes of the function values differ a lot. Here, we can make a similar observation. This unpleasant feature can be avoided by employing Algorithm 4 used in [47] as well. However, this is not mandatory. Note that the scaled objectives have the same optima than the original ones. Other possible scaling options valid with the improvement function are presented in [40].

Remark 2 In the case of the sophisticated constraint handling strategy, some modifications to the algorithms are needed. First of all, at the current iteration point $\boldsymbol{x}_h$ the modified objectives $\tilde{f}_i(\cdot, \boldsymbol{x}_h)$, $i \in I$ (6.12) are used instead of the original objectives f_i while the individual descent directions are calculated. Due to this, the

Algorithm 2 Individual search direction

Data: The index $i \in I$ of the objective, the descent parameter $m_1 \in (0, 1)$, the decrease parameters $r_1, r_2 \in (0, 1)$, the increase parameter $R > 1$, the stopping tolerance $\delta_1 \in (0, 1)$, the enlargement parameter $\theta > 0$, and the proximity measure $\varepsilon_1 > 0$.

Step 0. *(Criticality)* If $\|\boldsymbol{\xi}_i^1(\boldsymbol{x}_h) - \boldsymbol{\xi}_i^2(\boldsymbol{x}_h)\| < \delta_1$ then go to Step 3.

Step 1. *(Initialization)* Calculate $j^* = \arg\max_{j \in J_{i,h}^2} \left\{\|\boldsymbol{\xi}_{i,j}^2(\boldsymbol{y}_{i,j})\|\right\}$, set $\boldsymbol{\xi}_{\max}^2 = \boldsymbol{\xi}_{i,j^*}^2(\boldsymbol{y}_{i,j^*})$ and initialize the parameters

$$t_{\min} = r \cdot \frac{\theta}{2\left(\|\boldsymbol{\xi}_i^1(\boldsymbol{x}_h)\| + \|\boldsymbol{\xi}_{\max}^2\|\right)} \tag{6.15}$$

and $t_{\max} = R\, t_{\min}$. Choose the value $t \in [t_{\min}, t_{\max}]$.

Step 2. *(Search direction)* Compute the search direction $\boldsymbol{d}^*$ by solving problem (6.7). If $\|\boldsymbol{d}^*\| < \delta_1$ then go to Step 3 else go to Step 4.

Step 3. *(Clarke stationarity)* Execute Algorithm 3 for the point $\boldsymbol{x}_h$ with the objective f_i and the proximity measure ε_1 to obtain $\boldsymbol{x}^+$. Set $\boldsymbol{d}_i = \boldsymbol{x}^+ - \boldsymbol{x}_h$ and EXIT.

Step 4. *(Descent test)* Set $\boldsymbol{y} = \boldsymbol{x}_h + \boldsymbol{d}^*$. If $f_i(\boldsymbol{y}) - f_i(\boldsymbol{x}_h) \le m_1 \left(\hat{f}_{i,h}(\boldsymbol{x}_h + \boldsymbol{d}^*) - f_i(\boldsymbol{x}_h)\right)$ then set $\boldsymbol{d}_i = \boldsymbol{d}^*$ and EXIT.

Step 5. *(Bundle update)*

(a) If $f_i(\boldsymbol{y}) - f_i(\boldsymbol{x}_0) > 0$ and $\|\boldsymbol{d}^*\| > \theta$ then set $t = t - r_1(t - t_{\min})$ and go to Step 2.

(b) Otherwise, if $f_i(\boldsymbol{y}) - f_i(\boldsymbol{x}_h) \ge -m_1 \left(\hat{f}_{i,h}(\boldsymbol{x}_h + \boldsymbol{d}^*) - f_i(\boldsymbol{x}_h)\right)$ then set $t = t - r_2(t - t_{\min})$. Compute $\boldsymbol{\xi}^l \in \partial f_i^l(\boldsymbol{y})$ and set $\alpha^l = f_i^l(\boldsymbol{x}_h) - f_i^l(\boldsymbol{y}) + (\boldsymbol{\xi}^l)^T \boldsymbol{d}^*$ for $l = 1, 2$. Insert the element $(\boldsymbol{\xi}^l, \alpha^l)$ into $\mathcal{B}_{i,h}^l$ for $l = 1, 2$.

Step 6. *(Parameter update)* If $\|\boldsymbol{\xi}^2\| > \|\boldsymbol{\xi}_{\max}^2\|$ then set $\boldsymbol{\xi}_{\max}^2 = \boldsymbol{\xi}^2$ and update $t_{\min}$ using formula (6.15). Go to Step 2.

bundles $\mathcal{B}_{i,h}^1, i \in I$ for $\tilde{f}_i^1(\cdot, \boldsymbol{x}_h)$ are constructed differently. Since the cutting plane model for $\tilde{f}_i^1(\cdot, \boldsymbol{x}_h)$ is build by combining the separate models of the components inside the maximum in (6.13), we add a bundle element for each component to the bundle $\mathcal{B}_{i,h}^1$. Thus, whenever we update the bundle we add $|C| + 1$ elements to $\mathcal{B}_{i,h}^1$. Furthermore, whenever we get a new iteration point $\boldsymbol{x}_{h+1}$, the updated linearization errors for $\tilde{f}_i^1(\cdot, \boldsymbol{x}_{h+1})$ are obtained with the formula (22) in [47]. However, the use of modified objectives has no impact on $\mathcal{B}_{i,h}^2$.

In the sophisticated strategy, we scale the constraints. The scaling depends on the dimension of the problem. When $n \ge 250$, we include the constraint functions in Algorithm 4 together with the objectives. Basically, we think that the constraints $g_c, c \in C$ are objectives f_i, where $i = k + 1, \ldots, k + |C|$. If $n < 250$, the scaling is started by calculating $i^* = \text{argmax}\,\{|f_i(\boldsymbol{x}_0)| \,\big|\, i \in I\}$ and searching the value $\kappa \in \mathbb{Z}$ such that

$$10^{\kappa-1} \le |f_{i^*}(\boldsymbol{x}_0)| \le 10^{\kappa}.$$

The scaling factor $\omega_c = 10^{\eta_c}$, where $\eta_c \in \mathbb{Z}$, for $c \in C$ is determined such that

Algorithm 3 Escape procedure

Data: The point $\boldsymbol{x} \in \mathbb{R}^n$ under consideration, the descent parameter $m_2 \in (0, 1)$, the stopping tolerance $\delta_2 \in (0, 1)$, and the proximity measures $\varepsilon > 0$.

Step 0. *(Initialization)* Select $\boldsymbol{d}_0 \in \{\boldsymbol{d} \in \mathbb{R}^n \mid \|\boldsymbol{d}\| = 1\}$ and find $\bar{\boldsymbol{d}}_0 \in T_{DC}$ at $\boldsymbol{x}$ using $\boldsymbol{d}_0$. Compute $\boldsymbol{\xi}^1 \in G^1(\boldsymbol{x}; \bar{\boldsymbol{d}}_0)$ and $\boldsymbol{\xi}^2 \in G^2(\boldsymbol{x}; \bar{\boldsymbol{d}}_0)$. Set $U_0 = \{\boldsymbol{\xi}^1 - \boldsymbol{\xi}^2\}$, $\tilde{\boldsymbol{x}} = \boldsymbol{x}$ and $j = 0$.
Step 1. *(Clarke stationarity)* Find $\bar{\boldsymbol{u}}_j$ as the solution of the problem (6.11). If $\|\bar{\boldsymbol{u}}_j\| \leq \delta_2$, then approximate Clarke stationarity is obtained and EXIT with $\boldsymbol{x}^+ = \boldsymbol{x}$.
Step 2. *(Search direction)* Compute the search direction $\boldsymbol{d}_{j+1} = -\bar{\boldsymbol{u}}_j / \|\bar{\boldsymbol{u}}_j\|$.
Step 3. *(New subgradient)* Find $\bar{\boldsymbol{d}}_{j+1} \in T_{DC}$ at $\tilde{\boldsymbol{x}}$ using $\boldsymbol{d}_{j+1}$. Compute $\boldsymbol{\xi}^1 \in G^1(\tilde{\boldsymbol{x}}; \bar{\boldsymbol{d}}_{j+1})$ and $\boldsymbol{\xi}^2 \in G^2(\tilde{\boldsymbol{x}}; \bar{\boldsymbol{d}}_{j+1})$. Set $\bar{\boldsymbol{\xi}}_{j+1} = \boldsymbol{\xi}^1 - \boldsymbol{\xi}^2$. If $\boldsymbol{x} \neq \tilde{\boldsymbol{x}}$, then go to Step 5.
Step 4. *(Descent test)* If $(\bar{\boldsymbol{\xi}}_{j+1})^T \boldsymbol{d}_{j+1} \leq -m_2 \|\bar{\boldsymbol{u}}_j\|$, then go to Step 6.
Step 5. *(Update)* Set $U_{j+1} = \text{conv}\,\{U_j \bigcup \{\bar{\boldsymbol{\xi}}_{j+1}\}\}$, $\tilde{\boldsymbol{x}} = \boldsymbol{x}$ and $j = j + 1$. Go to Step 1.
Step 6. *(Step-length)* Determine a point $\hat{\boldsymbol{x}} = \boldsymbol{x} + \beta^* \boldsymbol{d}_{j+1}$ such that

$$\beta^* = \arg\max\{\beta > 0 \mid f(\boldsymbol{x} + \beta \boldsymbol{d}_{j+1}) - f(\boldsymbol{x}) \leq -m_2 \beta \|\bar{\boldsymbol{u}}_j\|\}.$$

If $\beta^* \geq \varepsilon$, then $\boldsymbol{x}^+ = \hat{\boldsymbol{x}}$, and EXIT. Otherwise, $\tilde{\boldsymbol{x}} = \hat{\boldsymbol{x}}$ and go to Step 3.

Algorithm 4 Scaling

Step 1. Calculate $i^* = \text{argmin}\,\big\{ |f_i(\boldsymbol{x}_0)| \,\big|\, i \in I \big\}$.
Step 2. For each $i \in I$, search $\kappa_i \in \mathbb{Z}$ such that $10^{\kappa_i - 1} \leq |f_i(\boldsymbol{x}_0)| \leq 10^{\kappa_i}$. If $\kappa_i < 0$, then $\kappa_i = 0$.
Step 3. For each $i \in I$, set $\nu_i = \kappa_{i^*} - \kappa_i$. If $\nu_i \leq -2$, then $\nu_i = \nu_i + 1$. Set $\omega_i = 10^{\nu_i}$ and $f_i = \omega_i f_i$ being the scaled objective function.

$$10^{\nu - 1} \leq \omega_c |g_c(\boldsymbol{x}_0)| \leq 10^{\nu},$$

where $\nu = \kappa$ if $n < 100$ and $\nu = \kappa + 1$ if $100 \leq n < 250$.

6.3.3 Convergence

In this section, the convergence of MB-DC is analysed. The first two theorems guarantee the finiteness of Algorithms 2 and 3. After that, the focus is on to prove that Algorithm 1 is finite and that it terminates at an approximate weakly Pareto stationary solution. Throughout the convergence analysis, the assumptions **A1** and **A2** are supposed to be valid.

First, we discuss the finiteness of Algorithm 2. Since Algorithm 2 corresponds to the main iteration of DBDC, we can utilize Theorem 5.8 from [26].

Theorem 5 ([26]) *Let* **A2** *be valid. For any* $\delta_1 \in (0, 1)$, *Algorithm 2 is stopped after a finite number of steps.*

Proof See Theorem 5.8 in [26]. □

Similarly, Algorithm 3 corresponds to the escape procedure in [26]. Thus, we recall Theorem 4.11 in [26] giving a finite upper bound for the number of iterations.

Theorem 6 ([26]) *Let **A1** be valid and the level set $\mathcal{F} = \{z \in \mathbb{R}^n \mid f(z) \leq f(\boldsymbol{x})\}$ be compact for $\boldsymbol{x} \in \mathbb{R}^n$. Algorithm 3 terminates after at most*

$$\left\lceil \frac{\ln\left(\frac{\delta_2^2}{K^2}\right)}{\ln\left(1 - \frac{(1-m_2)^2\delta_2^2}{8K^2}\right)} \right\rceil$$

iterations, where $\lceil\cdot\rceil$ is a ceiling of the number, $m_2, \delta_2 \in (0, 1)$, and $K > \delta_2$ is the Lipschitz constant of f at the point $\boldsymbol{x} \in \mathbb{R}^n$.

Proof See Theorem 4.11 in [26]. □

Next, we consider the finiteness of Algorithm 1. This proof is divided into two lemmas and one theorem. The first lemma shows that the new iteration point is obtained after a finite number of steps while the other lemma states that the sequence of solutions generated by the algorithm is finite.

Lemma 1 *Let **A1** and **A2** be valid. The iteration h of Algorithm 1 is finite and yields either the new iteration point $\boldsymbol{x}_{h+1}$ or stops the execution of the algorithm.*

Proof During the iteration h, individual search directions $\boldsymbol{d}_i$ at Step 1 are obtained after a finite number of steps due to Theorem 5. Additionally, the number of consecutive null steps is finite, since at most $n_{null}^{\max}$ consecutive null steps can be performed. Therefore, after a finite number of steps, we perform either Step 5 or 6. Both of these steps are finite since Algorithm 3 is finite as stated in Theorem 6. As a result, we have obtained a new iteration point $\boldsymbol{x}_{h+1}$ in Step 5c, 5e, or 6 or the execution of the algorithm is terminated in Step 5e. □

The second lemma applies ideas from the proofs of Theorem 5.9 in [26] and Theorem 5 in [47].

Lemma 2 *Let **A1** and **A2** be valid. For any stopping tolerance $\delta_2 \in (0, 1)$ and proximity measure $\varepsilon_2 > 0$ the sequence $\{\boldsymbol{x}_h\}$ generated by the Algorithm 1 is finite.*

Proof Assume that $\{\boldsymbol{x}_h\}$ is infinite and the stopping condition in Step 5e of Algorithm 1 is never met. Thus, Algorithm 3 for the improvement function cannot end at Step 1. Lemma 1 guarantees that the new iteration point $\boldsymbol{x}_{h+1}$ is obtained after a finite number of steps. In addition, there exists an iteration h' after which the DM has used all her/his options to steer the solution process. Therefore, $\boldsymbol{x}_{h+1}$ for $h > h'$ can be obtained only in Step 5e or 6 of Algorithm 1.

If $\boldsymbol{x}_{h+1}$ is found in Step 5e, then in Algorithm 3 we get a stepsize $\beta^* \geq \varepsilon_2$ such that

$$H(\boldsymbol{x}_{h+1}, \boldsymbol{x}_h) - H(\boldsymbol{x}_h, \boldsymbol{x}_h) \leq -m_2\varepsilon_2\delta_2 < 0. \tag{6.16}$$

This implies the following deduction for all $i \in I$. First, if (6.16) holds, then

$$f_i(\boldsymbol{x}_{h+1}) - f_i(\boldsymbol{x}_h) \leq -m_2\varepsilon_2\delta_2 < 0 \tag{6.17}$$

since $H(\boldsymbol{x}_h, \boldsymbol{x}_h) = 0$. Otherwise, $\boldsymbol{x}_{h+1}$ is obtained in Step 6 and

$$f_i(\boldsymbol{x}_{h+1}) - f_i(\boldsymbol{x}_h) \leq -\gamma < 0. \tag{6.18}$$

By combining (6.17) and (6.18), we know that at each iteration $h > h'$

$$f_i(\boldsymbol{x}_{h+1}) - f_i(\boldsymbol{x}_h) < -\eta < 0, \quad \text{where } \eta = \min\{m_2\varepsilon_2\delta_2, \gamma\} > 0.$$

Thus by doing $\bar{h}$ iterations after the iteration $h' + 1$ we have

$$f_i(\boldsymbol{x}_{h'+1+\bar{h}}) - f_i(\boldsymbol{x}_{h'+1}) < -\bar{h}\eta < 0.$$

The passage to the limit $\bar{h} \to \infty$ yields

$$\lim_{\bar{h}\to\infty} f_i(\boldsymbol{x}_{h'+1+\bar{h}}) - f_i(\boldsymbol{x}_{h'+1}) < -\infty.$$

By Remark 1 the point $\boldsymbol{x}_{h'+1} \in \mathcal{F}_{i,0}$ and we have a contradiction by **A2** and Lipschitz continuity since every f_i for all $i \in I$ must be bounded from below. □

Thus, by combining the results above covering all the cases where infinite cycles may happen, we are ready to claim the finite convergence of Algorithm 1 to an approximate weakly Pareto stationary solution.

Theorem 7 *Let **A1** and **A2** be valid. Algorithm 1 stops after a finite number of iterations with the solution $\boldsymbol{x}^*$ being an approximate weakly Pareto stationary solution for the problem* (6.1).

Proof Due to Lemmas 1 and 2, $\boldsymbol{x}^*$ is obtained after a finite number of steps. Thus, during the last iteration of Algorithm 1, the stopping condition in Step 5e is met. This means that Algorithm 3 for $H(\cdot, \boldsymbol{x}_h)$ is terminated in Step 1. Therefore, $\boldsymbol{x}^*$ satisfies the approximate Clarke stationary condition for the improvement function. From this we deduce that approximately $\boldsymbol{0} \in \partial H(\boldsymbol{x}^*, \boldsymbol{x}^*)$. Based on Theorem 2, $\boldsymbol{x}^*$ is an approximate weakly Pareto stationary solution for the problem (6.1). □

6.4 Numerical Experiments

In this section, we present the numerical results. We study the efficiency of MB-DC in the unconstrained case and compare it with two other methods. In addition, we compare two different strategies for constraint handling and exemplify the effect of

Table 6.1 Default parameters

$m_1 = 0.2 \quad m_2 = 0.01$	$\delta_1 = 10^{-5} \quad \delta_2 = 10^{-5}$
$\varepsilon_1 = 10^{-6}$ $\varepsilon_2 = \begin{cases} 10^{-5}, & \text{if } n < 10 \\ 10^{-4}, & \text{if } 10 \le n < 50 \\ 10^{-3}, & \text{if } n \ge 50 \end{cases}$	$r_1 = \begin{cases} 0.75, & \text{if } n < 10 \\ \frac{n}{n+5}, & \text{if } 10 \le n \le 300 \\ 0.99, & \text{if } n > 300 \end{cases}$ $r_2 = 0.1$
$\gamma = \begin{cases} 10^{-7}, & \text{if } n \le 10 \\ 10^{-6}, & \text{if } n > 10 \end{cases}$	$\tau_{tol} = \begin{cases} 10^{-2}, & \text{if } n \le 100 \\ 10^{-3}, & \text{if } n > 100 \end{cases}$
$R = \begin{cases} 10^{7}, & \text{if } n < 10 \\ 10^{5}, & \text{if } 10 \le n \le 50 \\ 10^{8}, & \text{if } 50 < n \le 100 \\ 10^{6}, & \text{if } n > 100 \end{cases}$	$\theta = 5 \cdot 10^{-5} \; n_{null}^{\max} = 5$

the DM. The MB-DC[1] is implemented with Fortran 95. The default parameters are given in Table 6.1 and the maximum size of bundles is $\min\{n+5, 50\}$ for $\mathcal{B}_{i,h}^1$, 3 for $\mathcal{B}_{i,h}^2$, and $\min\{(n+5)\cdot 2, 250\}$ for U_j in the escape procedure. However, MB-DC is sensitive for the selection of parameters, and thus, it may be beneficial to adjust parameters with respect to a problem.

6.4.1 Numerical Efficiency

First, we consider the numerical efficiency of MB-DC by solving unconstrained multiobjective DC test problems with two and three objectives. The test problems are selected to be a vast set of DC problems presented together with the starting points (for more details, see Tables 6.2 and 6.3 in [47]). The results are calculated without the interaction with the DM meaning that we set $n_{dm}^{\max} = 0$ and $L = 0$. The solutions are then compared with the solutions obtained with MDBDC[2] and MPB[3] in terms of function and subgradient evaluations and the CPU time. Similarly to MB-DC, MDBDC is designed for multiobjective nonsmooth DC optimization while MPB is a more general method aimed for nonconvex nonsmooth multiobjective problems. The implementations of these methods are described in [47] and [37], respectively. All of these three methods utilize the same quadratic programming solver by Lukšan [36]. We use the default values of the parameters for these methods if it is not mentioned otherwise. MDBDC is implemented with Fortran 95 and MPB with Fortran 77. Therefore, all tests are compiled with `gfortran` (the GNU Fortran

[1] Available at https://github.com/kjjoki/IMB-DC.

[2] Available at https://github.com/kjjoki/MDBDC.

[3] Available at http://napsu.karmitsa.fi/proxbundle/.

compiler) and performed on an Intel® Core™ i5-7200U CPU (2.50 GHz, 2.70 GHz) running on Windows 10.

The solutions of the bi-objective case are given in Table 6.2 and the case of three objectives in Table 6.3. There the solution $f(\boldsymbol{x}^*) = (f_1(\boldsymbol{x}^*), \ldots, f_k(\boldsymbol{x}^*))$ and the number of function and subgradient evaluations (n_f and n_ξ) together with the CPU time are given. Due to the nature of MB-DC, the function and subgradient evaluations are given for all the objectives separately. However, in MDBDC and MPB the number of function and subgradient evaluations are the same for each objective, and thus, only one occurrence is reported in Tables 6.2 and 6.3. Furthermore, for MPB $n_f = n_\xi$, and thus, only one with is reported.

At the first glance, the notable difference in the results is related to the problem 10 with $n = 500$. In this problem, both MDBDC and MPB fail but MB-DC can produce a solution. Then, while comparing the other test problems, we say that one solution is better than the other if all the objective values have a smaller value. In total, there are 39 cases, and in addition to problem 10 with $n = 500$, MB-DC obtains a better solution than MPB in three more cases, namely problems 2, 3, and 9 with $n = 100$. However, MDBDC obtains a better solution in one case, namely problem 10 with $n = 10$. Nevertheless, even if some solutions might be better in practice, all the solutions are still theoretically equally good. We see that all the methods produce mainly different solutions but in several cases the nature of the solutions seems to be more similar in MDBDC and MPB than in MB-DC. Additionally, the effect of the starting point can be seen in Problems 2 and 3 since in these problems the only difference is in the starting point.

Considering the computational side of the results, we notice that in the bi-objective case (Table 6.2), the average number of function and subgradient evaluations as well as the CPU time are greater for MB-DC than for MDBDC. Comparing MB-DC with MPB, the average number of function evaluations is greater but the number of subgradient evaluations is on the same magnitude as well as the CPU time. Note that here the problem 10 with $n = 500$ is excluded from the averages due to the failure in MDBDC and MPB. What is interesting is that when we move to the case of three objectives with MB-DC, the average function evaluations for objectives f_2 and f_3 are smaller than in the bi-objective case and the average subgradient evaluations is around half. For the objective function f_1, these are higher. Furthermore, the CPU time is similar to the bi-objective case. However, in the case of three objectives, all the computational measures for MDBDC and MPB grow significantly compared to the bi-objective case. This is probably due to the reason that in MB-DC the sizes of bundles do not depend on the number of objectives. Thus if we add new objectives to the problem the size of these bundles is not affected keeping the corresponding quadratic problems quite reasonable in size. However, in MDBDC and MPB at least one bundle depends on the number of objectives. Thus, each new objective makes the corresponding quadratic problem more and more difficult to solve.

The number of function and subgradient evaluations are quite extensive in some cases of MB-DC. This can be result of, for example, problems in finding an individual descent direction, the escape procedure while stopping, or the line search. Thus, the number of the function and subgradient evaluations are divided into different

Table 6.2 Numerical results for unconstrained bi-objective test problems

P	n	MB-DC						MDBDC				MPB		
		n_{f_1}	n_{f_2}	n_{ξ_1}	n_{ξ_2}	CPU	$f(\boldsymbol{x}^*)$	n_f	n_ξ	CPU	$f(\boldsymbol{x}^*)$	n_ξ	CPU	$f(\boldsymbol{x}^*)$
1.	2	520	481	103	102	0.01	(12.55, −1.29)	9	9	0.00	(16.86, −1.32)	6	0.00	(21.60, −1.34)
2.	2	32	46	10	24	0.01	(0.49, 0.62)	119	92	0.03	(0.50, 0.50)	21	0.00	(1.00, 1.00)
3.	2	51	59	14	27	0.00	(0.50, 0.50)	84	58	0.03	(0.50, 0.50)	28	0.02	(1.00, 1.00)
4.	2	388	386	75	76	0.01	(−1.54, 64.72)	9	9	0.02	(−1.69, 100.26)	6	0.00	(−1.69, 102.70)
5.	4	38	38	22	21	0.00	(0.00, 11.00)	30	26	0.02	(99.06, 5.40)	5	0.00	(267.70, 9.20)
6.	10	35	37	8	16	0.00	(−0.74, 0.00)	110	99	0.02	(−1.64, 0.00)	69	0.00	(−3.18, 2.28)
7.	10	323	240	59	57	0.02	(11.67, 39.81)	244	236	0.08	(42.28, 15.01)	129	0.03	(70.35, 8.69)
8.	10	170	154	38	35	0.00	(0.67, −2.95)	128	108	0.03	(0.24, −2.25)	19	0.00	(0.46, −0.57)
	50	676	318	219	92	0.56	(0.12, 1.70)	255	251	0.61	(0.96, −14.11)	291	0.17	(7.33, −15.22)*
	100	1229	672	988	451	6.11	(8.55, 87.18)	122	105	0.44	(10.00, −89.50)	74	0.05	(83.75, −7.45)*
	250	1624	225	1426	70	17.55	(1.92, 17.41)	229	109	0.92	(18.00, −231.50)	95	0.23	(193.00, −56.50)
	500	3003	240	2684	59	61.03	(1.63, 24.62)	94	84	1.11	(49.00, −450.50)	594	20.69	(51.00, −448.50)
9.	10	187	185	37	39	0.02	(−1.02, 31.59)	80	71	0.05	(−5.25, 35.01)	63	0.00	(−6.20, 64.02)
	50	712	719	116	123	0.09	(3.73 , 49.07)	156	145	0.17	(−33.72, 206.89)	1402	0.66	(−39.67, 263.23)**
	100	721	1626	390	1308	1.75	(−80.86, 629.84)	128	119	0.25	(−45.12, 535.88)	7	0.00	(−0.50, 995.00)
	250	1388	1342	271	282	1.08	(30.13, 445.80)	142	138	0.63	(−177.01, 1319.82)	1898	60.09	(−193.45, 1952.50)**
	500	2558	2574	450	470	4.14	(−219.89, 1517.60)	158	122	1.05	(−407.50, 2933.63)	9	0.02	(−2.50, 4992.03)
10.	10	133	150	36	49	0.00	(0.11, 1.95)	137	103	0.05	(0.10, 0.00)	332	0.02	(0.00, 22.29)
	50	115	214	21	135	0.19	(8.72, 0.04)	276	229	0.34	(2.35, 0.39)	151	0.14	(0.00, 65.87)
	100	257	2054	38	1869	8.84	(24.26, 0.87)	551	511	5.86	(3.45, 1.46)	375	1.53	(0.01, 173.44)*
	250	227	1609	52	1464	28.16	(31.47, 21.93)	581	539	8.86	(21.72, 419.19)	892	38.33	(0.01, 415.28)*
	500	436	2966	279	2814	208.41	(75.26, 48.56)	Fail				Fail		
Avg:		685	637	336	322	6.17		173	151	0.98		308	5.81	

* Locality measure is chosen to be 0.9 (default: 0.5) in MPB

** Locality measure is chosen to be 0.95 (default: 0.5) in MPB

Table 6.3 Numerical results for unconstrained test problems with three objectives

P	n	MB-DC								MDBDC				MPB		
		n_{f_1}	n_{f_2}	n_{f_3}	n_{ξ_1}	n_{ξ_2}	n_{ξ_3}	CPU	$f(\boldsymbol{x}^*)$	n_f	n_ξ	CPU	$f(\boldsymbol{x}^*)$	n_ξ	CPU	$f(\boldsymbol{x}^*)$
11.	2	503	518	491	103	102	106	0.00	(12.55, −1.29, 12.55)	9	9	0.00	(16.86, −1.32, 16.86)	6	0.00	(21.58, −1.34, 21.57)
12.	4	341	290	294	67	69	117	0.01	(320.17, 3.46, 1.95)	99	81	0.06	(177.00, 1.79, 3.90)	24	0.02	(119.06, 1.20, 5.40)
13.	10	223	213	198	51	48	48	0.00	(1.89, −3.79, 68.26)	171	156	0.05	(0.43, −3.18, 62.70)	61	0.08	(0.58, −2.05, 82.87)
	50	218	140	130	87	33	33	0.09	(1.97, −1.92, 247.40)	183	179	0.30	(17.77, 69.94, 211.57)	2828	8.77	(1.16, −34.16, 286.14)
	100	514	654	842	239	395	615	2.22	(35.90, 94.94, 646.18)	138	128	0.48	(25.56, −74.47, 865.41)	283	1.41	(35.48, −53.71, 758.41)
	250	1315	585	536	702	153	157	8.41	(9.21, 29.11, 974.49)	40	36	0.17	(70.90, −201.09, 1826.21)	4286	260.30	(71.19, −198.71, 1287.12)
	500	2903	625	625	1937	117	117	47.64	(11.27, 24.57, 3052.33)	237	112	1.23	(122.99, −399.49, 4347.55)	11533	4508.91	(36.56, −351.27, 3466.50)
14.	10	23	24	23	6	6	9	0.00	(4.00, −0.50, 0.00)	161	157	0.06	(1.32, −1.34, 0.00)	96	0.00	(2.24, 0.81, 0.00)
	50	209	163	163	67	44	45	0.02	(3.98, −0.68, 0.00)	17	15	0.01	(14.96, 6.45, 0.00)	39	0.02	(12.12, 1.22, 0.00)
	100	2517	991	370	2143	657	99	11.05	(2.35, 10.37, 46.22)	34	34	0.03	(48.38, −1.23, 0.00)	33	0.02	(68.98, −11.67, 10.09)
	250	740	111	101	645	43	38	7.49	(8.68, 1.97, 2.65)	114	112	0.70	(198.95, −50.91, 0.00)	83	0.16	(134.21, −50.90, 36.60)
	500	596	145	140	465	45	44	8.91	(35.34, −1.87, 0.00)	148	145	3.23	(157.79, −149.35, 0.00)	6652	3732.69	(96.44, −184.71, 1.72)
15.	10	210	202	192	43	44	44	0.01	(−3.32, 0.01, 13.95)	2666	2109	0.59	(−3.26, 0.00, 3.50)	230	0.02	(−3.29, 0.00, 8.02)
	50	539	489	463	94	99	109	0.09	(−0.50, 0.05, 41.48)	1116	966	1.66	(−6.78, 0.00, 96.77)	406	0.11	(1.58, 0.00, 176.73)
	100	363	142	239	239	38	127	0.44	(1.87, 0.09, 586.33)	411	409	4.61	(−68.14, 15.24, 66.65)	906	2.08	(10.40, 0.01, 444.10)
	250	4136	4092	2050	507	508	644	5.36	(8.01, 5.17, 339.08)	555	550	21.70	(−61.73, 0.00, 490.22)	6474	56.17	(96.70, 0.01, 1762.77)
	500	903	891	655	99	99	287	6.52	(22.01, 17.80, 472.84)	900	852	118.33	(−32.09, 0.02, 373053.52)	23714	1247.44	(757.32, 0.03, 112689.85)
Avg:		956	604	442	441	147	155	5.78		412	356	9.01		3391	577.54	

Table 6.4 Function and subgradient evaluations in MB-DC for unconstrained bi-objective test problems

P	n	Function evaluations									Subgradient evaluations						
		Algorithm 2		Line search		Escape procedure			Total		Algorithm 2		Escape procedure			Total	
		n_{f_1}	n_{f_2}	n_{f_1}	n_{f_2}	n_{f_1}	n_{f_2}	all	n_{f_1}	n_{f_2}	n_{ξ_1}	n_{ξ_2}	n_{ξ_1}	n_{ξ_2}	all	n_{ξ_1}	n_{ξ_2}
1.	2	125	156	74	4	0	0	321	520	481	43	42	0	0	60	103	102
2.	2	23	41	6	2	0	0	3	32	46	7	21	0	0	3	10	24
3.	2	36	52	12	4	0	0	3	51	59	11	24	0	0	3	14	27
4.	2	120	126	68	60	0	0	200	388	386	37	38	0	0	38	75	76
5.	4	24	32	5	1	4	0	5	38	38	12	16	5	0	5	22	21
6.	10	21	26	13	10	0	0	1	35	37	7	14	0	1	1	8	16
7.	10	129	127	146	65	0	0	48	323	240	49	47	0	0	10	59	57
8.	10	88	81	65	49	0	7	17	170	154	32	28	0	1	6	38	35
	50	347	220	325	94	0	0	4	676	318	217	90	0	0	2	219	92
	100	1051	504	53	43	0	0	125	1229	672	964	427	0	0	24	988	451
	250	1512	156	97	54	0	0	15	1624	225	1422	66	0	0	4	1426	70
	500	2782	157	217	79	0	0	4	3003	240	2682	57	0	0	2	2684	59
9.	10	83	85	87	83	0	0	17	187	185	31	33	0	0	6	37	39
	50	326	333	371	371	0	0	15	712	719	112	119	0	0	4	116	123
	100	430	1348	78	65	0	0	213	721	1626	350	1268	0	0	40	390	1308
	250	429	440	405	348	0	0	554	1388	1342	169	180	0	0	102	271	282
	500	604	624	630	626	0	0	1324	2558	2574	208	228	0	0	242	450	470
10.	10	51	69	29	23	0	5	53	133	150	21	30	0	4	15	36	49
	50	80	154	32	32	0	25	3	115	214	20	110	0	24	1	21	135
	100	216	1851	37	37	0	162	4	257	2054	36	1719	0	148	2	38	1869
	250	89	902	46	9	0	606	92	227	1609	34	849	0	597	18	52	1464
	500	88	1253	85	80	0	1370	263	436	2966	27	1211	0	1351	252	279	2814

categories in Tables 6.4 and 6.5. As we see, in some cases one of the objectives needs much more computational efforts than the others. In these cases, we may save some computational effort since we do not need to do the same computations for each objective unlike in MDBDC or MPB where all the objectives are calculated simultaneously. It is also noticed that the first objective has the most function evaluations in the line search but this is due to how the line search is implemented.

6.4.2 *Comparison Between Constraint Handling Strategies*

We have proposed two alternative ways to handle constraints, the default one and the sophisticated one. Before considering the numerical efficiency of these strategies, we give an example with a convex constraint. Due to convexity, we can omit the feasibility check in the line search by Theorem 4 if $\tau \in [0, 1]$.

We investigate Problem 1 having two objectives

Table 6.5 Function and subgradient evaluations in MB-DC for unconstrained test problems with three objectives

P	n	Function evaluations													Subgradient evaluations									
		Algorithm 2			Line search			Escape procedure				Total			Algorithm 2			Escape procedure				Total		
		n_{f_1}	n_{f_2}	n_{f_3}	n_{f_1}	n_{f_2}	n_{f_3}	n_{f_1}	n_{f_2}	n_{f_3}	all	n_{f_1}	n_{f_2}	n_{f_3}	n_{ξ_1}	n_{ξ_2}	n_{ξ_3}	n_{ξ_1}	n_{ξ_2}	n_{ξ_3}	all	n_{ξ_1}	n_{ξ_2}	n_{ξ_3}
11.	2	162	193	168	20	4	2	0	0	0	321	503	518	491	43	42	46	0	0	0	60	103	102	106
12.	4	162	111	155	43	43	3	0	0	0	136	341	290	294	30	32	80	0	0	0	37	67	69	117
13.	10	124	117	119	57	47	37	0	7	0	42	223	213	198	38	34	35	0	1	0	13	51	48	48
	50	150	96	96	63	39	29	0	0	0	5	218	140	130	84	30	30	0	0	0	3	87	33	33
	100	357	509	729	92	80	48	0	0	0	65	514	654	842	221	377	597	0	0	0	18	239	395	615
	250	940	391	395	308	127	74	0	0	0	67	1315	585	536	682	133	137	0	0	0	20	702	153	157
	500	2243	423	423	655	197	197	0	0	0	5	2903	625	625	1934	114	114	0	0	0	3	1937	117	117
14.	10	18	15	21	4	1	1	0	7	0	1	23	24	23	5	4	8	0	1	0	1	6	6	9
	50	100	77	78	32	9	8	0	0	0	77	209	163	163	46	23	24	0	0	0	21	67	44	45
	100	2278	779	197	83	56	17	0	0	0	156	2517	991	370	2115	629	71	0	0	0	28	2143	657	99
	250	693	91	86	42	15	10	0	0	0	5	740	111	101	642	40	35	0	0	0	3	645	43	38
	500	525	105	104	54	23	19	0	0	0	17	596	145	140	459	39	38	0	0	0	6	465	45	44
15.	10	101	102	102	60	51	41	0	0	0	49	210	202	192	32	33	33	0	0	0	11	43	44	44
	50	342	347	357	191	136	100	0	0	0	6	539	489	463	90	95	105	0	0	0	4	94	99	109
	100	313	97	216	33	28	4	0	0	2	17	363	142	239	233	32	120	0	0	1	6	239	38	127
	250	1102	1103	1242	2532	2487	306	0	0	0	502	4136	4092	2050	385	386	522	0	0	0	122	507	508	644
	500	235	235	425	532	520	94	0	0	0	136	903	891	655	73	73	261	0	0	0	26	99	99	287

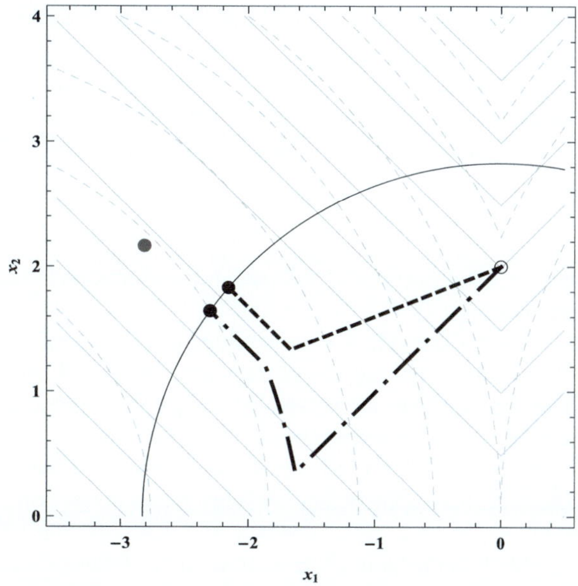

Fig. 6.1 Example of two constraint handling strategies with a convex constraint

$$
\begin{aligned}
f_1(\boldsymbol{x}) &= |x_1 - 1| + 200 \max\{0,\ |x_1| - x_2\} - 100(|x_1| - x_2)\\
f_2(\boldsymbol{x}) &= x_2 + 0.1(x_1^2 + x_2^2) + 10 \max\{0, -x_2\} - |x_1| - |x_2|.
\end{aligned}
$$

In this problem, the dimension n is 2. The convex constraint is of the form:

$$g(\boldsymbol{x}) = x_1^2 + x_2^2 - 8 \le 0.$$

Instead of $\boldsymbol{x}_0 = (-1.2, 1)^T$ used in Problem 1, we use here $\boldsymbol{x}_0 = (0, 2)^T$.

In Fig. 6.1, the gray curves describe the contours of f_1, the dashed gray curves the contours of f_2, and the black curve the contour of g. With the gray dot, we have marked the solution of the unconstrained problem, the black circle refers to $\boldsymbol{x}_0$, and the black dots refer to the solutions. The dot-dashed and dashed black lines represent the solution paths obtained with the default and sophisticated strategy, respectively. As we see, the strategy has an impact on the solution.

The computational efforts needed in the unconstrained case and in the constrained case with both constraint handling strategies are given in Table 6.6. The default strategy yields a solution having objective values close to the unconstrained solution while the sophisticated strategy provides more variety. Compared with the unconstrained case, both constrained cases need less function and subgradient evaluations. The CPU time is practically the same for every case. By comparing the constraint handling strategies, we see that the sophisticated strategy uses less than the third of function and subgradient evaluations of objectives. In addition, the number of function evaluations for the constraint is smaller. However, the number of subgradient evaluations for the sophisticated strategy is greater than for the default strategy,

Table 6.6 Numerical results for the example with the convex constraint

Strategy	n_{f_1}	n_{f_2}	n_g	n_{ξ_1}	n_{ξ_2}	n_{ξ_g}	CPU	$f(\boldsymbol{x}^*)$
Unconstrained	719	714		143	143		0.03	(68.62, −1.55)
Default	352	394	146	75	76	21	0.03	(68.89, −1.50)
Sophisticated	105	106	106	25	29	106	0.02	(35.66, −1.36)

but the total number of subgradient evaluations in the sophisticated strategy is still smaller.

Next we consider the constrained test problems from Table 4 in [47]. We have 14 test problems having two or three objectives and they all include one DC constraint. The test problems are solved by using both constraint handling strategies, MDBDC, and MPB. These results are given in Table 6.7.

As in Problems 1–15, we observe some better solutions between the solvers. MB-DC with the sophisticated strategy has a better solution than MDBDC or MPB in Problem 21, $n = 100$. However, the solution with MB-DC is worse than MDBDC and MPB in Problem 21, $n = 10$ with the default strategy and worse than MDBDC in Problem 17 with the sophisticated strategy. When the average computational effort is compared, MDBDC uses less effort, then comes MB-DC, and lastly MPB. Nevertheless, the average number of subgradient evaluations for constraint is equal with the default version of MB-DC and MDBDC.

By considering the differences between two constraint handling strategies, we notice that the CPU time is on the same magnitude even if the default strategy has a little bit smaller average value. The average numbers of function and subgradient evaluations in general are smaller with the default strategy, but the average function and subgradient evaluations are less or equal magnitude for f_2 and f_3 with the sophisticated strategy. It is interesting that while we compare the solutions, the default strategy yields a better solution in three cases in Problems 17 and 18 with $n = 10, 50$. In its turn, the sophisticated strategy yields a better solution in six cases in Problems 18 with $n = 250, 500$ and 21 with $n = 10, 50, 250, 500$. Therefore, it is justified to include possibility for both strategies such that a suitable selection for each situation can be obtained.

6.4.3 *Interactive Examples*

We illustrate with two examples the role of the DM in MB-DC. In these examples, we select $L = 500$ and $n_{dm}^{\max} = 3$ meaning that we can interact at most three times with the DM. As stated in Remark 1, the DM can select either one feasible individual descent direction to improve the corresponding objective or the escape procedure. If the escape procedure is selected, the interaction with the DM ends even if $n_{dm}^{\max}$

Table 6.7 Numerical results for constrained test problems

	P	n	MB-DC										MDBDC				MPB		
			n_{f_1}	n_{f_2}	n_{f_3}	n_g	n_{ξ_1}	n_{ξ_2}	n_{ξ_3}	n_{ξ_g}	CPU	$f(x^*)$	n_f	n_ξ	CPU	$f(x^*)$	n_ξ	CPU	$f(x^*)$
Default	16.	2	362	397		283	55	93		13	0.01	(0.71, 0.71)	44	38	0.01	(0.71, 0.71)	65	0.00	(0.71, 0.71)
	17.	4	865	901		593	187	247		98	0.03	(41.85, 8.34)	147	136	0.08	(33.73, 8.65)	12	0.00	(271.37, 9.20)
	18.	10	187	185		63	37	39		6	0.00	(−1.02, 31.59)	24	24	0.02	(−8.46, 82.24)	583	0.0312	(−8.45, 81.60)
		50	712	719		212	116	123		4	0.14	(3.73, 49.07)	65	63	0.09	(−47.00, 315.13)	2009	1.13	(−48.27, 422.61)
		100	721	1923		245	390	1605		40	2.25	(−80.86, 629.84)	48	48	0.11	(−97.88, 824.93)	3886	7.42	(−98.46, 951.47)
		250	1510	1343		1083	282	294		170	0.78	(29.43, 497.29)	62	52	0.34	(−241.39, 1900.22)	17916	377.70	(−247.85, 2215.82)
		500	2558	2574		1684	450	470		242	4.25	(−219.89, 1517.60)	122	64	0.72	(−491.29, 4014.57)	23590	1866.05	(−496.19, 4718.86)
	19.	2	200	199	189	114	44	43	45	21	0.02	(12.69, −1.24, 12.69)	12	12	0.00	(18.66, −1.28, 18.66)	11	0.00	(21.82, −1.29, 21.82)
	20.	4	296	260	227	144	61	64	67	38	0.00	(289.79, 3.24, 2.14)	54	38	0.03	(173.21, 1.75, 4.03)	20	0.00	(121.26, 1.22, 5.38)
	21.	10	68	64	64	29	19	18	18	2	0.00	(2.98, 1.94, 0.58)	22	20	0.02	(2.90, −0.59, 0.00)	68	0.00	(2.52, −0.93, 0.00)
		50	1037	1022	1022	894	103	102	102	42	0.05	(38.99, 12.10, 3.24)	54	54	0.03	(3.00, −46.50, 93.00)	52	0.02	(8.47, −32.30, 17.70)
		100	523	406	107	44	430	319	30	1	1.89	(42.53, 21.37, 58.62)	77	77	0.27	(3.87, −96.77, 161.61)	64	0.05	(7.25, −92.48, 178.88)
		250	355	319	318	254	60	33	32	5	0.13	(250.15, 60.06, 234.27)	44	26	0.06	(14.81, −235.13, 286.13)	96	0.31	(11.11, −238.50, 474.24)
		500	159	150	150	109	28	20	20	2	0.05	(472.06, 124.64, 24.39)	39	31	0.16	(2.35, −498.44, 974.06)	184	1.53	(14.94, −489.08, 873.94)
	Avg:		682	747	297	411	162	248	45	49	0.69		58	49	0.14		3468	161.02	
Sophisticated	16.	2	97	92		116	36	39		60	0.02	(0.71, 0.71)	44	38	0.01	(0.71, 0.71)	65	0.00	(0.71, 0.71)
	17.	4	188	205		220	51	109		131	0.00	(217.39, 9.20)	147	136	0.08	(33.73, 8.65)	12	0.00	(271.37, 9.20)
	18.	10	396	395		278	66	66		84	0.02	(8.39, 53.97)	24	24	0.02	(−8.46, 82.24)	583	0.0312	(−8.45, 81.60)
		50	532	532		440	54	54		65	0.05	(86.80, 248.76)	65	63	0.09	(−47.00, 315.13)	2009	1.13	(−48.27, 422.61)
		100	425	1546		1427	79	1216		1234	1.03	(−5.35, 340.69)	48	48	0.11	(−97.88, 824.93)	3886	7.42	(−98.46, 951.47)
		250	1662	1704		1307	256	299		314	1.17	(−157.78, 312.36)	62	52	0.34	(−241.39, 1900.22)	17916	377.70	(−247.85, 2215.82)
		500	2362	2391		1944	341	372		387	4.05	(−260.53, 744.82)	122	64	0.72	(−491.29, 4014.57)	23590	1866.05	(−496.19, 4718.86)
	19.	2	241	132	202	275	99	35	97	183	0.00	(2.46, −1.03, 2.46)	12	12	0.00	(18.66, −1.28, 18.66)	11	0.00	(21.82, −1.29, 21.82)
	20.	4	464	467	418	496	101	115	138	198	0.02	(312.09, 3.28, 2.05)	54	38	0.03	(173.21, 1.75, 4.03)	20	0.00	(121.26, 1.22, 5.38)
	21.	10	158	144	145	138	34	33	38	51	0.00	(2.79, 1.10, 0.00)	22	20	0.02	(2.90, −0.59, 0.00)	68	0.00	(2.52, −0.93, 0.00)
		50	81	80	81	68	14	13	16	17	0.00	(29.70, −3.95, 0.00)	54	54	0.03	(3.00, −46.50, 93.00)	52	0.02	(8.47, −32.30, 17.70)
		100	5584	267	242	5569	5347	75	56	5392	9.83	(2.20, −97.47, 158.46)	77	77	0.27	(3.87, −96.77, 161.61)	64	0.05	(7.25, −92.48, 178.88)
		250	98	90	92	84	22	16	24	32	0.02	(158.79, −53.56, 0.00)	44	26	0.06	(14.81, −235.13, 286.13)	96	0.31	(11.11, −238.50, 474.24)
		500	315	280	283	269	58	53	58	73	0.16	(399.37, −49.55, 0.00)	39	31	0.16	(2.35, −498.44, 974.06)	184	1.53	(14.94, −489.08, 873.94)
	Avg:		900	595	209	902	468	178	61	587	1.17		58	49	0.14		3468	161.02	

Table 6.8 The information given to the DM in the first example

Round	Option	Change in value of f_1	Change in value of f_2
8	0	Escape procedure improving both objectives or stop	
	1	from 13.06 to 2.01	from −0.96 to −0.81
	2	from 13.06 to 377.47	from −0.96 to −2.40
9	0	Escape procedure improving both objectives or stop	
	1	from 2.01 to 1.48	from −0.81 to −0.44
	2	from 2.01 to 372.43	from −0.81 to −2.36

would not be met yet. Throughout the examples, one round considers the steps from Step 1 of Algorithm 1 to the next visit in Step 1 of Algorithm 1.

First, we consider Problem 1 and the communication is described in Table 6.8. The first possibility to communicate with the DM is in the round 8. From the options, the DM selects the option 1 in order to gain improvement for f_1 even if f_2 deteriorates but the deterioration is on an acceptable level. When the second interaction happens in the round 9, the DM is happy with the current state of the solution process. Thus, (s)he selects the escape procedure and lets the algorithm to continue until the stopping condition is met. In this case, the algorithm stops after the escape procedure meaning that the solution in round 9 shown to the DM is an approximate weakly Pareto stationary. Thus, the final solution is $f(\boldsymbol{x}^*) = (2.01, -0.81)$ and the solution obtained in Table 6.2 without the DM is $f(\boldsymbol{x}) = (12.55, -1.29)$. In total, this solution needs 62 function evaluations and 18 subgradient evaluations for f_1 and 65 function evaluations and 17 subgradient evaluations for f_2.

The second example is Problem 13 with $n = 250$ and the alternatives for the communication are given in Table 6.9. In round 18, the DM selects the option 3 in order to improve f_3. Even if the increase in the other two objectives is quite large, the DM is willing to take a risk, since the impaired objectives are likely to be improved while the solution process continues. However, if the current iteration point is already weakly Pareto stationary, we have still interactions left to improve f_1 and f_2. The next chance for interaction occurs in the round 29. In this case, the DM is already satisfied with the current level of f_3 even if the improvement for it could be significant. Instead of that, the DM selects f_1 to be improved, and meanwhile, f_3 decreases a little. The third and final interaction occurs in the round 37. The DM is content with the solution and selects 0 meaning the execution of the escape procedure. The final solution is obtained after a total of 50 rounds and it is $f(\boldsymbol{x}^*) = (4.99, 82.86, 139.99)$. The solution obtained without the DM in Table 1 is $f(\boldsymbol{x}^*) = (9.21, 29.11, 974.49)$. As we see, this solution improves f_2 more while the DM does not consider this as important and does not select f_2 to be improved even if the major decrease is available at other objectives' expense. The computational efforts for f_1, f_2, and f_3 are 1184, 431, and 449 function evaluations, respectively, and 850, 97, and 115 subgradient evaluations, respectively.

Table 6.9 The information given to the DM in the second example

Round	Option	Change in value of f_1	Change in value of f_2	Change in value of f_3
18	0	Escape procedure improving all objectives or stop		
	1	from 24.53 to 16.24	from 31.76 to 34.71	from 1005.66 to 1008.54
	2	from 24.53 to 160.99	from 31.76 to −91.34	from 1005.66 to 1888.04
	3	from 24.53 to 58.05	from 31.76 to 87.29	from 1005.66 to 289.86
29	0	Escape procedure improving all objectives or stop		
	1	from 13.85 to 10.07	from 77.23 to 84.27	from 172.39 to 150.02
	2	from 13.85 to 179.36	from 77.23 to −113.33	from 172.39 to 1849.31
	3	from 13.85 to 7.95	from 77.23 to 92.88	from 172.39 to 52.05
37	0	Escape procedure improving all objectives or stop		
	1	from 7.08 to 5.16	from 83.10 to 85.89	from 141.63 to 141.75
	2	from 7.08 to 180.49	from 83.10 to −112.19	from 141.63 to 1832.23
	3	from 7.08 to 21.04	from 83.10 to 89.49	from 141.63 to 97.13

6.5 Conclusions

The paper describes a new descent type method for constrained nonsmooth multiobjective DC optimization. The method finds a common descent direction by using descent directions of the individual objectives obtained by using a special bundle method. In order to guarantee the stationarity, the escape procedure with the improvement function is applied [47]. The proposed MB-DC method has an ability to modify the solution process based on the DM's wishes. Additionally, two different strategies for constraint handling are introduced. The method is proven to be finitely convergent to an approximate weakly Pareto stationary solution.

The numerical results show that MB-DC works well in practice. The new method is compared with one multiobjective DC method [47] and one multiobjective nonconvex method [39]. Both of these reference methods use the improvement function as a tool to obtain a common descent direction. Based on the results, MB-DC succeeds in solving a problem in which two other improvement function based methods in the comparison fail. In general, the nature of the solutions obtained with the reference methods seems to be more similar when compared to each other than to the solutions obtained with MB-DC, even if all the methods produce mainly different solutions. Another interesting observation is that when the number of objective functions is increased, the computational effort of MB-DC is on the same magnitude while the reference methods need more computational effort compared to the bi-objective case.

The future development includes the usage of the parallel computing in the implementation of MB-DC during the determination of the individual search directions. Another way to improve the numerical behaviour of MB-DC would be a more efficient line search strategy. The similar interactivity and constraint handling techniques than used in this paper could be applied to other methods using the same core idea for the direction finding like [15, 49, 55]. Furthermore, the outline for the direction finding could be combined to other bundle methods to make more variants for problems of different types.

Acknowledgements The authors are grateful to Prof. Napsu Karmitsa for her valuable comments during the preparation of this paper. The authors would like to extend their thank to Prof. Marko Mäkelä for his continuous support.

Funding The research is financially supported by Academy of Finland Projects No. 319274 led by Prof. Napsu Karmitsa and No. 310507 led by Prof. Tero Aittokallio and University of Turku.

Competing Interests The authors have no conflicts of interest to declare that are relevant to the content of this chapter.

References

1. van Ackooij W, de Oliveira W (2019) Non-smooth DC-constrained optimization: constraint qualification and minimizing methodologies. Optim Methods Softw 34(4):890–920
2. Astorino A, Fuduli A, Giallombardo G, Miglionico G (2019) SVM-based multiple instance classification via DC optimization. Algorithms 12(12):249
3. Astorino A, Miglionico G (2016) Optimizing sensor cover energy via DC programming. Optim Lett 10(2):355–368
4. Bagirov A, Karmitsa N, Mäkelä MM (2014) Introduction to nonsmooth optimization: theory, practice and software. Springer, Cham, Heidelberg
5. Bagirov AM, Karmitsa N, Taheri S (2020) Partitional clustering via nonsmooth optimization: clustering via optimization. Springer, Cham
6. Bello Cruz JY, Iusem AN (2011) A strongly convergent method for nonsmooth convex minimization in Hilbert spaces. Numer Funct Anal Optim 32(10):1009–1018
7. Bonnel H, Iusem AN, Svaiter BF (2005) Proximal methods in vector optimization. SIAM J Optim 15(4):953–970
8. Bosman PAN, de Jong ED (2005) Exploiting gradient information in numerical multi–objective evolutionary optimization. In: Proceedings of the 7th annual conference on genetic and evolutionary computation, GECCO '05, pp 755–762
9. Buchanan J, Gardiner L (2003) A comparison of two reference point methods in multiple objective mathematical programming. Eur J Oper Res 149(1):17–34
10. Carrizosa E, Guerrero V, Romero Morales D (2018) Visualizing data as objects by DC (difference of convex) optimization. Math Program 169:119–140
11. de Carvalho Bento G, Bitar S, da Cruz Neto J, Soubeyran A, de Oliveira Souza J (2020) A proximal point method for difference of convex functions in multi-objective optimization with application to group dynamic problems. Comput Optim Appl 75:263–290
12. Clarke FH (1983) Optimization and nonsmooth analysis. Wiley, New York

13. Craft D, Halabi T, Shih HA, Bortfeld T (2007) An approach for practical multiobjective IMRT treatment planning. Int J Radiat Oncol Biol Phys 69(5):1600–1607
14. Désidéri JA (2009) Multiple-gradient descent algorithm (MGDA). Technical Report 6953, INRIA Research Report
15. Désidéri JA (2012) Multiple-gradient descent algorithm (MGDA) for multiobjective optimization. C. R. Acad Sci Paris Sér. I. **350**, 313–318
16. Fuduli A, Gaudioso M, Giallombardo G (2004) A DC piecewise affine model and bundling technique in nonconvex nonsmooth minimization. Optim Methods Softw 19(1):89–102
17. Gaudioso M, Giallombardo G, Miglionico G, Vocaturo E (2020) Classification in the multiple instance learning framework via spherical separation. Soft Comput 24:5071–5077
18. Hiriart-Urruty JB (1985) Generalized differentiability, duality and optimization for problems dealing with differences of convex functions. Lect Notes Econ Math Syst 256:37–70
19. Horst R, Thoai NV (1999) DC programming: overview. J Optim Theory Appl 103(1):1–43
20. Hoseini Monjezi N, Nobakhtian S (2020) An inexact multiple proximal bundle algorithm for nonsmooth nonconvex multiobjective optimization problems. Ann Oper Res. https://doi.org/10.1007/s10479-020-03808-0
21. Hutterer A, Jahn J (2003) On the location of antennas for treatment planning in hyperthermia. OR Spectr 25(3):397–412
22. Ji Y, Goh M, de Souza R (2016) Proximal point algorithms for multi-criteria optimization with the difference of convex objective functions. J Optim Theory Appl 169(1):280–289
23. Ji Y, Qu S (2017) Proximal point algorithms for vector DC programming with applications to probabilistic lot sizing with service levels. Discret Dyn Nat Soc. https://doi.org/10.1155/2017/5675183
24. Jin Y (2006) Multi-objective machine learning. Studies in computational intelligence, vol 16. Springer, Berlin, Heidelberg
25. Joki K, Bagirov A, Karmitsa N, Mäkelä MM (2016) A proximal bundle method for nonsmooth DC optimization utilizing nonconvex cutting planes. J Glob Optim 68(3):501–535
26. Joki K, Bagirov A, Karmitsa N, Mäkelä MM, Taheri S (2018) Double bundle method for finding Clarke stationary points in nonsmooth DC programming. SIAM J Optim 28(2):1892–1919
27. Joki K, Bagirov A, Karmitsa N, Mäkelä MM, Taheri S (2020) Clusterwise support vector linear regression. Eur J Oper Res 287(1):19–35
28. Joki K, Bagirov AM (2020) Bundle methods for nonsmooth DC optimization. In: Bagirov A, Gaudioso M, Karmitsa N, Mäkelä MM, Taheri S (eds) Numerical nonsmooth optimization: state of the art algorithms. Springer, Cham, pp 263–296
29. Karas E, Ribeiro A, Sagastizábal C, Solodov M (2009) A bundle-filter method for nonsmooth convex constrained optimization. Math Program 116(1):297–320
30. Karmitsa N, Bagirov A, Taheri S (2017) New diagonal bundle method for clustering problems in large data sets. Eur J Oper Res 263(2):367–379
31. Kiwiel KC (1985) A descent method for nonsmooth convex multiobjective minimization. Large Scale Syst 8(2):119–129
32. Kiwiel KC (1985) Methods of descent for nondifferentiable optimization. Springer, Berlin
33. Le Thi HA, Pham Dinh T (1997) Solving a class of linearly constrained indefinite quadratic problems by d.c. algorithms. J. Glob. Optim. **11**(3), 253–285
34. Le Thi HA, Pham Dinh T (2005) The DC (difference of convex functions) programming and DCA revisited with DC models of real world nonconvex optimization problems. Ann Oper Res 133(1):23–46
35. Lemaréchal C, Nemirovskii A, Nesterov Y (1995) New variants of bundle methods. Math Program 69:111–147
36. Lukšan L (1984) Dual method for solving a special problem of quadratic programming as a subproblem at linearly constrained nonlinear minimax approximation. Kybernetika 20:445–457
37. Mäkelä MM (2003) Multiobjective proximal bundle method for nonconvex nonsmooth optimization: Fortran subroutine MPBNGC 2.0. Technical Report B 13/2003, Reports of the Department of Mathematical Information Technology, Series B, Scientific Computing, University of Jyväskylä, Jyväskylä

38. Mäkelä MM, Eronen VP, Karmitsa N (2014) On nonsmooth multiobjective optimality conditions with generalized convexities. In: Rassias TM, Floudas CA, Butenko S (eds) Optim Sci Eng. Springer, New York, pp 333–357
39. Mäkelä MM, Karmitsa N, Wilppu O (2016) Proximal bundle method for nonsmooth and nonconvex multiobjective optimization. In: Tuovinen T, Repin S, Neittaanmäki P (eds) Mathematical modeling and optimization of complex structures, Computational methods in applied sciences, vol 40. Springer, Cham, pp 191–204
40. Mäkelä MM, Montonen O (2020) New multiobjective proximal bundle method with scaled improvement function. In: Bagirov A, Gaudioso M, Karmitsa N, Mäkelä MM, Taheri S (eds) Numerical nonsmooth optimization: state of the art algorithms. Springer, Cham, pp 461–479
41. Mäkelä MM, Neittaanmäki P (1992) Nonsmooth optimization: analysis and algorithms with applications to optimal control. World Scientific Publishing Co., Singapore
42. Miettinen K (1999) Nonlinear multiobjective optimization. Kluwer Academic Publishers, Boston
43. Miettinen K, Eskelinen P, Ruiz F, Luque M (2010) Nautilus method: An interactive technique in multiobjective optimization based on the nadir point. Eur J Oper Res 206(2):426–434
44. Miettinen K, Mäkelä MM (1995) Interactive bundle-based method for nondifferentiable multiobjective optimization: NIMBUS. Optimization 34(3):231–246
45. Mifflin R (1977) An algorithm for constrained optimization with semi-smooth functions. Math Oper Res 2(2):191–207
46. Montonen O, Eronen VP, Ranta T, Huttunen JAS, Mäkelä MM (2020) Multiobjective mixed integer nonlinear model to plan the schedule for the final disposal of the spent nuclear fuel in Finland. Mathematics 8(4):528
47. Montonen O, Joki K (2018) Bundle-based descent method for nonsmooth multiobjective DC optimization with inequality constraints. J Glob Optim 72(3):403–429
48. Montonen O, Joki K (2020) Multiobjective double bundle method for DC optimization. In: Bagirov A, Gaudioso M, Karmitsa N, Mäkelä MM, Taheri S (eds) Numerical nonsmooth optimization: state of the art algorithms. Springer, Cham, pp 481–497
49. Montonen O, Karmitsa N, Mäkelä MM (2018) Multiple subgradient descent bundle method for convex nonsmooth multiobjective optimization. Optimization 67(1):139–158
50. Montonen O, Ranta T, Mäkelä MM (2019) Planning the schedule for the disposal of the spent nuclear fuel with interactive multiobjective optimization. Algorithms 12(12):252
51. Mukai H (1979) Algorithms for multicriterion optimization. IEEE Trans Autom Control AC-25 177–186
52. de Oliveira W (2019) Proximal bundle methods for nonsmooth DC programming. J Glob Optim 75:523–563
53. Outrata J, Koĉvara M, Zowe J (1998) Nonsmooth approach to optimization problems with equilibrium constraints. Theory, applications and numerical results. Kluwer Academic Publishers, Dordrecht
54. Pham Dinh T, Le Thi HA (1997) Convex analysis approach to DC programming: theory, algorithms and applications. Acta Math Vietnam 22(1):289–355
55. Poirion F, Mercier Q, Désidéri JA (2017) Descent algorithm for nonsmooth stochastic multiobjective optimization. Comput Optim Appl 68(2):317–331
56. Purshouse RC, Deb K, Mansor MM, Mostaghim S, Wang R (2014) A review of hybrid evolutionary multiple criteria decision making methods. In: 2014 IEEE congress on evolutionary computation (CEC), pp 1147–1154
57. Qu S, Goh M, Wu SY, De Souza R (2014) Multiobjective DC programs with infinite convex constraints. J Glob Optim 59(1):41–58
58. Rangaiah GP (2009) Multi-objective optimization: techniques and applications in chemical engineering. Advances in process systems engineering. World Scientific, Singapore
59. Sagastizábal C, Solodov M (2005) An infeasible bundle method for nonsmooth convex constrained optimization without a penalty function or a filter. SIAM J Optim 16(1):146–169
60. Sun WY, Sampaio RJB, Candido MAB (2003) Proximal point algorithm for minimization of DC functions. J Comput Math 21(4):451–462

61. Wang H, Olhofer M, Jin Y (2017) A mini-review on preference modeling and articulation in multi-objective optimization: current status and challenges. Complex Intell Syst 3:233–245
62. Wang S (1989) Algorithms for multiobjective and nonsmooth optimization. In: Kleinschmidt P, Radermacher F, Sweitzer W, Wildermann H (eds) Methods of operations research, 58. Athenaum Verlag, Hanstein, pp 131–142
63. Wierzbicki AP (1982) A mathematical basis for satisficing decision making. Math Model 3(5):391–405

Chapter 7
Variational Study of Mesh and Shape Optimization

Alain Dervieux, Frédéric Alauzet, and Didier Chargy

Abstract This paper focuses on the combination of design optimization with anisotropic mesh adaptation. The key components are: unstructured mesh for CFD, use of adjoint state for design, use of metric-based mesh adaptation.

Keywords Anisotropic mesh adaptation · Goal-oriented optimization · Adjoint-based sensitivity analysis · Shape optimization in CFD · High-fidelity simulation

7.1 Motivations

In order to design the new transport devices answering to the climatic challenge, the needs in High-Fidelity CFD simulation are increasing. High-Fidelity CFD simulation demands:

- high-fidelity models like RANS and hybrid RANS/LES,
- high-fidelity approximations with a strong control of the approximation error.

This paper focuses on the combination of design with anisotropic mesh adaptation. The key components are: unstructured mesh for CFD, use of adjoint state for design, use of mesh adaptation.

A. Dervieux (✉)
Lemma, Biot, France, and University Côte d'Azur/INRIA Projet Ecuador, Sophia-Antipolis, France
e-mail: alain.dervieux@inria.fr

F. Alauzet
GammaO, Équipe INRIA-ONERA-Centre INRIA de Saclay, University Paris-Saclay, Palaiseau, France
e-mail: frederic.alauzet@inria.fr

D. Chargy
Lemma, Biot, Sophia-Antipolis, France
e-mail: didier.chargy@lemma-ing.com

T. Tuovinen et al. (eds.), *Challenges in Design Methods, Numerical Tools and Technologies for Sustainable Aviation, Transport and Industry*, Computational Methods in Applied Sciences 17, https://doi.org/10.1007/978-3-031-98675-8_7

The adjoint state was popularized by J. L. Lions in his book "Optimal Control" [1] in 60's. It was demonstrated by O. Pironneau for CFD design based on the Full Potential model [2] and then by A. Jameson [3] for the Euler model. Unstructured-mesh CFD was strongly supported and developed in France by Dassault Aviation (J. Périaux, P. Perrier, program HERMES). Adjoint methods use intensively code differentiation, strongly supported in France by Dassault Aviation (J. Périaux, B. Stoufflet, program GENIE). A cooperation on metric-based mesh adaptation started very early between INRIA, Paris 6 and Dassault (J. Périaux, B. Stoufflet).

It appears today that with the new algorithms for mesh adaptation, new efficient and accurate automatical tools are available for maximizing the quality of high-fidelity prediction. The advantages of these tools will probably in the near future make their use mandatory in the industrial practice. Automatic tools are already available for design optimization. Today these tools rely on a generation of non-adaptative or poorly adaptative CFD. It is then important to propose for the very near future a composite algorithm which combines in best conditions the novel mesh adaptation technology with most recent or novel design optimization loops.

The first part of this paper gives some state-of-the-art of the anisotropic goal-oriented mesh adaptation algorithms. The second part examines the integration of these algorithms in a design loop.

7.2 Anisotropic Mesh Adaptation

The ingredients of anisotropic mesh adaptation are in short:

- Spalart-Allmaras RANS and DDES vertex-centered FEM-FVM using tetrahedra,
- anisotropic metric-based local remesher.

They are described and discussed in details in [4, 5]. The Goal Oriented (GO) version minimizes the error on a prescribed scalar output. The error analysis relies on an adjoint state. The double loop mesh adaptation algorithm (Fig. 7.1) contains an *inner loop* which iteratively adapts the mesh and recomputes the flow, the adaptation being performed while maintaining the total number of nodes in the mesh. The double loop mesh adaptation algorithm contains also an *outer loop* of refinement, that is responsible of increasing the total number of nodes in such a way that the double loop organizes the convergence of the discrete solution to the exact one.

7.2.1 Earlier Higher-Order Convergence and Important Details of Flow

Anisotropic adaptation results in *earlier* high-order convergence.

We illustrate this fact with a case of the AIAA High Lift Prediction Workshop [6, 11] see Fig. 7.2. In the right figure, the convergence of the lift coefficient is depicted.

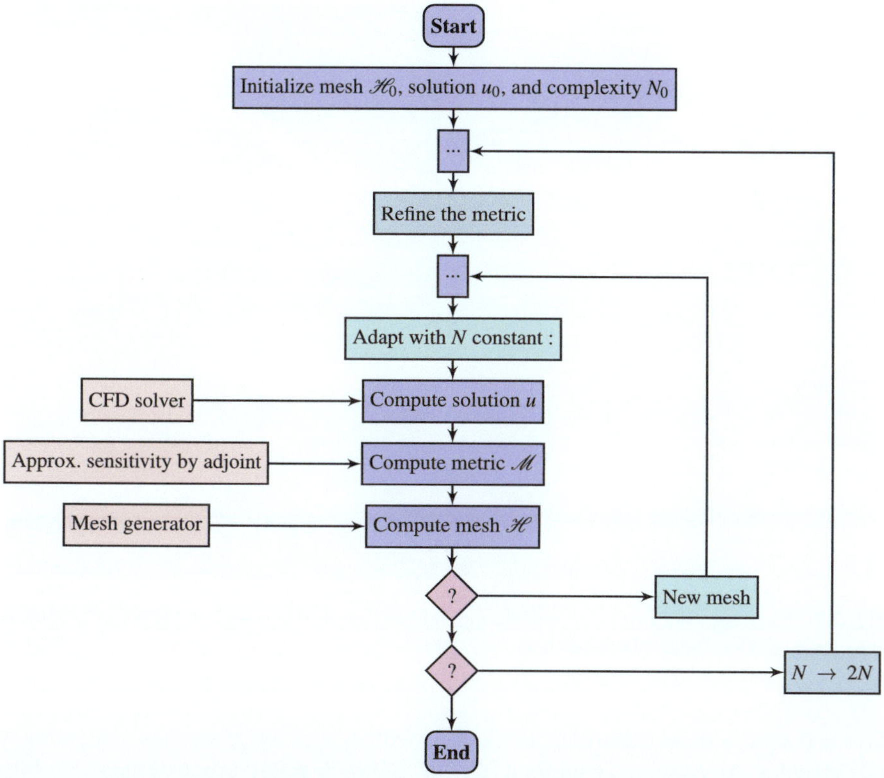

Fig. 7.1 Anisotropic double loop Goal Oriented mesh adaptation

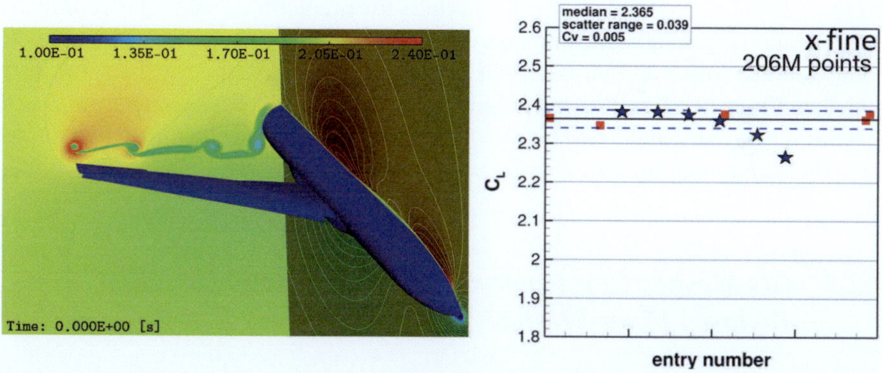

Fig. 7.2 High-lift workshop. *Left*: Goal Oriented adaptation with lift as functional. Local Mach number computed with an adapted mesh of 5.41M nodes. *Right*: lift predictions against different computations. ■: five computations with the workshop mesh of 206M cells. ★ (right to left) are produced by a Goal Oriented adaptation loop with lift as functional, with successive meshes of (right to left) 0.68M, 1.35M, 2.73M, 5.41M, 10.5M, and 20.0M nodes

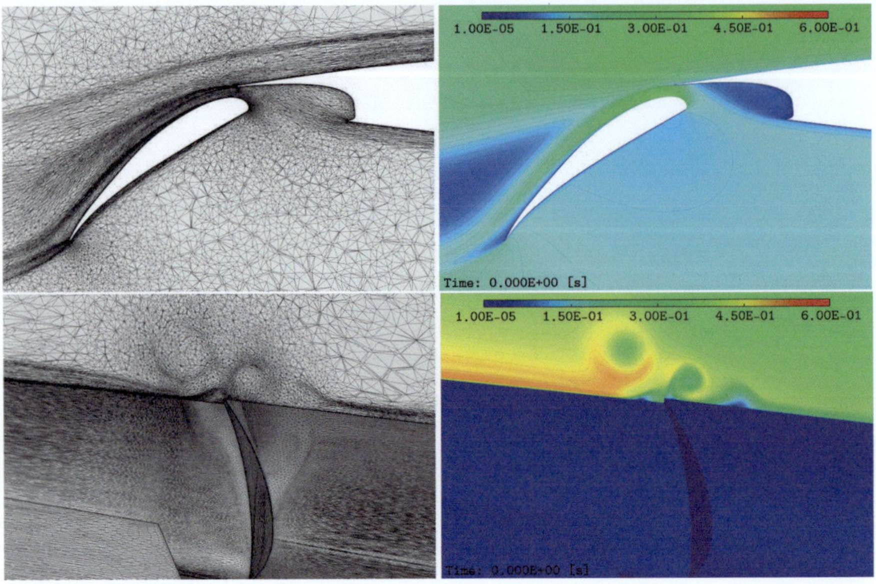

Fig. 7.3 Same high-lift workshop calculation, adapted mesh composed of 5.41M cells and the associated solution (local Mach number)

Five red square show different calculations with best practice meshes composed of 205M nodes. An confidence interval $[C_L = 2.365 \pm 0.05]$ is then derived. Six blue stars are also depicted, produced by the application of the GO mesh adaptation double loop (values of lift at end of outer loops). We observe that meshes with 2.73M, 5.41M, 10.5M, 20.0M nodes of the process produce lift predictions inside the confidence interval, showing an accuracy comparable to the accuracy of the five 205 M nodes computations. Since adjoint-based/GO approaches are designed to find the mesh which minimizes the error on the target scalar output, this method *concentrate mesh in regions which are important for the output accuracy*. This is illustrated by pictures from the above computation (Fig. 7.3).

7.2.2 GO Meshes Can Produce Much Better Results Than Taylored Finer Meshes

This statement is illustrated by computations from the 2nd AIAA Sonic-Boom Prediction Workshop, [7]. The geometry, depicted in Fig. 7.4, is a shape for a supersonic aircraft with mitigated sonic boom. The pressure perturbation is measured in a horizontal plane at some distance under the aircraft. A non-adapted and an adapted calculation have been performed with a post-processing [8] showing the local error

Fig. 7.4 View of the 2nd AIAA Sonic-Boom prediction workshop shape used for comparison between Goal Oriented mesh and workshop Taylored mesh

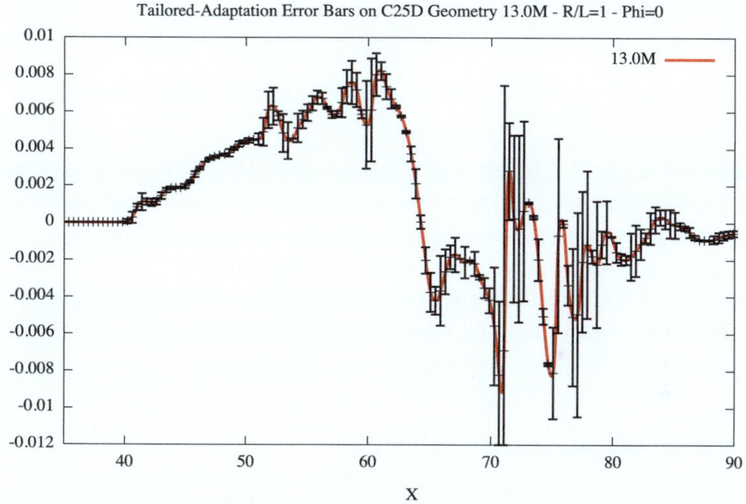

Fig. 7.5 13M Taylored mesh $dp/pinf$ on an axis under the aircfaft. Black vertical intervals are estimates of local approximation uncertainty and indicate very large uncertainties for $60 \leq X \leq 80$

(black vertical intervals). It appears that the computation with a 13M nodes taylored mesh (aligned with main shock direction) still show large local errors. In contrast, the GO approach with 5.9M nodes shows small error intervals almost everywhere (Figs. 7.5 and 7.6).

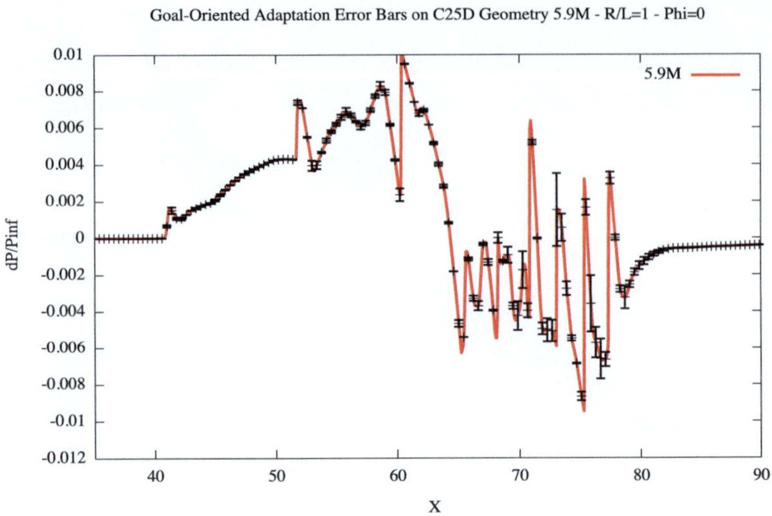

Fig. 7.6 5.9M adapted mesh $dp/pinf$ on an axis under the aircfaft. Black vertical intervals are estimates of local approximation uncertainty. Local approximation uncertainty is much smaller (by an order of magnitude) than in Fig. 7.5

7.2.3 Some Remarks on the Goal-Oriented Adaptation

As illustrated in Fig. 7.2, a rather neat convergence of the *functional* is obtained. We emphasize that the GO algorithm does *not* provide a mesh-converged *flow*. Indeed, the error which is minimized is weighted by the derivative of the adjoint state. For example, regions where adjoint is constant are not refined.

Further, the GO algorithm relies on an adjoint flow while *not* having any evident mechanism for ensuring that the mesh is sufficient for computing this *adjoint flow*.

7.3 Mesh Adaptation for Design

Let us make precise the kind of tool we are developing. We focus on the *last phase* of design:

- main characteristics of the shape are chosen, defining the initial shape,
- an initial flow is computed as accurately as possible on this initial shape with GO mesh adaptation, maximizing the accuracy on the functional, for example the drag,
- it remains to slightly modify the shape in order to get the *last attainable improvements by design*.

The central question is:

How to continue towards drag minimization while still enjoying the accuracy of mesh adaptation.

7.3.1 *Theoretical Issues in Mesh Adaptation for Design*

7.3.1.1 Single Adjoint Approach

It is remarkable that:

- the GO adjoint state W^* computed for minimizing the *error* δj *on the functional* j w.r.t. metric $\mathcal{M}$ and,
- the design adjoint state for minimizing the functional j itself w.r.t. shape parameter γ_h, are strictly the same:

$$\min_{\mathcal{M}} |j - j_h| \quad \Rightarrow \quad W^*$$
$$\min_{\gamma_h} j_h \quad \Rightarrow \quad W^*$$

It is then natural to combine the two minimization loops using the same adjoint. This *single-adjoint* strategy will be illustrated in the numerical examples.

7.3.2 *Double Adjoint Approach*

The purpose of the design is to approximate the continuous optimal shape γ_{opt}.

$$\gamma_{opt} = Arg \min_{\gamma} j.$$

Then a more rational mesh adaptation criterion should concentrate on the accuracy for the optimal parameter:

Find a metric $\mathcal{M}_{opt}$ and an approximate optimal shape $\gamma_{opt,h}$

$$\gamma_{opt,h} = Arg \min_{\gamma_h} j_{(h)}.$$

such that the deviation $||\gamma_{opt,h} - \gamma_{opt}||$ is minimized (in some norm):

$$\min_{\mathcal{M}} ||\gamma_{opt,h} - \gamma_{opt}||. \square$$

In the above statement, index h holds for the discretization based on $\mathcal{M}_{opt}$. To address that latter problem, Becker et al. [9] proposed the following analysis:

$$0 = j'(\gamma_{opt}) - j_h'(\gamma_{opt,h}) = j'(\gamma_{opt}) - j_h'(\gamma_{opt}) + j_h'(\gamma_{opt}) - j_h'(\gamma_{opt,h}).$$

Thus:

$$j_h'(\gamma_{opt}) - j_h'(\gamma_{opt,h}) = j_h'(\gamma_{opt}) - j'(\gamma_{opt}).$$

Now:

$$\begin{aligned} j_h'(\gamma_{opt}) - j_h'(\gamma_{opt,h}) &= j_h''(\gamma_{opt})(\gamma_{opt} - \gamma_{opt,h}) + R \\ &= j_h''(\gamma_{opt,h})(\gamma_{opt} - \gamma_{opt,h}) + R'. \end{aligned}$$

Then:

$$\gamma_{opt} - \gamma_{opt,h} \approx \left[j_h''(\gamma_{opt})\right]^{-1}\Big(j_h'(\gamma_{opt}) - j'(\gamma_{opt})\Big). \tag{7.1}$$

Then Becker et al. [9] propose to simplify the square Hessian matrix $j_h''(\gamma_{opt})$ by a scalar number times Identity. In fact, using Automatic Differentiation for example, the Hessian matrix $j_h''(\gamma_{opt})$ is certainly complex but not impossible to evaluate. Instead, let us simplify the initial minimal norm problem into the following Goal Oriented formulation:

$$\text{Minimize the error on} \quad (g_0, \gamma_{opt} - \gamma_{opt,h}).$$

We simplify it further by putting:

$$g = \left[j_h''(\gamma_{opt})\right] g_0$$

and analyzing the *GO minimization* of

$$(g,\ j_h'(\gamma_{opt}) - j'(\gamma_{opt})) \equiv (g,\ j_h'(\gamma_{opt}))$$

for an arbitrary g.

$\big(g,\ j_h'(\gamma_h)\big)$ is a function θ_h of (γ_h, W_h, W_h^*):

$$\min_{\mathcal{M}}\ \theta_h\big(\gamma_h, W_h, W_h^*\big) \ \text{ under the constraint } \ K(\gamma_h, W_h, W_h^*) = 0$$

with $\theta_h\big(\gamma_h, W_h, W_h^*\big) = \Big(g, \Big(W_h^*, \frac{\partial \Psi_h}{\partial u}(\gamma_h, W_h)\Big) + \frac{\partial J_h}{\partial u}(\gamma_h, W_h)\Big)$ and where the *previous* state and adjoint state are the unknowns of a *new* state equation

$$K(\gamma_h, W_h, W_h^*) = 0$$

with:

$$\begin{aligned} K &= (K_1, K_2) \\ K_1(\gamma_h, W_h, W_h^*) &= \Psi_h(\gamma_h, W_h), \\ K_2(\gamma_h, W_h, W_h^*) &= \frac{\partial \Psi_h}{\partial W_h}^*(\gamma_h, W_h) W_h^* + \frac{\partial J_h}{\partial W_h}(\gamma_h, W_h). \end{aligned}$$

This time, the adaptation accounts for both state and adjoint flows local errors, weighted by a new adjoint $K'^{-1*}\partial\theta_h/\partial(W, W^*)$.

It remains to express the residuals K_1 and K_2 in terms of interpolation errors of state (W, W^*) in order to build a GO mesh adaptation loop for the optimal shape. □

The double-adjoint approach presented here does not take into account the constraints which are introduced on shape parameters in quasi all practical problems. Next section present an implementation of the single adjoint approach for which introducing constraints is trivial.

7.3.3 Implementation of Single Adjoint Approach

In the last phase of industrial design, the shape optimization has to rely on:

- a sufficiently accurate *shape parametrization* in order to be able to get the best improvement,
- a highly accurate *simulation*, in order to better identify a local minimum,
- an accurate algorithm for searching the optimum.

We focus on the combination of:

- adjoint-based sensitivity to shape,
- Sequential Quadratic Programming for shape optimization, and
- anisotropic mesh adaptation for accurate functional evaluation.

An option would be to consider the mesh-adaptive solver as the solver to be installed inside the design loop. But the function which maps the *Shape* onto the *Mesh-adapted flow* is *not a differentiable mapping* due to the discontinuous changes in the underlying mesh performed by adaptation and SQP cannot be applied efficiently. It is possible to work on the coupling between SPQ and BFGS from one side with the adaptation. An example of this coupling for a simpler gradient loop as optimizer is discussed in [10]. In the present work, we install the double mesh-adaptation loop as an external loop with respect to the optimization loop. This is developed in Fig. 7.7. Two applications are then presented.

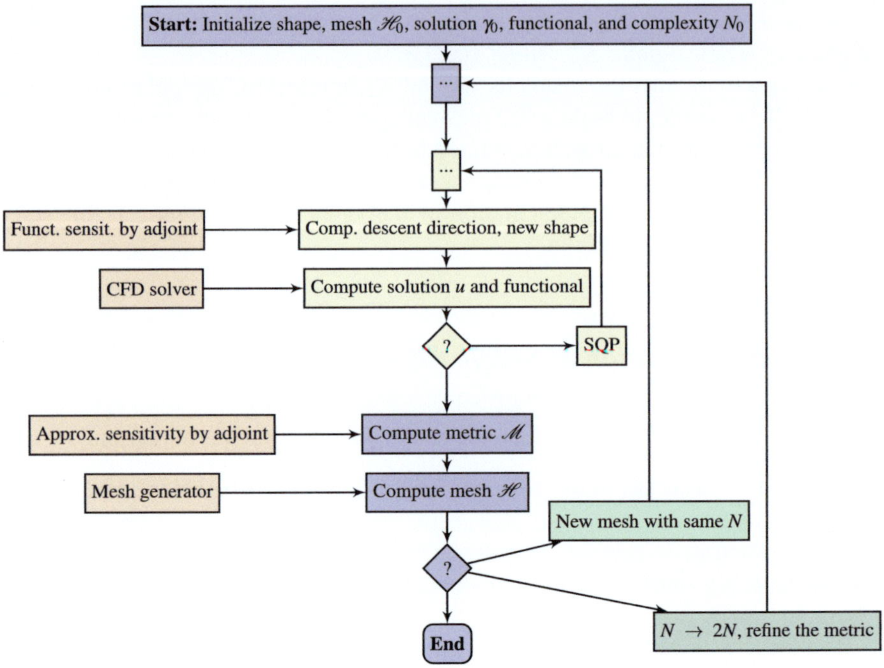

Fig. 7.7 Single-adjoint triple loop for adaptation, refinement and design

7.3.4 Application to Hydraulic Design, 1, Bulbous Bow Optimization

In this first example, the above loop is applied without the mesh adaptation sequence. This will permit to evaluate the efficiency of the association of SQP, shape parametrization and volumic mesh deformation. We consider the well known David Taylor Model Basin 5415, a historical post-second war shape for a combattant ship, see the shape in Fig. 7.8. We consider the optimization of the front bulb. The flow model involves the free surface. The mesh size is of 849709 nodes. An idea of the initial flow is given in Fig. 7.9, top. in which are presented the free surface (indicated by the variation of fluid density) and the total pressure. Applying the SQP algorithm, we

Fig. 7.8 Initial shape: David Taylor model Basin 5415

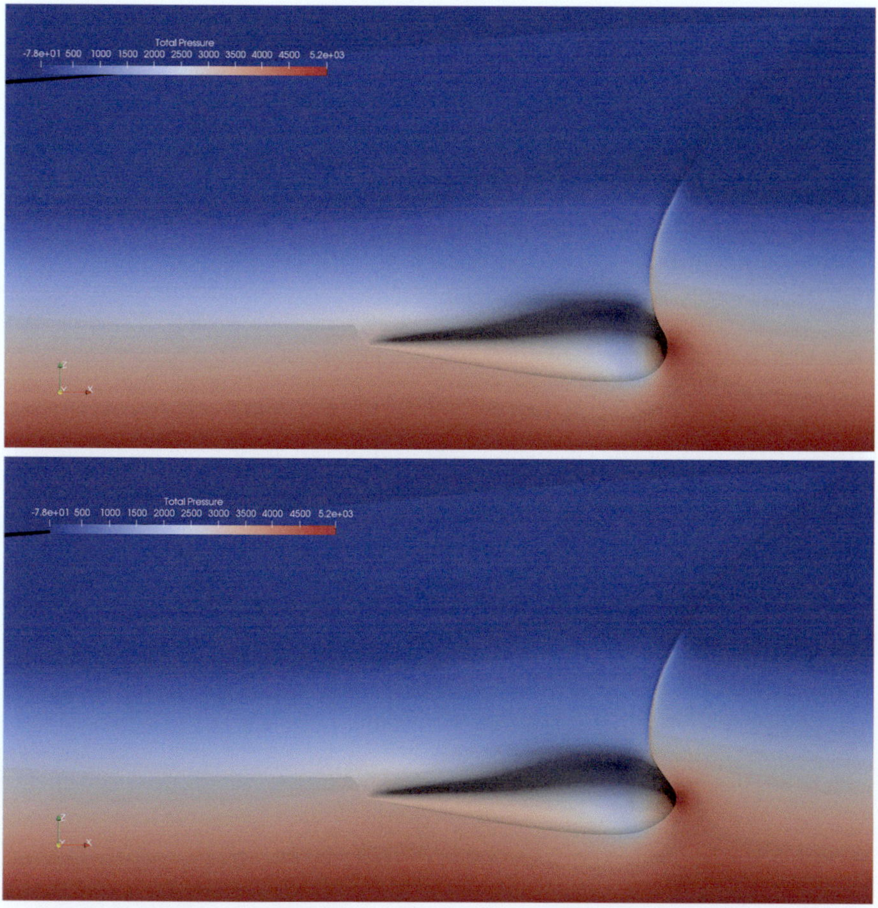

Fig. 7.9 David Taylor model Basin 5415. Top: density before and after optimization. Bottom: pressure before and after optimization

Table 7.1 Optimisation of DTMB bulb

Mesh size	Optimization steps	Initial drag	Final drag
849709	5	10.9	10.1 (−9%)

observe that with solely 5 SQP minimization iterations, this shape can be improved in such a way that the ship drag is reduced by 9%, Table 7.1. The modification of bulb shape (less flat at stagnation point) is slightly perceptible when examining the flow obtained at the end of optimization (Fig. 7.9, bottom)).

7.3.5 Applications to Hydraulic Design, 2, VP1304 Propeller

The last example combines mesh adaptation and shape optimization. The geometry is the ship VP1304 propeller model proposed by Postdam university and depicted in Fig. 7.10. A first mesh adaptive double loop predicts accurately the flow, with a final mesh of 2.5M nodes. Then 100 shape degrees of freedom are introduced for optimizing the blades. Then the optimization process combined with mesh adaptation is started. The purpose is to increase the so-called *propeller efficiency*, ratio of propulsive power to absorbed power, which is then chosen as the scalar output, while the cost functional to minimize is chosen as the inverse of propeller efficiency. With this medium number of shape parameters, the combination of adaptation and optimization uses 20 mesh adaptations and a total of 80 SQP iterations. Initial and final meshes are shown in Fig. 7.11. Initial and final flows are presented in Fig. 7.12. Differences seem very small. The blades modifications are more evident, although not very large (Figs. 7.13 and 7.14) This computation permitted to increase the propeller efficiency by 4.5% (Figs. 7.13, 7.14 and 7.15).

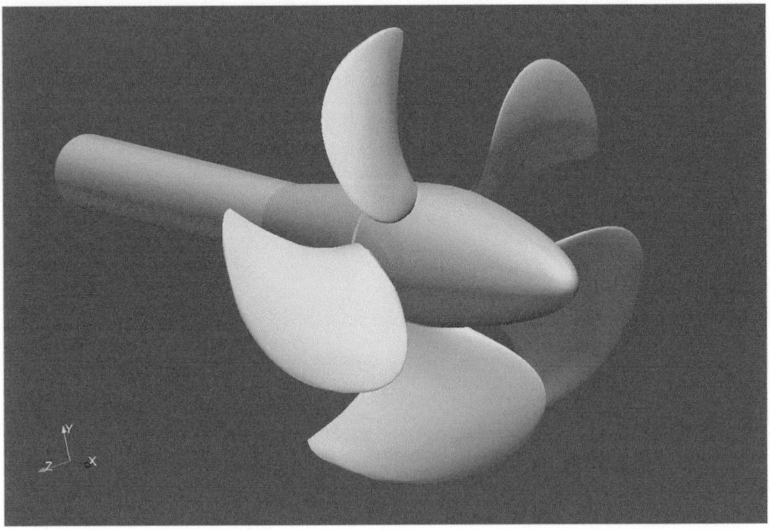

Fig. 7.10 VP1304 propeller: initial geometry

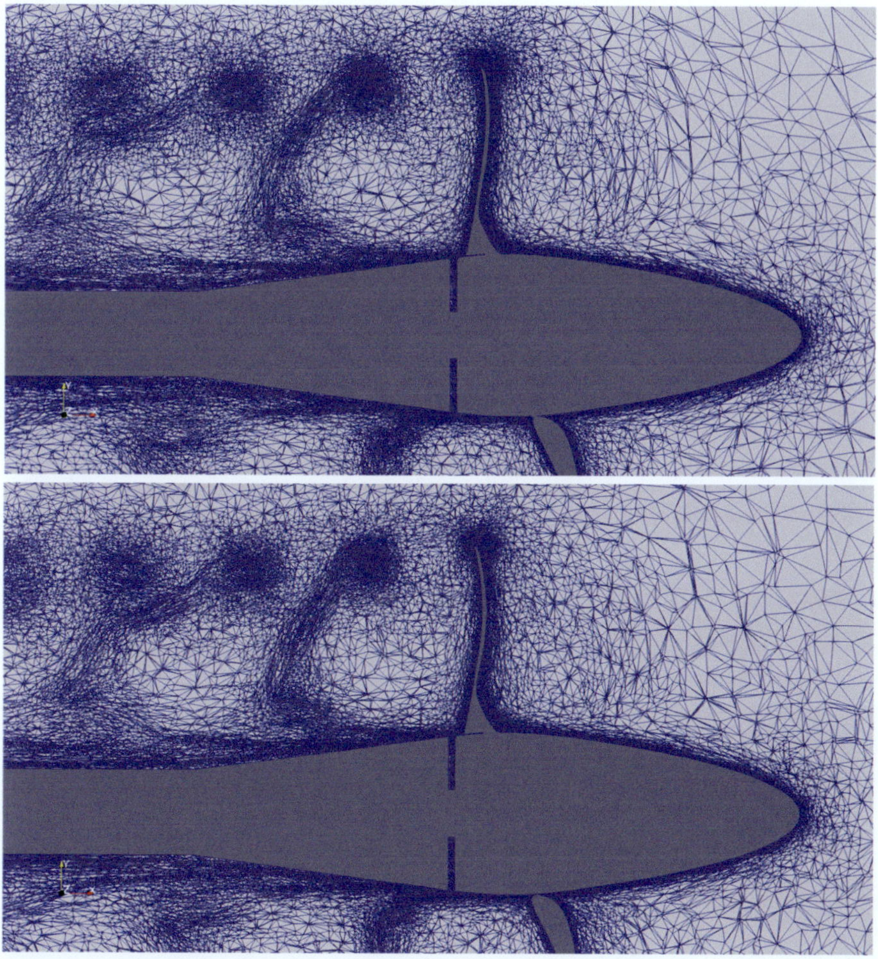

Fig. 7.11 VP1304 propeller: initial and final meshes

7.4 Concluding Remarks

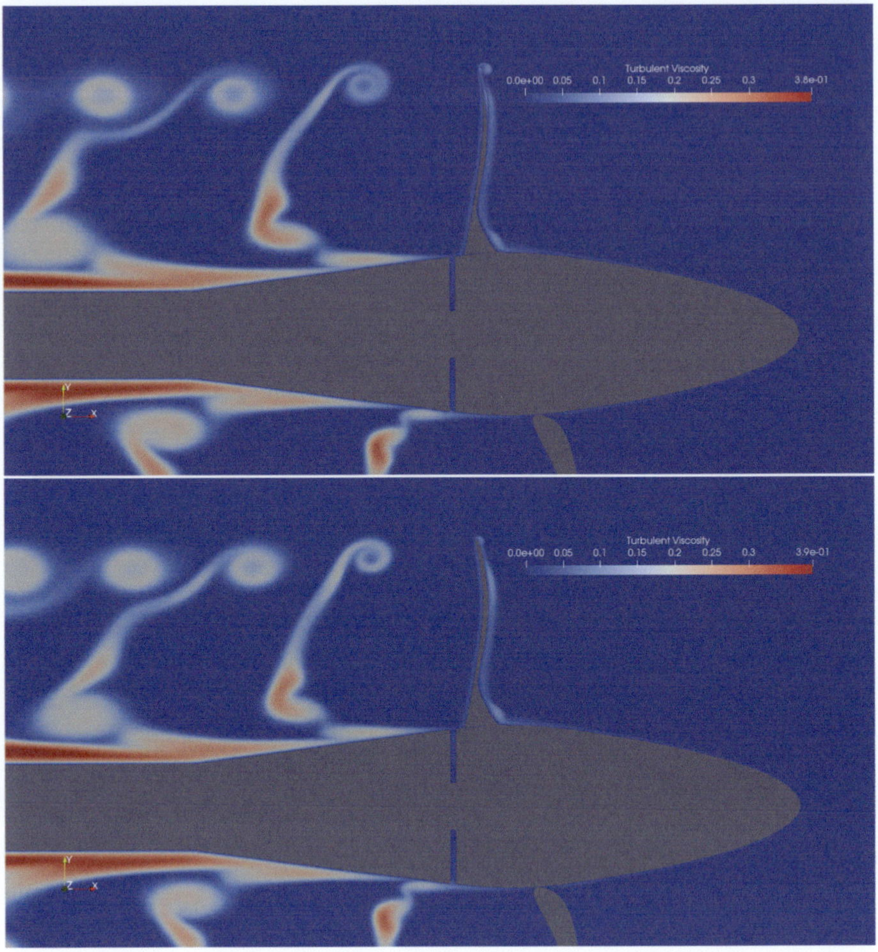

Fig. 7.12 VP1304 propeller: initial and final flows

This short paper first shows that the new anisotropic mesh adaptation algorithms are efficient to get a more accurate and more efficient solution. As concerns GO adaptation we observe that:

– a medium (3M) mesh GO-adaptation provides results that are equivalent to usual fine (200M) mesh computations.
– It results that when a massively parallel fine (> 30M nodes) mesh GO-adaptation is available, the quality of the results will be hardly affordable when using a traditional non-adaptive approach.

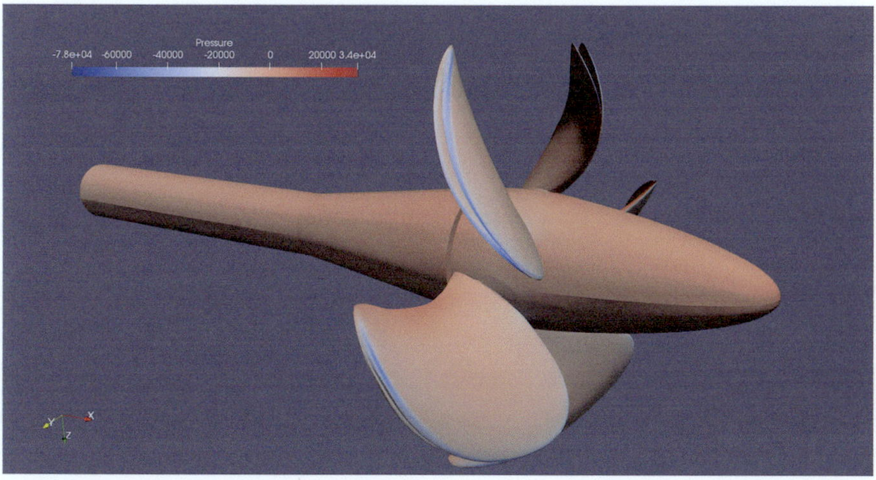

Fig. 7.13 VP1304 propeller: initial and final shapes

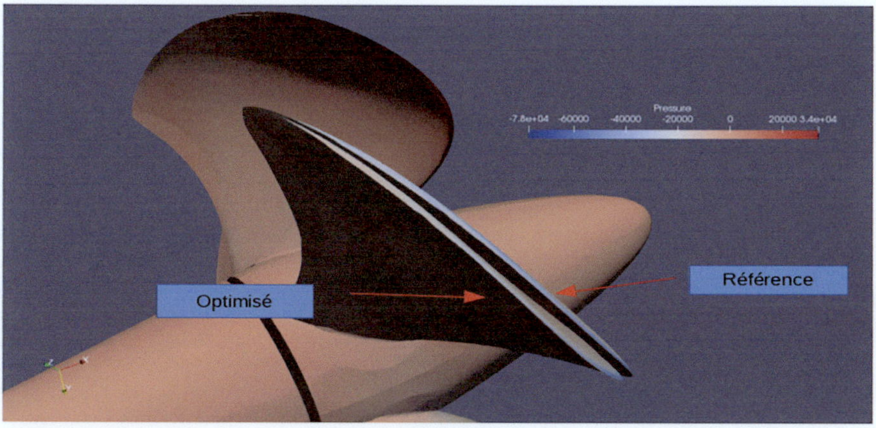

Fig. 7.14 VP1304 propeller: initial and final shapes

These properties are important for initializing and performing a final design phase. Starting from a medium GO-adapted solution, the last example shows that

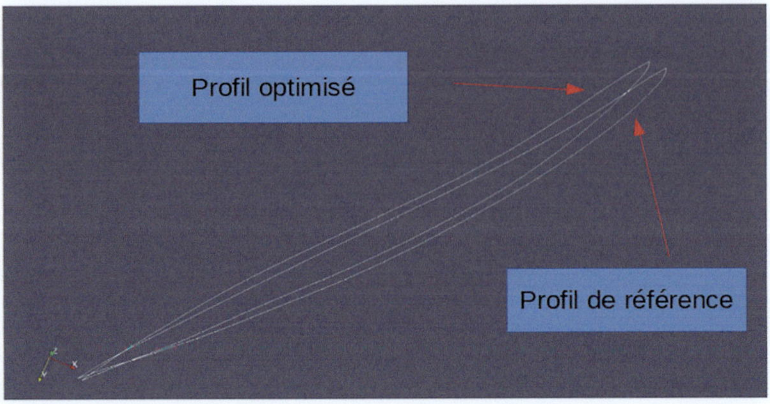

Fig. 7.15 VP1304 propeller: initial and final shapes

the single-adjoint composite design + adaptation loop, performing the *simultaneous* GO-adaptation and shape improvement, provides a High-Fidelity shape improvement.

Acknowledgements Lemma's NiceFlow-Design project is supported by France 2030 and Next Generation EU and uses HPC resources from GENCI-CINES (Grant 2023-A0142A14070).

Competing Interests The authors have no conflicts of interest to declare that are relevant to the content of this chapter.

Ethics Approval No ethics approval was required.

References

1. Lions JL (1971) Optimal control of systems governed by partial differential equations. Springer-Verlag
2. Pironneau O (1994) Optimal shape design for aerodynamics. In: AGARD REPORT, p 803
3. Jameson A (1988) Aerodynamic design via control theory. J Sci Comput 3:233–260
4. Dervieux A, Alauzet F, Loseille A, Koobus B (2023) Mesh adaptation for computational fluid dynamics, tome 1, 1st ed. ISTE Ltd and Wiley, New York. ISBN: 9-781-78630-832-0
5. Dervieux A, Alauzet F, Loseille A, Koobus B (2023) Mesh adaptation for computational fluid dynamics, tome 2, 1st ed. ISTE Ltd and Wiley, New York. ISBN: 9-781-78630-832-0
6. Rumsey CL, Slotnick JP (2015) Overview and summary of the second AIAA high lift prediction workshop. J Aircr 52(4):1006–1025
7. Park MA, Nemec M (2017) Near field summary and statistical analysis of the second AIAA sonic boom prediction workshop. In: 23th AIAA computational fluid dynamics conference, AIAA Paper 2017-3256, Denver, CO, USA
8. Frazza L, Loseille A, Dervieux A, Alauzet F (2019) Nonlinear corrector for Reynolds averaged Navier Stokes equations. Int J Numer Meth Fluids 91(11):567–586
9. Becker R, Braack M, Meidner D, Rannacher R, Vexler B (2006) Adaptive finite element methods for PDE-constrained optimal control problems. In: Reactive flows, diffusion and transport. Springer

10. Mesri Y, Alauzet F, Dervieux A (2007) Coupling optimum shape design and mesh adaptation for sonic boom reduction. In: 42th CAA-AAAF, Sophia-Antipolis (F), March 2007
11. Alauzet F, Frazza L (2020) 3D RANS anisotropic mesh adaptation on the high-lift version of NASA's common research model (HL-CRM). AIAA paper 2019-2947

Chapter 8
Early Ideas and Innovations in Bayesian and Model-Assisted Multiobjective Optimization

Michael T. M. Emmerich

Abstract The early 2000s saw rapid growth in multiobjective design optimization, fueled by advances in evolutionary algorithms, Bayesian optimization, and surrogate models such as neural networks, Kriging, and Gaussian processes. These methods enabled efficient exploration of complex design spaces while reducing the cost of expensive function evaluations. Model-Assisted Multiobjective Optimization transformed the field by integrating surrogate-assisted techniques to enhance both efficiency and solution quality. This chapter reviews its historical development, emphasizing key contributions from European research efforts—including the European Community on Computational Methods in Applied Sciences and the INGENET network. We highlight foundational methodologies such as Pareto Efficient Global Optimization, Expected Hypervolume Improvement, and confidence-bound and interval-based pre-selection in Bayesian optimization. The chapter also examines the impact of these approaches on engineering applications, particularly in computational fluid dynamics and finite element method simulations, offering a comparative analysis of core concepts and their roles in shaping the landscape of modern multiobjective optimization.

Keywords Multiobjective optimization · Surrogate modeling · Hypervolume improvement · Engineering design · Bayesian optimization · Evolutionary algorithms

8.1 Introduction

The early 2000s marked an explosion of ideas in multiobjective design optimization, fueled by the convergence of several key advancements: the emergence of multiobjective evolutionary algorithms (MOEAs) [8, 49], the increasing adoption of Bayesian optimization for black-box problems [28, 41], and the computational maturity of surrogate models—such as response surface models based on neural networks

M. T. M. Emmerich (✉)
Faculty of Information Technology, University of Jyväskylä, Jyväskylä, Finland
e-mail: michael.t.m.emmerich@jyu.fi

T. Tuovinen et al. (eds.), *Challenges in Design Methods, Numerical Tools and Technologies for Sustainable Aviation, Transport and Industry*, Computational Methods in Applied Sciences 17, https://doi.org/10.1007/978-3-031-98675-8_8

and Kriging (or Gaussian Process Regression) [32, 45]—that could handle multidimensional and multimodal functions. Together, these innovations transformed the approach to complex optimization problems, shifting the paradigm from exhaustive brute-force methods to intelligent, model-assisted strategies.

In many engineering and scientific applications, optimization problems involve multiple conflicting objectives. Traditional solution methods relied on either exhaustive simulation-based approaches or heuristic search techniques, both of which are computationally intensive [26]. As simulation models grew more complex, particularly in areas like computational fluid dynamics (CFD) and finite element method (FEM) simulations [19], conventional optimization techniques struggled to cope with the high cost of function evaluations. This challenge led to the development of *Model-Assisted Multiobjective Optimization* (MAMO), a framework that integrates surrogate models to approximate expensive functions and efficiently explore the Pareto-optimal set [14].

By making use of data-driven approximations, MAMO reduces the number of costly function evaluations while maintaining high-quality approximations of Pareto-optimal solutions. Early research demonstrated that surrogate-assisted optimization could accelerate convergence and yield robust solutions across various engineering disciplines [36]. The integration of Kriging-based Gaussian processes [45], neural network surrogates, and Bayesian sampling strategies [37] allowed optimization algorithms to learn from previous evaluations, predict promising regions of the search space, and iteratively refine solutions.

The paper is organized as follows. In Sect. 8.2, we provide historical context by discussing early research initiatives that addressed computationally expensive design problems in CFD and FEM simulations [19]. Section 8.3 introduces key methodological foundations in optimization, covering single-objective and multiobjective optimization as well as surrogate modeling techniques, including Radial Basis Function Networks (RBFNs) [39] and Gaussian Process Regression (GPR) [45]. Section 8.4 surveys surrogate models in single-objective optimization.

The next three sections present foundational contributions to surrogate-assisted multiobjective optimization: Sect. 8.5 reviews the RBFN-based approach introduced by Giotis and Giannakoglou [22], Sect. 8.6 details the ParEGO method that employs a scalarization-based strategy with Gaussian Process models [30], and Sect. 8.7 discusses the Expected Hypervolume Improvement (EHVI) framework that targets Pareto front enhancement directly [44].

Sections 8.8, 8.9, and 8.10 discuss alternative infill criteria and early methods for using surrogate models: Section 8.8 examines non-dominated sorting of expected improvements (following the work of Jeong and Obayashi [25]), Sect. 8.9 describes the SMS-EGO framework that exploits lower confidence bounds, and Sect. 8.10 introduces interval filters that compare two-sided confidence intervals for candidate selection.

Section 8.11 provides a comparative analysis of these early contributions, highlighting their respective advantages and computational trade-offs. Finally, Sect. 8.12 concludes with a summary of findings and an outlook on future research directions.

An appendix is provided with detailed examples, including a comprehensive Kriging prediction with confidence intervals.

8.2 Historical Context: European Initiatives in Multi-objective Design Optimization

The early 2000s were transformative for multi-objective optimization in engineering design, driven in part by systematic efforts within the *European Community on Computational Methods in Applied Sciences* (ECCOMAS) and the *European Thematic Network on Intelligent Optimization and Inverse Problems in Engineering* (INGENET) [10, 21]. These communities recognized that real-world engineering challenges—especially those involving CFD and FEM simulations—required optimization methods capable of handling computationally expensive, simulation-based black-box models.

A central objective of these efforts was to develop systematic benchmark suites that not only advanced optimization algorithms but also provided a common testing ground for evaluating methodologies in realistic engineering contexts. The INGENET network, in particular, played a key role in curating benchmark problems (see Appendix for two examples) that embodied the trade-offs inherent in multiobjective design, spanning areas such as aerodynamics, structural mechanics, and thermal systems [35]. These initiatives bridged applied mathematics and computational engineering, fostering interdisciplinary collaboration and accelerating innovation in real-world optimization.

During this period, other significant research forums emerged. The **Evolutionary Multi-Criterion Optimization** (EMO) conference series (inaugurated in 2001) consolidated both theoretical and applied advances in MOEAs, while the **Adaptive Design and Manufacture** (ACDM) conferences facilitated knowledge exchange on engineering design optimization, including pioneering work in multiobjective formulations. The **EUROGEN** conference series, organized under ECCOMAS, further served as a publication channel for early advances in surrogate-assisted methods and their practical applications.

A breakthrough concerning the enablement of multiobjective optimization in design engineering was the emergence of *Model-Assisted Multiobjective Optimization* (MAMO), which employed surrogate models to significantly reduce the computational burden of expensive multiobjective tasks. By constructing data-driven approximations of the objective functions, MAMO reduced the number of costly simulations required while still delivering high-quality approximations of the Pareto front. These early contributions laid the foundation for today's efficient, surrogate-based optimization techniques and have continued to influence applications ranging from digital twin technology to sustainability-driven engineering design.

8.3 Preliminaries on Optimization and Surrogate Models

Subsequently, we provide a brief introduction of the fundamental concepts that form the basis for the discussions in the following sections.

8.3.1 *Single- and Multiobjective Optimization*

Optimization involves finding the best solution from a set of feasible alternatives. In **single-objective optimization**, one seeks the optimal solution for a function $f(\mathbf{x})$:

$$\min_{\mathbf{x} \in \mathcal{X}} f(\mathbf{x}), \tag{8.1}$$

where $\mathcal{X}$ is the feasible search space and the optimal solution $\mathbf{x}^*$ satisfies

$$f(\mathbf{x}^*) \leq f(\mathbf{x}), \quad \forall \mathbf{x} \in \mathcal{X}. \tag{8.2}$$

A special case is direct or black-box optimization, where function evaluations are expensive (e.g., due to CFD or FEM simulations). In such settings, surrogate models (such as RBFNs) provide efficient approximations of these expensive evaluations and guide the search toward optimal solutions.

In **multiobjective optimization**, several conflicting objectives

$$f_1(\mathbf{x}), \ f_2(\mathbf{x}), \ \ldots, \ f_m(\mathbf{x})$$

are optimized simultaneously:

$$\min_{\mathbf{x} \in \mathcal{X}} \mathbf{f}(\mathbf{x}) = \big(f_1(\mathbf{x}), f_2(\mathbf{x}), \ldots, f_m(\mathbf{x})\big), \tag{8.3}$$

with no single solution simultaneously optimizing all objectives. Instead, one seeks the set of Pareto-optimal solutions, where a solution $\mathbf{x}^*$ is Pareto-optimal if there is no other $\mathbf{x} \in \mathcal{X}$ that improves one objective without worsening at least one other.

Radial Basis Function Networks (RBFNs) Radial Basis Function Networks (RBFNs) [3] are a class of artificial neural networks commonly used for function approximation. They typically have three layers: an input layer, a hidden layer where a nonlinear mapping is applied, and a linear output layer. For simplicity, we first consider the case of a single output ($l = 1$).

For the class of RBFN suggested by Giotis and Giannakoglou [22] the number of RBF centers equals the number of training points. During the training phase, the network learns to reproduce exactly the function values at known points. Unlike polynomial regression, it can reproduce exact data points from multimodal functions. When predicting the response at a new point $\mathbf{x} \in \mathbb{R}^n$, the network uses a weighted

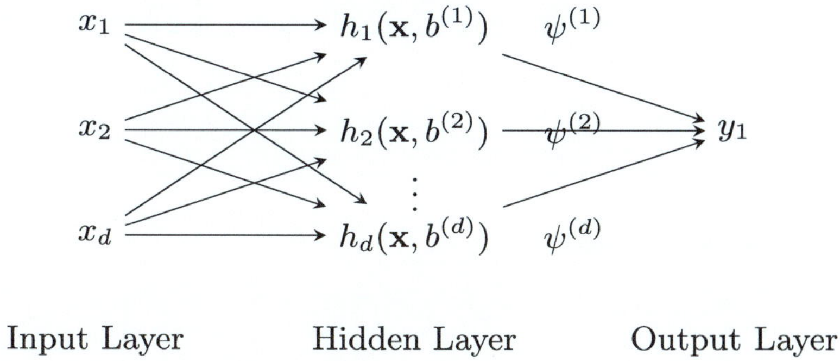

Fig. 8.1 Architecture of a Radial Basis Function Network (reproduced from [11])

sum of radial basis functions centered at the training points. Emmerich [11] showed that this method produces equations much like those in Simple Kriging, which we will cover later (Fig. 8.1).

RBFN Architecture Let $\mathbf{x}^{(1)}, \ldots, \mathbf{x}^{(m)}$ denote the training points and let $y^{(i)} = y(\mathbf{x}^{(i)})$. For each training point, the RBF center is defined as:

$$\mathbf{b}^{(i)} := \mathbf{x}^{(i)}, \quad i = 1, \ldots, m. \tag{8.4}$$

Using a norm $|\cdot|$ on $\mathbb{R}^n$ and a positive definite function $r : \mathbb{R}_0^+ \to \mathbb{R}_0^+$, the activation function of the ith hidden unit is defined by:

$$h(\mathbf{x}, \mathbf{b}^{(i)}) := r\big(|\mathbf{x} - \mathbf{b}^{(i)}|\big), \quad i = 1, \ldots, m. \tag{8.5}$$

Giotis and Giannakoglou [22] suggest using

$$r(\mathbf{x}) = \exp\big(-|\mathbf{x} - \mathbf{x}'|^q\big) \quad \text{with } q = 2. \tag{8.6}$$

Alternatively, a weighted distance measure

$$|\mathbf{x} - \mathbf{x}'|_\theta = \sum_{i=1}^{n} \theta_i |x_i - x_i'|$$

can be used, with weights θ_i provided by the user or estimated from local gradients.

Input-Output Mapping The output of the RBFN is a linear combination of the hidden layer activations:

$$\hat{y}(\mathbf{x}) = \sum_{i=1}^{m} \psi^{(i)} h(\mathbf{x}, \mathbf{b}^{(i)}). \tag{8.7}$$

The weights $\psi^{(i)}$ are determined during training by requiring that the network exactly interpolates the training data:

$$\sum_{i=1}^{m} \psi^{(i)}\, h(\mathbf{x}^{(j)}, \mathbf{x}^{(i)}) = y^{(j)}, \quad j = 1, \ldots, m. \tag{8.8}$$

In matrix form, this is written as:

$$\underbrace{\begin{bmatrix} h(\mathbf{x}^{(1)}, \mathbf{b}^{(1)}) & \cdots & h(\mathbf{x}^{(1)}, \mathbf{b}^{(m)}) \\ \vdots & \ddots & \vdots \\ h(\mathbf{x}^{(m)}, \mathbf{b}^{(1)}) & \cdots & h(\mathbf{x}^{(m)}, \mathbf{b}^{(m)}) \end{bmatrix}}_{\mathbf{H}} \underbrace{\begin{bmatrix} \psi^{(1)} \\ \vdots \\ \psi^{(m)} \end{bmatrix}}_{\boldsymbol{\psi}} = \underbrace{\begin{bmatrix} y^{(1)} \\ \vdots \\ y^{(m)} \end{bmatrix}}_{\mathbf{y}}. \tag{8.9}$$

Assuming no duplicate training points and positive definite r, the solution is

$$\boldsymbol{\psi} = \mathbf{H}^{-1}\mathbf{y}. \tag{8.10}$$

8.3.2 Kriging and Gaussian Process Regression

Kriging is a statistical interpolation method used for spatial prediction and function approximation. It was developed by Matheron and was inspired by the initial ideas of Krige, the mining engineer [32, 36].

The Kriging method constructs an estimate of an unknown function at a new point by forming a weighted combination of known values while taking spatial correlation into account. In addition to providing a linear predictor, Kriging also supplies an uncertainty measure, often expressed as mean squared error.

For a set of sample points $\mathbf{x}^{(1)}, \mathbf{x}^{(2)}, \ldots, \mathbf{x}^{(m)}$ with corresponding values $y^{(1)}$, $y^{(2)}, \ldots, y^{(m)}$, the prediction at a new point $\mathbf{x}'$ is given by:

$$\hat{y}(\mathbf{x}') = \beta + \sum_{i=1}^{m} \lambda^{(i)}\, c\big(\mathbf{x}', \mathbf{x}^{(i)}\big), \tag{8.11}$$

where:

- β is the known mean in Simple Kriging,
- $\lambda^{(i)}$ are weights determined from the covariance function,
- $c(\mathbf{x}', \mathbf{x}^{(i)})$ is the covariance (or correlation) between $\mathbf{x}'$ and $\mathbf{x}^{(i)}$.

The weights $\lambda^{(i)}$ are computed by solving

$$\mathbf{C}\boldsymbol{\lambda} = \mathbf{c}(\mathbf{x}'), \tag{8.12}$$

with $\mathbf{C}$ being the covariance matrix whose entries are $C_{ij} = s(\mathbf{x}^{(i)}, \mathbf{x}^{(j)})$ and $\mathbf{c}(\mathbf{x}')$ the covariance vector between $\mathbf{x}'$ and each training point.

It is important to note that

- the covariance function is estimated from the data.
- The statistical model quantifies prediction uncertainty using variance $\hat{s}(\mathbf{x}')$, which depends on proximity to data points and can be interpreted as the prediction's mean squared error (MSE) via the BLUP.

In the appendix we provide an example with a more details.

Simple Kriging In Simple Kriging, the mean β is assumed known and constant over the domain, simplifying the prediction to:

$$\hat{y}(\mathbf{x}') = \beta + \sum_i \lambda_i \, (y^{(i)} - \beta). \tag{8.13}$$

While this assumption works well when the mean is indeed constant, it is less flexible if the mean varies spatially.

Gaussian Process Regression (GPR) Gaussian Process Regression (GPR) [45] is closely related to Kriging. In GPR, the prediction at a new point is given by the mean of the predictive posterior distribution, and the associated variance quantifies uncertainty. Although both methods yield a predictor and an uncertainty measure, GPR is embedded in a full probabilistic framework that includes priors and Bayesian inference.

8.3.3 Relation Between Kriging and RBFNs

RBFNs and Simple Kriging share a common idea: both construct predictions as a linear combination of basis functions centered at the training points. Table 8.1 summarizes the correspondences between the two approaches.

Table 8.1 Comparison simple kriging versus radial basis function Networks

Simple kriging	RBFN
Prediction: $\hat{y}(\mathbf{x}') = \beta + \sum \lambda_i \, c\big(\mathbf{x}', \mathbf{x}^{(i)}\big)$	Output: $\hat{y}(\mathbf{x}') = m + \sum w_i \, h\big(\mathbf{x}', \mathbf{x}^{(i)}\big)$
Spatial indices: $\mathbf{x}^{(i)}$	RBF centers: $\mathbf{x}^{(i)}$
Best linear unbiased predictor (BLUP)	Predicted output
Covariance function $c(\mathbf{x}', \mathbf{x}^{(i)})$	Activation function $h\big(\mathbf{x}', \mathbf{x}^{(i)}\big)$
Weights λ_i	Weights w_i
Variance $\hat{s}$	No uncertainty quantification
Correlation parameters adapted to data	Parameters of activation function fixed

The activation function in RBFNs corresponds to the positive definite correlation function in Kriging. Thus, when RBFNs are used for interpolation, the resulting equations are equivalent to those of Simple Kriging.

8.4 Surrogate Models in Single-Objective Optimization

The use of surrogate models in single-objective optimization has a long history, with many ideas later generalized to multiobjective settings (e.g., Bayesian optimization and expected improvement [37, 38]). In particular, the work of Yaochu Jin on data-driven approximations [26] greatly influenced single-objective evolutionary optimization. We briefly introduce two archetypal algorithms.

8.4.1 Bayesian Optimization for a Single-Objective Problem

Bayesian optimization, originally introduced by Mockus et al. [37, 38], is a model-based approach designed for optimizing expensive black-box functions. It is widely used in hyperparameter tuning, engineering design, and scientific optimization problems where function evaluations are costly. The key idea is to maintain a probabilistic surrogate model (often a Gaussian Process) that provides both predictions and uncertainty estimates. An acquisition function is then used to balance exploration and exploitation by deciding where to sample the next point.

$$EI(\mathbf{x}) = \begin{cases} (\hat{y}^* - \hat{y}(\mathbf{x}))\,\Phi\Big(\dfrac{\hat{y}^* - \hat{y}(\mathbf{x})}{\hat{s}(\mathbf{x})}\Big) + \hat{s}(\mathbf{x})\,\phi\Big(\dfrac{\hat{y}^* - \hat{y}(\mathbf{x})}{\hat{s}(\mathbf{x})}\Big), & \hat{s}(\mathbf{x}) > 0, \\ 0, & \hat{s}(\mathbf{x}) = 0. \end{cases} \tag{8.14}$$

This value can be interpreted as the expected progress in the direction of the global optimum, given the observed function evaluation are a sample of a random field indexed by the input variable vector.

The algorithm, summarized in the pseudocode of Algorithm 1, works as follows: In the beginning one selects a surrogate model (for example, a Gaussian Process) and evaluates the objective function at some initial points. Then one fit the surrogate model S on the current dataset $\mathcal{D}$ to train it well. Next, one uses an acquisition function (for example, Expected Improvement or Upper Confidence Bound) to choose the next point to evaluate, which is very important to guide where to search next. After that, one evaluates the objective function at the selected point and update the dataset $\mathcal{D}$ accordingly. Finally, the process continues until a stopping criterion, like a budget or convergence, is met, and when no further improvement is observed, one stops.

This approach was later popularized under the name "Efficient Global Optimization (EGO)" after a publication by Jones et al. [28] which revived interest in the

Algorithm 1 Bayesian Optimization for a Single-Objective Problem

Initialize: Choose a surrogate model S (e.g., a Gaussian Process).
Collect an initial dataset $\mathcal{D} = \{(\mathbf{x}_i, y_i)\}$ by evaluating $f(\mathbf{x})$ at a set of initial points.
for $t = 1$ to max_iterations **do**
  Train the surrogate model S using $\mathcal{D}$.
  Select the next evaluation point:

$$\mathbf{x}_{\text{new}} = \arg\max_{\mathbf{x}} \text{AcquisitionFunction}(S, \mathcal{D})$$

  Evaluate the expensive objective: $y_{\text{new}} = f(\mathbf{x}_{\text{new}})$
  Update the dataset: $\mathcal{D} \leftarrow \mathcal{D} \cup \{(\mathbf{x}_{\text{new}}, y_{\text{new}})\}$
  if stopping criterion is met **then**
    Break
  end if
end for
return Best solution found.

approach—a term that, while widely used, is somewhat of a misnomer since it does not guarantee efficiency if the evaluation budget is known a priori [48].

Note that the term "Bayesian Optimization" is also used to refer to estimation of distribution algorithms, which model populations in evolutionary algorithms through stochastic distributions [34]. However, these methods are not specifically designed for problems with expensive function evaluations.

8.4.2 Surrogate-Assisted Evolutionary Algorithm with Pre-selection

Evolutionary algorithms generate new offspring by applying mutation and recombination operators to a population of candidate solutions. In surrogate-assisted variants [19], machine learning models are used to reduce the number of expensive function evaluations. In pre-selection strategies, offspring candidates are first evaluated using the surrogate model, and only the most promising ones are subsequently evaluated with the true objective function.

Early approaches ranged from simple neural network predictions [27] to sophisticated methods incorporating uncertainty quantification via Kriging. Notable examples include metamodel-assisted evolution strategies using lower confidence bounds [15], the parameterless probability of improvement [43], and variants using Expected Improvement together with constraint-handling strategies [11, 14].

The main loop of a model-assisted evolutionary algorithm is shown in Algorithm 2. It begins with the initialization of a population by randomly sampling candidate solutions. A surrogate model is then trained using a limited number of true evaluations to approximate the objective function. Offspring are generated through mutation and crossover, creating new candidate solutions. To reduce computational costs, a pre-selection step utilizes the surrogate model to identify promising candidates. These

Algorithm 2 Model-Assisted Evolutionary Algorithm with Pre-Selection

Initialize: Generate an initial population P of random individuals.
Train a surrogate model S using initial evaluations.
for $t = 1$ to max_generations **do**
  Generate an offspring population Q via variation operators (mutation/crossover).
  Evaluate each $\mathbf{x} \in Q$ using the surrogate model:

$$\{\hat{y}(\mathbf{x}), \hat{s}(\mathbf{x}) \mid \mathbf{x} \in Q\}$$

  Select a subset $Q' \subseteq Q$ of promising candidates based on $\hat{y}$.
  Evaluate the true objective for individuals in Q'.
  Update the surrogate model S with the new evaluations.
  Form the next generation population P via selection
   (e.g., comma-selection or plus-selection).
  if stopping criterion is met **then**
    Break
  end if
end for
return Best solution found.

selected candidates are then evaluated using the actual objective function, and the surrogate model is updated accordingly. Finally, a selection mechanism determines the next generation, ensuring iterative refinement of the population over successive iterations.

Evolutionary algorithms were quite popular in the early attempts to utilize surrogate models, but seem to recently receive less attention. However, there is notable work on the introduction of multiobjective optimization based on decomposition [5, 23].

A conceptually similar alternative to model-assisted evolutionary optimization is model-assisted pattern search [42]. Unlike evolutionary algorithms that rely on stochastic sampling, model-assisted pattern search employs a deterministic sampling strategy. This approach enables the establishment of local convergence guarantees.

8.5 Giotis and Giannakoglou's RBFN-Based Method

Giotis and Giannakoglou [20, 22] were early pioneers in the integration of surrogate models with evolutionary algorithms for multi-objective optimization. Focusing on aerodynamic design optimization, where objective evaluations involved costly CFD simulations, they employed response surfaces generated via polynomial regression and RBF-based neural networks. By integrating these surrogates with evolutionary algorithms, their work demonstrated that surrogate-supported optimization could effectively address complex engineering challenges. A special feature of this approach is the computation of importance factors as an interpretation of

RBF weights. In hindsight, this technique might be viewed as an early eXplainable artificial intelligence (XAI) approach [2].

Key Features

Giotis and Giannakoglou pioneered the integration of surrogate models with evolutionary algorithms for multi-objective aerodynamic design optimization by approximating costly CFD simulations with response surfaces generated through RBF-based neural networks. Their approach not only reduced computational costs in complex engineering problems but also featured the computation of importance factors derived from RBF weights, offering an early demonstration of model interpretability that anticipated modern explainable artificial intelligence.

8.6 ParEGO: Pareto-Based Efficient Global Optimization

ParEGO, introduced by Knowles in 2006 [30, 31], transformed multiobjective optimization into a series of scalarized single-objective problems. By systematically varying the weights in a Chebyshev aggregation function, ParEGO explores trade-offs among objectives. A Gaussian Process (or Kriging) model is used to quantify uncertainty, and the Expected Improvement acquisition function is applied to the scalarized objective.

The weighted Chebyshev scalarization function with the ideal point as reference is defined as:

$$g(\mathbf{x}, \mathbf{w}, \mathbf{z}^*) = \max_{i=1,\dots,m} w_i \left| f_i(\mathbf{x}) - z_i^* \right|,$$

where:

- $\mathbf{x} \in \mathbb{X}$ is the decision vector,
- $\mathbf{w} = (w_1, \dots, w_m)$ is a weight vector with $w_i > 0$ and $\sum_{i=1}^m w_i = 1$,
- $\mathbf{f}(\mathbf{x}) = \left(f_1(\mathbf{x}), \dots, f_m(\mathbf{x})\right)$ is the objective vector,
- $\mathbf{z}^* = \left(z_1^*, \dots, z_m^*\right)$ is the ideal point with $z_i^* = \min_{\mathbf{x} \in \mathbb{X}} f_i(\mathbf{x})$.

This scalarization converts the multiobjective problem into a single-objective one:

$$\min_{\mathbf{x} \in \mathbb{X}} g(\mathbf{x}, \mathbf{w}, \mathbf{z}^*),$$

allowing ParEGO to approximate the Pareto front by varying the weight vector $\mathbf{w}$. Notably, the Chebyshev scalarization can handle concave Pareto fronts more effectively than additive or linear scalarizations.

Key Features

ParEGO uses Gaussian Process models for surrogate modeling, iteratively updating with new samples to efficiently explore and converge to the Pareto-optimal set. It transforms the problem into single-objective tasks via weighted Chebyshev scalarization and calculates expected improvement. Its computational efficiency and broad applicability have made it widely adopted.

8.7 Expected Hypervolume Improvement (EHVI)

The Expected Hypervolume Improvement (EHVI) criterion extends the concept of expected improvement from single-objective to multiobjective settings by assessing the improvement of an approximation set toward the Pareto front as a whole. EHVI interprets improvement as an multidimensional integral the non-dominated space, where the weighting is derived from a probability density function. (see Fig. 8.2) Emmerich et al. [14] proposed an EHVI framework that directly targets improvements in the hypervolume indicator, which measures the dominated volume in the objective space. Although EHVI was initially presented alongside other acquisition functions, it later gained prominence, especially after its successful application in Bayesian optimization schemes [18, 33, 44].

Early implementations of EHVI were computationally demanding due to expensive integral evaluations and reliance on Monte Carlo approximations. These challenges led to the development of more efficient exact evaluation schemes [6, 24], culminating in algorithms with asymptotically optimal $O(n \log n)$ runtime for two and three objectives [18, 46]. These improvements have made EHVI a practical and powerful tool for high-dimensional MAMO problems.

Note that an earlier approach to computing the expected improvement for Pareto front approximation with two objective functions was proposed by Keane. This method calculates the center of gravity over the dominated region using a measure that rewards filling gaps in the Pareto front approximation [29]. In [44] it has been compared to the EHVI.

Key Features

EHVI directly rewards improvements in the hypervolume indicator, thus explicitly targeting the Pareto front. Its reliance on a Lipschitz continuous, Pareto-compliant indicator helps ensure convergence, and recent algorithmic advances have mitigated its initial computational drawbacks.

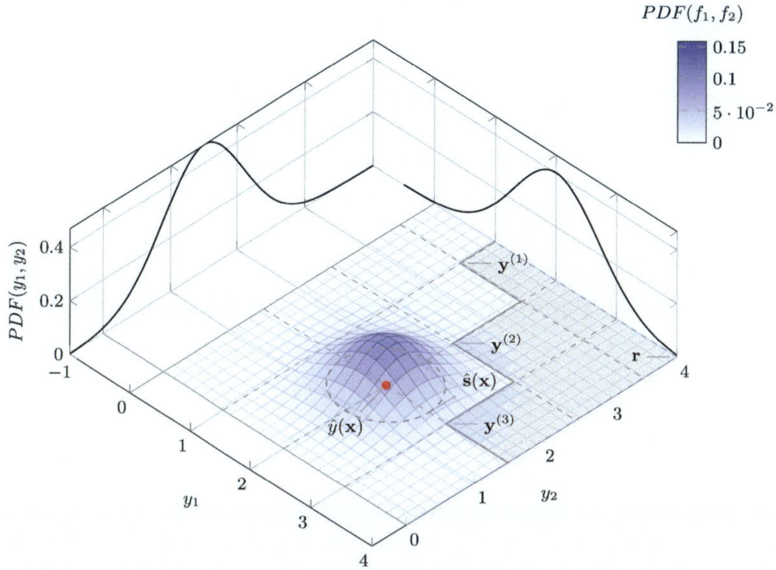

Fig. 8.2 Pareto front approximation and 2-D predictive distribution. The incumbent Pareto front is provided by $\{\mathbf{y}^{(1)}, \mathbf{y}^{(2)}, \mathbf{y}^{(3)}\}$. The grey area can be interpreted as the hypervolume indicator. The red point marks $\hat{y}(\mathbf{x})$ (mean), and the dashed ellipse represents $\hat{\mathbf{s}}(\mathbf{x})$ (variance) below the predictive surface

8.8 Non-dominated Sorting of Expected Improvement (Jeong and Obayashi)

Jeong and Obayashi [25] extended the Expected Improvement (EI) concept to multi-objective optimization by applying non-dominated sorting to the EI values computed for individual objectives. Instead of aggregating objectives through scalarization or hypervolume measures, their method computes the EI for each objective separately and ranks candidate solutions based on Pareto dominance.

This approach naturally produced a well-distributed set of solutions and was computationally efficient owing to the closed-form expression of the single-objective EI. However, challenges such as correlations between EI values and efficient classification hindered their early adoption. Recent advances in multiobjective Bayesian optimization have revisited these principles, incorporating faster sorting techniques and uncertainty-aware strategies.

Key Features

By computing the EI separately for each objective and using non-dominated sorting, this approach avoids the pitfalls of scalarization and hypervolume computation. Its main advantage is computational efficiency, although the early versions did not explicitly address crowding distances or correlations among objectives.

8.9 SMS-EGO: Exploiting the Lower Confidence Bound

The lower confidence bound of a Kriging or GP prediction

$$\mathrm{lb}_{\omega}(\mathbf{x}) = \hat{y}(\mathbf{x}) - \omega \cdot \hat{s}(\mathbf{x}) \tag{8.15}$$

has been used in single objective optimization as a infill criterion [7], but can be generalized in a straightforward way to multiobjective optimization by using a vector of independent predictions of objective function values. The factor ω can be used to scale between exploration and exploitation.

The idea of using hypervolume improvement based on a lower confidence bound within the model-assisted evolutionary algorithm was introduced in the seminal work on the SMS-EMOA multiobjective optimization algorithm. The hypervolume indicator offered a straightforward way to define improvement of a Pareto front approximation [13]. Applied initially to two problems from the INGENET benchmark suite (e.g., the NACA RAE 2822 drag minimization test case), the method showed promise for surrogate-assisted multi-objective optimization.

Within the Bayesian Optimization scheme, the approach of using lower bounds was successfully introduced by Ponweiser et al. [40]. A limitation noted by Ponweiser et al. [40] was that the method struggled to detect improvements in dominated regions. To address this, the concept of negative hypervolume improvement was introduced and later refined in [44]. The resulting SMS-EGO framework combines Bayesian optimization with hypervolume-based selection and draws inspiration from both SMS-EMOA and EGO. Although the initial computational advantage of SMS-EGO over EHVI has diminished as exact EHVI computation methods improved, the idea of incorporating negative hypervolume improvement continues to influence recent developments, such as gradient-based hypervolume optimization [9].

Key Features

SMS-EGO leverages a lower confidence bound to extend single-objective strategies to a multiobjective context by applying independent predictions for each objective. The parameter ω balances exploration and exploitation, while the use of hypervolume improvement—including its negative variant—provides a direct measure for advancing Pareto front approximations, particularly in dominated regions. By combining Bayesian optimization with hypervolume-based selection, SMS-EGO integrates the strengths of SMS-EMOA and traditional EGO into a robust framework for surrogate-assisted multiobjective optimization.

8.10 Interval Filters

Thus far, the methods described have relied on scalar acquisition criteria. In contrast, interval filters compare multi-dimensional two-sided confidence intervals assigned to each candidate solution. These filters use possibilistic logic to sort, accept, or reject points for evaluation based on the entire confidence interval rather than a single scalar value.

Emmerich et al. [16] proposed defining lower and upper confidence bounds for each solution as:

$$\mathrm{lb}_\omega(\mathbf{x}) = \hat{y}(\mathbf{x}) - \omega \cdot \hat{s}(\mathbf{x}), \quad \mathrm{ub}_\omega(\mathbf{x}) = \hat{y}(\mathbf{x}) + \omega \cdot \hat{s}(\mathbf{x}). \tag{8.16}$$

Here, $\hat{y}(\mathbf{x})$ is the predicted value, $\hat{s}(\mathbf{x})$ is the associated uncertainty, and ω is a confidence parameter controlling the interval width. Under the Gaussian assumption, the probability that the true value falls within this interval is given by

$$p_\alpha = \Phi(\omega) - \Phi(-\omega), \tag{8.17}$$

with the convention that if $\hat{s}(\mathbf{x}) = 0$, then $\mathrm{lb}_\omega(\mathbf{x}) = \mathrm{ub}_\omega(\mathbf{x}) = \hat{y}(\mathbf{x})$.

When selecting candidate solutions for evolutionary algorithms, two types of errors may occur:

(A) Selecting solutions that are not among the top μ candidates.
(B) Rejecting solutions that belong to the top μ candidates.

A filter with high *precision* minimizes error (A), while one with high *recall* minimizes error (B). In multiobjective settings (minimization), the confidence intervals are depicted as orthogonal ranges in the objective and constraint spaces (see Fig. 8.3). A solution is *possibly non-dominated* if its lower bound is not dominated with respect to the upper bounds of all other solutions (we can create a witness realization then where the solution is non-dominated) and *certainly dominated* if its lower bound is dominated by a upper bound of at least one other solution. A detailed discussion of interval filters can be found in [11].

These methods can also be extended to constrained optimization, as shown in [17] for a three-objective airfoil optimization test case. By training a surrogate model for the constraint boundary, lower and upper bounds for the constraint can be obtained, which then inform the acceptance or rejection of candidate solutions.

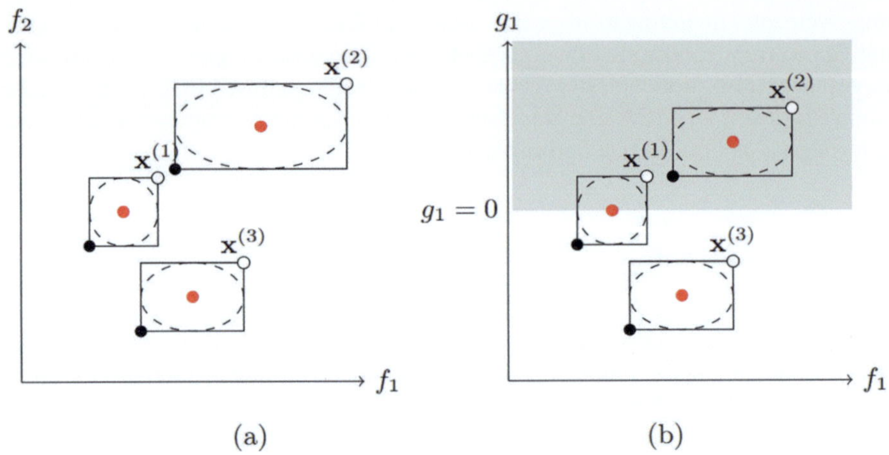

Fig. 8.3 Interval Filters for multiple objectives (**a**) and for implicit constraints (**b**), see also: [17]

Key Features

Interval filters assess candidate solutions using multi-dimensional two-sided confidence intervals rather than single scalar values. Each solution is assigned lower and upper confidence bounds—derived from its predicted value and uncertainty with a tunable width parameter—providing a probabilistic guarantee (under a Gaussian assumption) that the true value lies within the interval. This method effectively balances the risk of including inferior solutions and excluding top candidates, and in multiobjective settings, it differentiates possibly non-dominated solutions from those that are certainly dominated. Moreover, the approach extends naturally to constrained optimization by integrating surrogate models for constraint boundaries.

8.11 Comparison of Early Contributions and Ideas

Tables 8.2 and 8.3 compare the key aspects of these foundational methods. It becomes apparent that Kriging models (or Gaussian process models) became the prevalent choice, when it comes to surrogate models. It is interesting to note that Lipschitz models might be an interesting alternative approach, as outlined in a new article by Žilinskas [47].

Moreover, many ideas for acquisition functions can be used in both Evolutionary and Bayesian Optimization, and have counterparts from single-objective optimization. The only approach that does not make use of acquisition functions is the interval filters.

Table 8.2 Comparison of selected early contributions to MAMO

Aspect	Giotis and Giannakoglou	ParEGO	EHVI
Surrogate model	Radial basis functions	Kriging/GP (scalarized)	Kriging/GP (per objective)
Optimization strategy	Evolutionary algorithms	Bayesian optimization	Indicator-based EA, Bayesian optimization
Uncertainty quantification	–	Variances of scalarized improvement	Variance of hypervolume improvement
Applications	CFD optimization	Broad MAMO problems	Diverse multiobjective problems

Table 8.3 Comparison of additional early contributions to MAMO

Aspect	Jeong and Obayashi	SMS-EGO	Interval filters
Surrogate model	Kriging/GP	Kriging/GP	Kriging/GP (per objective)
Optimization strategy	NSGA-II	Bayesian optimization	Evolutionary algorithms
Uncertainty quantification	Single-objective EI	Lower confidence bound	Upper and lower confidence bounds
Applications	CFD/airfoil optimization	Broad MAMO problems	CFD/airfoil optimization

8.12 Conclusions and Outlook

The foundational contributions to Model-Assisted Multiobjective Optimization (MAMO) have paved the way for contemporary surrogate-assisted techniques. This paper has reviewed key early methods—including those by Giotis and Giannakoglou, ParEGO, EHVI, SMS-EGO, and interval-based filters—that introduced diverse strategies for balancing exploration and exploitation and reducing computational costs in multiobjective settings.

The pioneering integration of RBFNs with evolutionary algorithms by Giotis and Giannakoglou demonstrated the feasibility of surrogate-assisted optimization for aerodynamic design. ParEGO's scalarization approach provided an effective framework using Gaussian Process models, while EHVI focused on direct hypervolume improvement of the Pareto front. Subsequent methods, such as SMS-EGO and non-dominated sorting of EI, further refined surrogate-based selection. Interval-based methods introduced robust uncertainty handling, which influenced subsequent developments in Bayesian optimization.

The field of MAMO has been advancing a lot over recent years and parallel computing (batch evaluation) possibilities as well as advances in machine learning were driving forces. Moreover, infill criteria have been revised and/or thoroughly tested and integrated into programming libraries. For more recent overview papers see [1, 4].

Looking ahead, MAMO continues to evolve with advances in deep learning surrogates, transfer learning, and adaptive sampling strategies. These developments promise to further enhance the applicability of surrogate-assisted optimization in engineering, environmental planning, and even combinatorial fields such as model-driven drug discovery. The integration of scalable, model-aware optimization frameworks will be essential in addressing increasingly complex multi-objective problems.

A Kriging Prediction with Confidence Intervals: A Detailed Example

This appendix illustrates the computational process for using Kriging to predict values at unobserved locations. Although the example is for one-dimensional inputs, the method extends naturally to higher dimensions by adjusting the correlation function.

A.1 Mathematical Background

The Kriging prediction at a new point x is given by:

$$\hat{y}(x) = \beta + \mathbf{k}(x)^T \mathbf{K}^{-1}(\mathbf{y} - \beta \mathbf{1}),$$

where β is the global mean of the observed values, $\mathbf{y}$ is the vector of observations, and $\mathbf{1}$ is a vector of ones.

The variance of the prediction at x is:

$$\sigma^2(x) = \sigma_0^2 \left(1 - \mathbf{k}(x)^T \mathbf{K}^{-1} \mathbf{k}(x) + \frac{\left(1 - \mathbf{1}^T \mathbf{K}^{-1} \mathbf{k}(x)\right)^2}{\mathbf{1}^T \mathbf{K}^{-1} \mathbf{1}} \right),$$

with σ_0^2 representing the process variance.

A.2 Example with Three Data Points

Consider data points $x_1 = 1, x_2 = 2, x_3 = 3$ with observed values $y_1 = 1.0$, $y_2 = 2.0$, $y_3 = 1.5$. Assume a correlation function with parameter $\theta = 1$ and exponent $q = 2$.

Correlation Matrix The correlation matrix $\mathbf{K}$ is

$$\mathbf{K} = \begin{bmatrix} 1 & e^{-1} & e^{-4} \\ e^{-1} & 1 & e^{-1} \\ e^{-4} & e^{-1} & 1 \end{bmatrix}.$$

Its inverse $\mathbf{K}^{-1}$ is computed numerically.

Correlation Vector for $x = 2.5$ The correlation vector $\mathbf{k}(2.5)$ is

$$\mathbf{k}(2.5) = \begin{bmatrix} e^{-2.25} \\ e^{-0.25} \\ e^{-0.25} \end{bmatrix}.$$

Prediction and Variance The predicted value at $x = 2.5$ is

$$\hat{y}(2.5) = \beta + \mathbf{k}(2.5)^T \mathbf{K}^{-1}(\mathbf{y} - \beta \mathbf{1}),$$

with β typically set as the mean of the observations (here, 1.5). The prediction variance is computed as

$$\sigma^2(2.5) = \sigma_0^2 \left(1 - \mathbf{k}(2.5)^T \mathbf{K}^{-1} \mathbf{k}(2.5) + \frac{\left(1 - \mathbf{1}^T \mathbf{K}^{-1} \mathbf{k}(2.5)\right)^2}{\mathbf{1}^T \mathbf{K}^{-1} \mathbf{1}}\right).$$

A.3 Python Implementation

A Python implementation (see [12]) uses these formulas to calculate the predictions and uncertainties. By changing parameters (e.g., setting $q = 1$ instead of $q = 2$), one can study the impact on predictions. The figure below shows an example output with confidence intervals (Fig. 8.4).

B Key Terminology

- **Bayesian Optimization**: An optimization method using a statistical model-based acquisition function to guide evaluations of the objective function. The term is also used as a synonym for estimation of distribution algorithms in evolutionary computation.
- **CFD Optimization**: The use of optimization techniques to improve aerodynamic performance.
- **Confidence Intervals in Surrogates**: Quantification of uncertainty in surrogate model predictions.

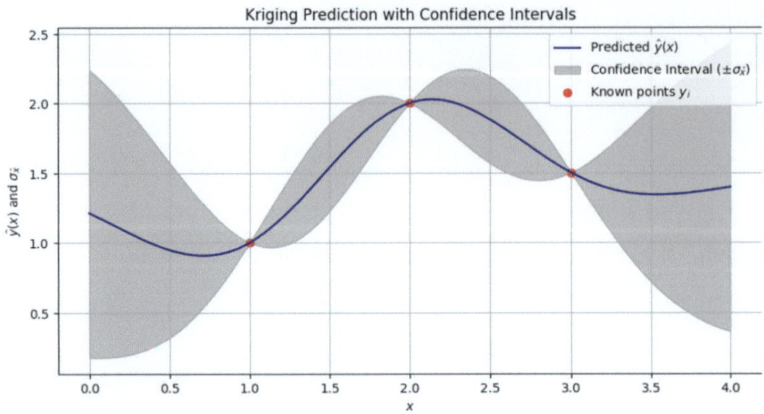

Fig. 8.4 Kriging prediction with confidence intervals

- **EHVI**: Expected Hypervolume Improvement, an acquisition function in Bayesian multiobjective optimization.
- **Expected Improvement (EI)**: An acquisition function that quantifies potential improvement in Bayesian optimization.
- **FEM**: Finite Element Method, a numerical technique in engineering design.
- **Gaussian Process Regression (GPR)**: A nonparametric regression method with strong conceptual similarities with Kriging, but focusing more on Bayesian interpretation of the prediction and variance and not less on linear estimator theory.
- **Hypervolume Indicator**: A metric that measures the volume dominated by a Pareto front.
- **Hypervolume-Based Infill Criterion**: A strategy that uses hypervolume improvement for candidate selection.
- **INGENET**: A European network focused on intelligent optimization methods.
- **Interval Filters**: Methods that compare confidence intervals to decide on candidate evaluation.
- **Kriging**: A surrogate modeling technique that provides predictions with uncertainty estimates.
- **Lower Confidence Bound (LCB)**: A statistic used to guide exploration in Bayesian optimization.
- **Metamodeling**: The use of surrogate models to approximate expensive simulations.
- **MAMO**: Model-Assisted Multiobjective Optimization.
- **Model-assisted Pattern Search**:
- **Multiobjective Optimization (MOO)**: The simultaneous optimization of multiple conflicting objectives.
- **Non-dominated Sorting**: A method for ranking solutions based on Pareto dominance.
- **NSGA-II**: A popular evolutionary algorithm for multiobjective optimization.

- **Pareto Front**: The set of non-dominated solutions in multiobjective optimization.
- **Radial Basis Function Networks (RBFN)**: Neural networks used for regression and multivariate function interpolation.
- **Scalarization**: The conversion of a multiobjective problem into a single-objective one.
- **Simple Kriging**: A form of Kriging with a known constant mean.
- **SMS-EGO**: A framework combining Bayesian optimization with hypervolume-based selection.
- **Surrogate-Assisted Evolutionary Algorithm (SA-EA)**: An evolutionary algorithm augmented with surrogate models.

C Summary of Airfoil Redesign Test Cases

Two airfoil redesign test cases from the INGENET database [10] served as benchmarks for early MAMO methods.

NACA Test Case

The goal is to compare an optimized airfoil with **NACA 0012** and **NACA 4412** under two flow conditions: take-off (high lift) and cruise (low drag). The objective functions are based on differences in the pressure distribution, and the test case features 18 degrees of freedom (Table 8.4).

RAE Drag Minimization Test Case

For the RAE 2822 airfoil, the goal is to minimize drag coefficients over three flow conditions while satisfying lift and pitch moment constraints. Geometric constraints ensure appropriate airfoil thickness, leading-edge radius, and trailing-edge angle (Table 8.5).

Table 8.4 Flow conditions for the NACA test case

Condition	Take-off	Cruise
Mach number	0.20	0.77
Reynolds number	5×10^6	10^7
Angle of attack (°)	10.8	1.0
Transition	3%	3%

Table 8.5 Flow conditions for the RAE test case

Condition	Cruise	Off-design 1	Off-design 2
Mach number	0.734	0.754	0.680
Reynolds number	6.5×10^6	6.2×10^6	5.7×10^6
Angle of attack (°)	2.8	2.8	1.8
Transition	3%	3%	11%

In the RAE test case, the objectives are to optimize the drag coefficient while maintaining sufficient lift and keeping the pitching moment within a 2% range.

References

1. Allmendinger R, Emmerich M, Hakanen J, Jin Y, Rigoni E (2017) Surrogate-assisted multicriteria optimization: complexities, prospective solutions, and business case. J Multi-Criter Dec Anal 24(1–2):5–24
2. Biggers LR, Bocovich C, Capshaw R, Eddy BP, Etzkorn LH, Kraft NA et al (2014) Explainable artificial intelligence: understanding, visualizing and interpreting deep learning models. Empirical Softw Eng 19(3)
3. Broomhead DS, Lowe D (1988) Multivariable functional interpolation and adaptive networks. Complex Syst 2(3):321–355
4. Chugh T, Jin Y, Miettinen K, Hakanen J, Sindhya K (2016) A surrogate-assisted evolutionary algorithm for multiobjective optimization. IEEE Trans Evol Comput 21(5):739–752
5. Chugh T, Jin Y, Miettinen K, Hakanen J, Sindhya K (2016) A surrogate-assisted reference vector guided evolutionary algorithm for computationally expensive many-objective optimization. IEEE Trans Evol Comput 22(1):129–142
6. Couckuyt I, Deschrijver D, Dhaene T (2014) Fast calculation of multiobjective probability of improvement and expected improvement criteria for pareto optimization. J Global Optim 60:575–594
7. Cox DD, John S (1992) A statistical method for global optimization. In: Proceedings of the 1992 IEEE international conference on systems, man, and cybernetics, pp 1241–1246. IEEE
8. Deb K, Pratap A, Agarwal S, Meyarivan T (2002) A fast and elitist multiobjective genetic algorithm: Nsga-ii. IEEE Trans Evol Comput 6(2):182–197
9. Deist TM, Maree SC, Alderliesten T, Bosman PAN (2020) Multi-objective optimization by uncrowded hypervolume gradient ascent. In: Parallel problem solving from nature–PPSN XVI: 16th international conference, PPSN 2020, Leiden, The Netherlands, September 5–9, 2020, Proceedings, Part II 16, pp 186–200. Springer
10. Désidéri J-A, Marco-Blaszka N (2000) INGENET test-case database: structure and examples of optimization problems. In: European congress on computational methods in applied sciences and engineering (ECCOMAS 2000), pp 1–20, Barcelona, Spain. Accessed 11–14 Sept 2000
11. Emmerich M (2005) Single- and multi-objective evolutionary design optimization assisted by Gaussian random field metamodels. PhD thesis, University of Dortmund
12. Emmerich MTM (2025) Kriging-based surrogate modeling: a simple 1-D example with uncertainty quantification and visualization in Python (Version 1.0) [Software]. Zenodo. https://doi.org/10.5281/zenodo.14776563
13. Emmerich M, Beume N, Naujoks B (2005, March) An EMO algorithm using the hypervolume measure as selection criterion. In: Proc. EMO, pp 62–76. Springer

14. Emmerich M, Giannakoglou KC, Naujoks B (2006) Single-and multiobjective evolutionary optimization assisted by gaussian random field metamodels. IEEE Trans Evol Comput 10(4):421–439
15. Emmerich M, Giotis A, Özdemir M, Bäck T, Giannakoglou K (2002) Metamodel—assisted evolution strategies. In: International conference on parallel problem solving from nature, pp. 361–370. Springer
16. Emmerich M, Naujoks B (2004) Metamodel assisted multiobjective optimisation strategies and their application in airfoil design. In: Adaptive computing in design and manufacture VI, pp 249–260. Springer
17. Emmerich M, Naujoks B (2004) Metamodel-assisted multiobjective optimization with implicit constraints and its application in airfoil design. In: International conference & advanced course ERCOFTAC, Athens, Greece
18. Emmerich M, Yang K, Deutz A, Wang H, Fonseca CM (2016) A multicriteria generalization of bayesian global optimization. Advances in stochastic and deterministic global optimization, pp 229–242
19. Galván B, Greiner D, Periaux J, Sefrioui M, Winter G (2003) Parallel evolutionary computation for solving complex cfd optimization problems: a review and some nozzle applications. Parallel Comput Fluid Dyn 2002:573–604
20. Giannakoglou KC (2001) Optimization and inverse design in aeronautics: how to couple genetic algorithms with radial basis function networks. Innov Tools Sci Comput Aeronaut Eng
21. Giannakoglou KC (2002) Ingenet: A european thematic network for design optimization using computational intelligence. Eng Optim 34(6):647–656
22. Giotis AP, Giannakoglou KC (1999) Evolutionary algorithms with surrogate models in multi-objective aerodynamic optimization. Comput Methods Appl Mech Eng 175(3):331–347
23. Habib A, Singh HK, Chugh T, Ray T, Miettinen K (2019) A multiple surrogate assisted decomposition-based evolutionary algorithm for expensive multi/many-objective optimization. IEEE Trans Evol Comput 23(6):1000–1014
24. Hupkens I, Deutz A, Yang K, Emmerich M (2015) Faster exact algorithms for computing expected hypervolume improvement. In: International conference on evolutionary multi-criterion optimization, pp 65–79. Springer
25. Jeong S, Obayashi S (2005) Efficient global optimization (ego) for multi-objective problem and data mining. In: 2005 IEEE congress on evolutionary computation, vol 3, pp 2138–2145. IEEE
26. Jin Y (2005) A comprehensive survey of fitness approximation in evolutionary computation. Soft Comput 9(1):3–12
27. Y. Jin, M. Olhofer, and B. Sendhoff. On evolutionary optimization with approximate fitness functions. In *Proceedings of the 2nd Annual Conference on Genetic and Evolutionary Computation*, pages 786–793, 2000
28. Jones DR, Schonlau M, Welch WJ (1998) Efficient global optimization of expensive black-box functions. J Global Optim 13(4):455–492
29. Keane AJ (2000) Statistical improvement criteria for surrogate-assisted multiobjective optimization. AIAA J 38(2):211–219
30. Knowles J (2006) Parego: a hybrid algorithm with on-line landscape approximation for expensive multiobjective optimization problems. In: IEEE congress on evolutionary computation, pp 325–332
31. Knowles J (2006) Parego: a hybrid algorithm with on-line learning for multiobjective optimization problems. In: Proceedings of the international conference on evolutionary multi-criterion optimization (EMO 2006), pp 394–409, Heidelberg. Springer
32. Krige DG (1951) A statistical approach to some basic mine valuation problems on the witwatersrand. J Chem Metall Min Soc S Afr 52(6):119–139
33. Łaniewski-Wołłk Ł, Obayashi S, Jeong S (2010) Development of expected improvement for multi-objective problem. In: Proceedings of the 42nd fluid dynamics conference/aerospace numerical simulation symposium

34. Laumanns M, Ocenasek J (2002) Bayesian optimization algorithms for multi-objective optimization. In: International conference on parallel problem solving from nature, pp 298–307. Springer
35. Mäkinen RAE, Périaux J, Toivanen J (1999) Multidisciplinary shape optimization in aerodynamics and electromagnetics using genetic algorithms. Int J Numer Meth Fluids 30(2):149–159
36. Matheron G (1963) Principles of geostatistics. In: Economic Geology, pp 1246–1266. Society of Economic Geologists
37. Mockus J (1974) On bayesian methods for seeking the extremum. In: Proceedings of the IFIP technical conference, pp 400–404
38. Mockus J, Tiesis V, Žilinskas A (1978) Bayesian methods for seeking the extremum. In: Toward global optimization. Elsevier
39. Myers DE (1992) Kriging, cokriging, radial basis functions and the role of positive definiteness. Comput & Math Appl 24(12):139–148
40. Ponweiser W, Wagner T, Biermann D, Vincze M (2008) Multiobjective optimization on a limited budget of evaluations using model-assisted-metric selection. In: International conference on parallel problem solving from nature, pp 784–794. Springer
41. Sacks J, Welch WJ, Mitchell TJ, Wynn HP (1989) Design and analysis of computer experiments. Stat Sci 4(4):409–423
42. Siefert CM, Model-assisted pattern search. Undergraduate Honors Thesis, College of William & Mary, Virginia, USA, (2000) Advisors: Virginia J. Trosset, and Stephen K. Park, Torczon, Michael W
43. Ulmer H, Streichert F, Zell A (2005) Model assisted evolution strategies. In: Knowledge incorporation in evolutionary computation, pp 333–355. Springer
44. Wagner T, Emmerich M, Deutz A, Ponweiser W (2010) On expected-improvement criteria for model-based multi-objective optimization. In: PPSN XI, pp 718–727. Springer
45. Williams C, Rasmussen C (1995) Gaussian processes for regression. Advances in neural information processing systems, vol 8
46. Yang K, Emmerich M, Deutz A, Fonseca CM (2017) Computing 3-d expected hypervolume improvement and related integrals in asymptotically optimal time. In: International conference on evolutionary multi-criterion optimization, pp 685–700. Springer
47. Žilinskas A (2025) On similiarities between two global optimization algorithms based on different (bayesian and lipschitzian) approaches. Springer optimization and its applications, pp 233–237
48. Žilinskas A, Žilinskas J (2015) Adaptation of a one-step worst-case optimal univariate algorithm of bi-objective Lipschitz optimization to multidimensional problems. Commun Nonlinear Sci Numer Simul 21(1–3):89–98
49. Zitzler E, Thiele L (1999) Multiobjective evolutionary algorithms: a comparative case study and the strength pareto approach. IEEE Trans Evol Comput 3(4):257–271

Chapter 9
Uncertainty Quantification Based on Multi-fidelity Metamodels and Reduced Order Models

Alberto Clarich, Luca Battaglia, Carlo Poloni, and Livia Trambaiolo

Abstract Optimization under uncertainties is achieving more and more agreement in the industrial design community. In fact, most of the industrial processes are permeated by uncertainties, including for instance dimensional tolerances and fluctuations in the operating conditions. These uncertainties are commonly transferred to the performance of the system, which cannot be determined by a single value, but rather by a statistical distribution of results. To deal with industrial problems characterized by a large number of uncertainties and expensive simulation time, it is particularly important to develop methodologies to obtain accurate results, and rely on a reduced number of sample evaluations at the same time. In this paper two different methodologies, applied to aeronautical test cases, are presented. The first one takes advantage of multi-fidelity metamodels, consisting of combining few high-fidelity simulations (HF) with low-fidelity ones (LF), to evaluate the propagation of the uncertainties using Cokriging algorithm. The method is particularly efficient for large number of uncertainties since it provides quite a linear correlation of the samples with the uncertainties number, while other methods, such as the ones based on Polynomial Chaos Expansion, may require at least a quadratic correlation. The second methodology is instead based on the UQ application of Reduced Order Models (ROM), that define an equivalent CFD model in function of a given set of parameters, by interpolating a sampling series of CFD snapshots. Surrogate models like Deep Learning can be used for the interpolation, or as alternative POD (Proper

---

A. Clarich (✉) · L. Battaglia
ESTECO SpA, Trieste, Italy
e-mail: clarich@esteco.com

L. Battaglia
e-mail: battaglia@esteco.com

C. Poloni
University of Trieste, Trieste, Italy
e-mail: poloni@units.it

L. Trambaiolo
Imperial College of London, London, UK
e-mail: livia.trambaiolo19@imperial.ac.uk

T. Tuovinen et al. (eds.), *Challenges in Design Methods, Numerical Tools and Technologies for Sustainable Aviation, Transport and Industry*, Computational Methods in Applied Sciences 17, https://doi.org/10.1007/978-3-031-98675-8_9

Orthogonal Decomposition) methods which are instead based on the interpolation of a reduced number of modes (the principal components). In both cases, the ROM model can be used to instantly evaluate a Monte Carlo DOE at the variation of the uncertain parameters, allowing the computation of the uncertainty propagation of the vectorial field of interest.

9.1 Introduction

Robust Design Optimization in industrial practice is generally characterized by a large number of uncertainties and expensive simulation time. Therefore, there is a need to develop methodologies targeting accuracy, while relying on a reduced number of sample evaluations.

In literature there are several examples of numerical methods to study stochastic phenomena, which can be generally divided into two categories: intrusive methods, which consist in resolution of a coupled system of deterministic equations, and non-intrusive methods, which consist in resolution of a decoupled system of deterministic equations. An example of the first group is the intrusive Chaos Polynomials [1] and in general the Adjoint formulation of optimization problems [2], while in the second group there are several methodologies, such as Monte Carlo, Stochastic Collocation [3], Tensorial-expanded Chaos Collocation [4] and methods based on Response Surfaces, such as DACE [5], or POD [6].

The complexity of designing an efficient intrusive solver is evident, both because of computational cost and of the obvious handicap to imply an internal modification of the deterministic solver. A remarkable advantage of non-intrusive methodologies is that they have a simpler computational management and the deterministic solver is represented by a black-box with no need to modify it.

Among the non-intrusive methods, Polynomial Chaos Expansion (PCE) is one of the most efficient methodologies to accurately manage uncertainties in industrial design [7], since stochastic moments are computed in a deterministic way, and they are not affected by approximation errors that may occur when applying meta-models. It can be proved that the convergence to the exact moments grows exponentially with the number of sampling points [8], while a Monte Carlo sampling converges linearly.

However, this methodology requires a minimum number of samples which heavily increases with the number of uncertainties, therefore a typical industrial optimization design with expensive simulation times and many uncertainties can hardly be treated as a feasible task.

In this paper, two methods are proposed for Uncertainty Quantification (UQ) to overcome these issues.

The first method is based on multi-fidelity metamodels. Using Cokriging algorithm [9] and [10], the purpose is to build a scalar field meta-model, using two or more series of samples of different fidelity, which are respectively naturally nested (LF or low-fidelity samples are used to determine where to evaluate the HF or high-fidelity samples). Multi-fidelity methodologies [11] have in fact been introduced to combine

HF simulations with middle or low-fidelity ones (MF/LF), to train efficiently metamodels with a significantly lower computational cost. LF models can be obtained for instance from a model grid that is coarser than the HF one, or from a simulation model which introduces a higher level of approximation keeping a good accuracy in the results. From the multi-fidelity metamodel, it is then possible to evaluate how the uncertainties of the system are propagated to the performances, by the evaluation of a large Monte Carlo surrogate DOE, that can be generated in few seconds.

The second method illustrated in this paper is instead based on the application of ROM or Reduced Order Model [12] to a vector field, that allows the definition of an equivalent CFD model as a function of the principal components of the original field (POD technique), or as an interpolation by Deep Learning NN (Neural Network) of the CFD field results in each point of the grid. Also in this case, by using the ROM model it is possible to compute the propagation of the uncertainties of the system by the evaluation of a large Monte Carlo DOE.

9.2 UQ by Multi-fidelity Metamodel

9.2.1 Cokriging Algorithm

Cokriging is a multi-fidelity surrogate model based on Bayesian theory. It was originally proposed for enhancing the accuracy of a high-fidelity surrogate using supplementary observations of low-fidelity models. In essence, it aims at exploiting the cross-correlation between two or more Gaussian process surrogates through a stochastic auto-regressive scheme, and it yields a predictive posterior distribution for the high-fidelity model output that encodes the contribution of lower fidelity levels with quantified uncertainty (Fig. 9.1).

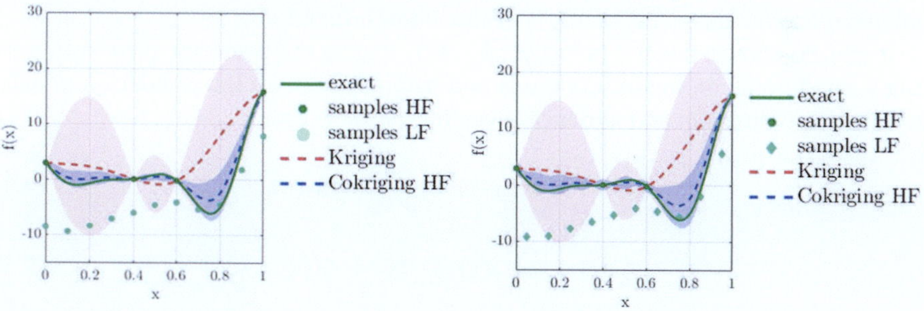

Fig. 9.1 Cokriging by nested (left) and not nested (right) sampling points

Cokriging was first proposed by Kennedy and O'Hagan in the landmark paper [4], and since then it has been used in many different applications, such as surrogate-based optimization [9]. The algorithm described in [4] has an unfavorable cubic scaling with the number of observed data and the number of surrogate models.

To overcome this issue, Le Gratièt [10] proposed a recursive version of the Cokriging algorithm, which allows to break up the computation of posterior distribution into a sequence of distinct inferences of smaller dimensions. The advantage of this approach is clear: the inversion of the small covariance matrices associated with the sequence of inferences is less expensive and better conditioned than the inversion of the large covariance matrix associated with the full inference problem.

To describe the recursive Cokriging approach, suppose we have available multiple models, say s-models $z_t(\mathbf{x})$ (with t = 1,...,s), providing an estimate of the real-valued scalar field,

$$z_t : \mathcal{X} \in \mathbb{R}^d \to \mathbb{R} \tag{9.1}$$

$$\mathbf{x} \to z_t(\mathbf{x}) \tag{9.2}$$

sorted by increasing order of fidelity, being $z_1(\mathbf{x})$ the model with the lowest level of fidelity, and modeled by Gaussian processes $(Z_t(\mathbf{x}))_{t=1,\ldots,s}$.

Such Gaussian processes are related to each other by the autoregressive model:

$$\begin{cases} Z_t(\mathbf{x}) = \rho_{t-1}(\mathbf{x})\tilde{Z}_{t-1}(\mathbf{x}) + \delta_t(\mathbf{x}) \\ \tilde{Z}_{t-1}(\mathbf{x}) \perp \delta_t(\mathbf{x}) \\ \rho_{t-1}(\mathbf{x}) = \mathbf{g}_{t-1}^T(\mathbf{x})\boldsymbol{\beta}_{\rho_{t-1}} \end{cases} \tag{9.3}$$

where $\tilde{Z}_{t-1}(\mathbf{x})$ is a Gaussian process with distribution $\left(Z_{t-1}(\mathbf{x})|\mathbf{z}_{t-1}, \boldsymbol{\beta}_{t-1}, \boldsymbol{\beta}_{\rho_{t-1}}, \boldsymbol{\theta}_{t-1}\right)$, $\delta_t(\mathbf{x}) \sim \mathcal{GP}\left(\mathbf{f}_t^T(\mathbf{x})\boldsymbol{\beta}_t, k_t\left(\mathbf{x}, \mathbf{x}'; \boldsymbol{\theta}_t\right)\right)$ and $Z_1(\mathbf{x}) \sim \mathcal{GP}\left(\mathbf{f}_1^T(\mathbf{x})\boldsymbol{\beta}_1, k_1\left(\mathbf{x}, \mathbf{x}'; \boldsymbol{\theta}_1\right)\right)$. k_t is the covariance function depending on hyper-parameters $\boldsymbol{\theta}_t$, $\boldsymbol{\beta}_t$ and $\boldsymbol{\beta}_{\rho_{t-1}}$ are vectors of unknown coefficients, $\mathbf{f}_t(\mathbf{x})$ and $\mathbf{g}_{t-1}(\mathbf{x})$ are chosen functions.

It can be shown that $\left(Z_t(\mathbf{x})|\mathbf{z}_t, \boldsymbol{\beta}_t, \boldsymbol{\beta}_{\rho_t}, \boldsymbol{\theta}_t\right)$, that is the posterior distribution of $Z_t(\mathbf{x})$, conditional on the observations and parameters of all lower-fidelity models, is Gaussian with mean and variance given by:

$$\hat{m}_{Z_t}(\mathbf{x}) = \rho_{t-1}(\mathbf{x})\hat{m}_{Z_{t-1}}(\mathbf{x}) + A_t(\mathbf{x}) \tag{9.4}$$

$$\hat{s}^2_{Z_t}(\mathbf{x}) = \rho^2_{t-1}(\mathbf{x})\hat{s}^2_{Z_{t-1}}(\mathbf{x}) + B_t(\mathbf{x}) \tag{9.5}$$

where $A_t(\mathbf{x})$ and $B_t(\mathbf{x})$ are functions that can be explicitly computed based on univariate GPR results at fidelity levels t and t−1 [4].

This is very convenient from a computational viewpoint. In Le Gratièt [10] a fully Bayesian formulation of recursive Cokriging which incorporates prior information in the maximum likelihood estimation of the model hyper-parameters is given. This allows to speed up the estimation process by leveraging on analytical expressions.

In the Cokriging modeling process, high-fidelity (HF) and low-fidelity (LF) sampling points must be nested to satisfy the Markov property. Thanks to the recursive formulation of the multi-fidelity Cokriging model, the method can be easily adapted also to the cases where the design sets are not nested [10]. The nested property of the design sets is not necessary to build the model, but it allows for a simple estimation of the model parameters.

In Fig. 9.1 there is an example of Cokriging models obtained by nested sampling points (left) and not nested sampling points (right).

The models are obtained with the same HF samples, but different LF samples. Shaded areas correspond to confidence intervals $\hat{m}_{Z_t}(x) \pm 3\hat{s}_{Z_t}(x)$ with $t = 1,2$ where z_1 is the LF model and z_2 is the HF model.

9.2.2 *Application of Multi-fidelity UQ to an Aeronautical Test Case*

To prove the efficiency of the multi-fidelity methodology for UQ, we propose in this section an application to an aeronautical test case.

A NACA0012 airfoil, at Mach (M) = 0.6, Re = 1.35E7, and angle of attack (AoA) of 4°, has been simulated using ANSYS Fluent software, using two grids of different quality (Fig. 9.2): the coarser grid (LF) has 30,000 elements while the refined one (HF) has 150,000 elements.

The aerodynamic coefficients obtained from the simulations for the baseline have been validated accordingly to literature results [13] for the selected flight conditions, so that the results (Fig. 9.3) with both grids differ from the experimental one by less than 1%.

The baseline airfoil has therefore been parameterized by adding two parametric Bezier curves of 4 control points on the upper and lower baseline profiles (Fig. 9.4), with the goal to train a multi-fidelity metamodel able predict the aerodynamic coefficients in function of the 8 parameters (coordinates of the Bezier control points).

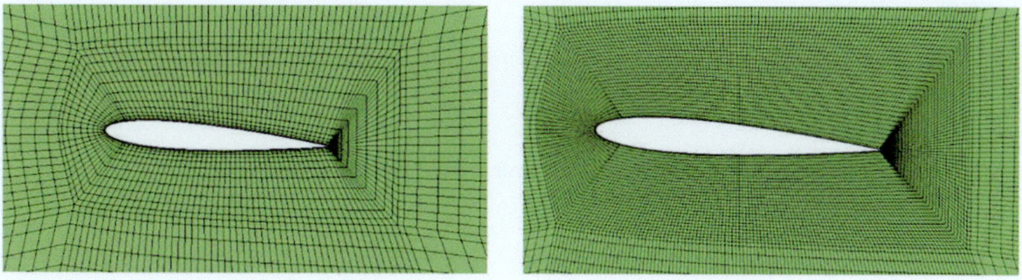

Fig. 9.2 NACA0012 grid: LF (30,000 elements), left—HF (150,000), right

Fig. 9.3 Baseline (M = 0.6, Re = 1.35E7, α = 4°) pressure field

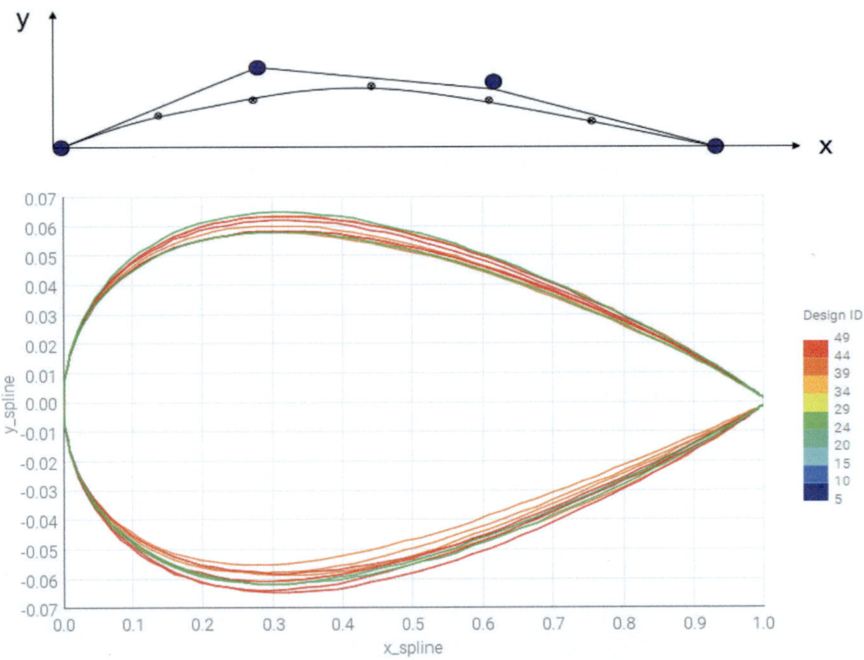

Fig. 9.4 Bezier control points (above) and parameterized shapes (below)

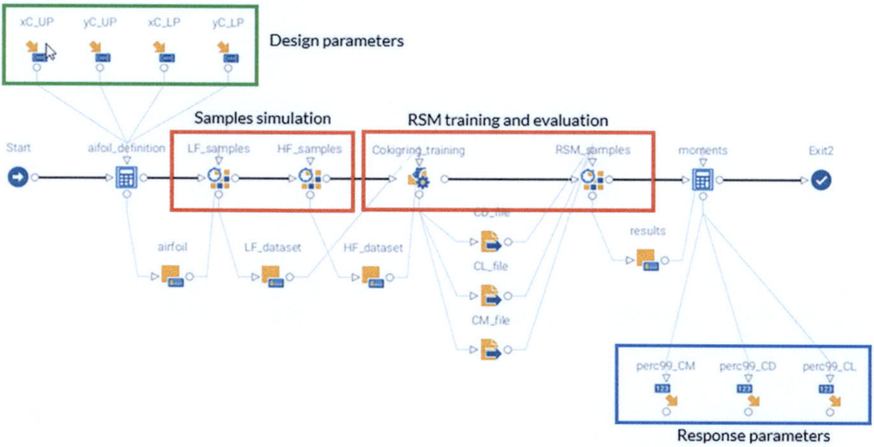

Fig. 9.5 Process automation (modeFRONTIER) to evaluate HF/LF samples

The two operational parameters are defined by an uncertainty distribution which is Normal, with standard deviation equal to 5E-3 for M and 0.1 for AoA. Accordingly to the distributions two series of samples have been evaluated for the baseline, using a Latin Hypercube DOE algorithm.

Figure 9.5 illustrates the process automation workflow that has been setup in modeFRONTIER software to evaluate the two series of samples.

In the first part of the workflow the baseline airfoil is parameterized, by adding the Bezier curves defined by the Bezier curve in function of the design parameters (Bezier control points coordinates). In the central part, two Scheduling nodes are used to evaluate the HF and LF DOEs, by changing M and AoA parameters according to the defined distributions, and by launching in batch Ansys ICEM and Fluent CFD solvers (mesh update and CFD analysis). The result data obtained is used by the RSM Trainer node that evaluate, by a Cokriging algorithm coded in Python, the response surface (RSM) for the three aerodynamic coefficients. Using the RSM, another Scheduling node is then used to evaluate a large Monte Carlo DOE (10,000 designs) in few seconds, according to the uncertain parameters distribution. It is then possible to evaluate the moments of the output parameters distribution (mean, standard deviation) and percentiles. In particular, the focus is on evaluating the 99% percentile of Cd and Cm, and the 0.1% percentile of Cl, since the optimization is carried out to minimize the first two values (minimizing the "max" values) and maximize the latter one (maximizing the "min" value).

It has been found that by using 10 LF and 5 HF samples, the error in the prediction of the aerodynamic coefficients is in the order of 0–1%, if compared to the same coefficients computed by (PCE) Polynomial Chaos Expansion [3] using 10 HF samples, which is the most accurate result (Table 9.1, first row).

Accepting this level of approximation, a Robust Design optimization has been executed, by considering the three objectives of the aerodynamics coefficients

Table 9.1 Baseline and optimized coefficients

Design	Cd_99% Cokriging	Cl_1% Cokriging	Cm_99% Cokriging	Cd_99% HF	Cl_1% HF	Cm_99% HF
Baseline	1.22E−2	0.505	6.81E−3	1.24E−2	0.507	6.82E−3
Optimal 1	1.09E−2	0.537	7.85E−3	1.26E−2	0.527	1.34E−3
Optimal 2	1.11E−2	0.546	7.31E−3	1.29E−2	0.537	0.73E−3

as defined above. The results of the optimization, applying the proprietary algorithm PilOPT of modeFRONTIER (an automatic combination of Gas, gradient and surrogates-based algorithms), are reported in Fig. 9.6.

The optimal designs selected for validation from the Pareto front (highlighted in green in Fig. 9.6), have been validated by using a PCE trained on 10 HF samples: the results are reported in Table 9.1 (second and third row).

It is important to note that Cl and Cm percentiles have been improved significantly by the optimization, while Cd has remained practically unchanged: this result may be due to the fact that Cd prediction by Cokriging is less accurate than the other two coefficients. Nonetheless, the UQ methodology based on Cokriging RSM trained on multi-fidelity metamodels has been proved to be efficient and a valid alternative to PCE methods, in particular if the simulation time for each design is important.

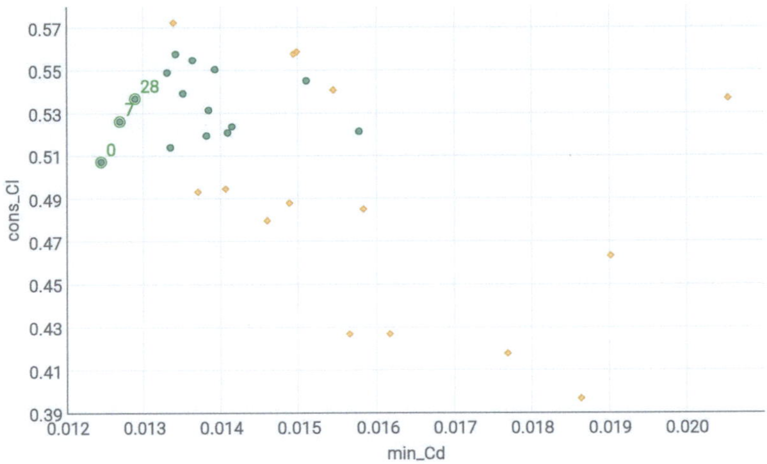

Fig. 9.6 Robust design optimization by Cokriging: Cd 99%ile versus Cl 1%ile

9.2.3 *Consideration About Number of Samples Versus Uncertainties*

An additional expected advantage of the Cokriging methodology is the rate of convergence versus the number of uncertainties. In the aeronautical test case illustrated in the previous section unfortunately it is not possible to scale drastically the number of parameters to prove this assumption. Thus, for testing purposes, an analytical case, based on the deflection computation of a cantilever beam (Fig. 9.7) in function of a scalable number of section thickness parameters t_i, was studied.

Considering the t_i parameters as uncertainties, it is easy to compare the efficiency of the method based on Kriging RSM prediction of the deflection (here multi-fidelity models are not used, therefore the Cokriging algorithm is equivalent to a single fidelity Kriging) versus the PCE computation of the results.

As expected (Fig. 9.8), the PCE polynomial expansion of order 2 requires a number of samples that is proportional at the 2nd order with the number of uncertainties, to obtain an error in the prediction of the results of less than1%. To obtain the same accuracy, the Kriging RSM prediction shows a linear relationship with the number of parameters. This mean that, if until 20–30 uncertain parameters the two methodologies shows the same level of accuracy, for a larger number of parameters the (Co)Kriging methodology is more convenient from the computational point of view.

9.3 UQ by Reduced Order Models

The second approach presented in this paper based on the application of Reduced Order Models (ROM) to evaluate how the parameters uncertainties are propagated to the vector field of interest.

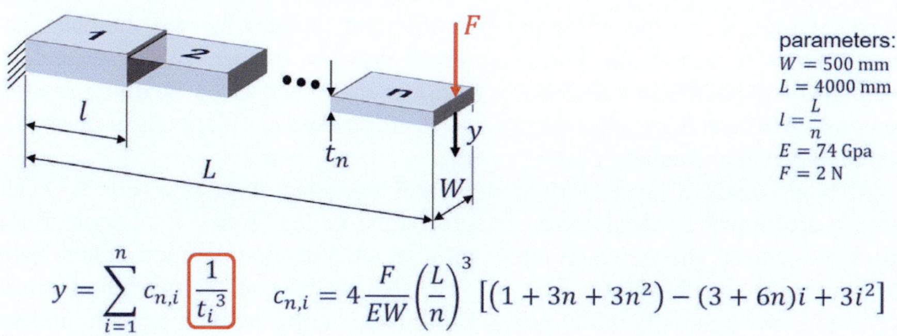

Fig. 9.7 Scalable cantilever beam; deflection as function of sections thickness

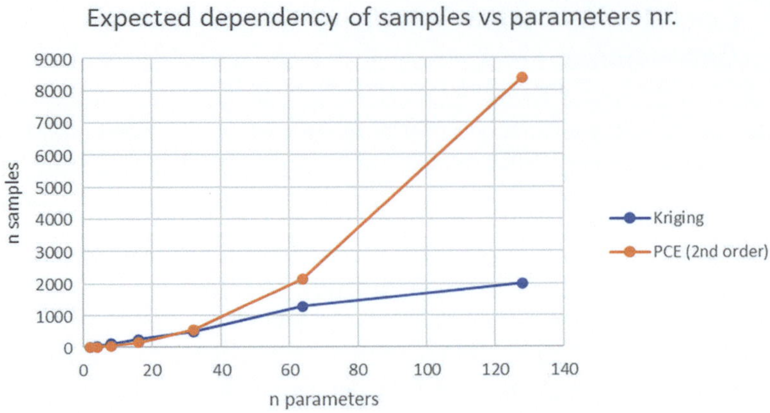

Fig. 9.8 Cantilever beam scalable test: samples nr. versus Uncertainties nr.

Two different ROM methods are presented (the first one based on Deep Learning and the second one on Proper Orthogonal Decomposition) and applied to the same baseline airfoil test case defined in Sect. 2.2.

In both cases, a training set of design samples is used in the training process of each methodology, and the ROM model is used to evaluate in a few seconds by a large Monte Carlo DOE (over than 10,000 designs) different vector fields, from which stochastic moments such as mean and standard deviation of the output of interest can be computed in each cell of the grid.

A comparison in terms of accuracy of the two methods in function of the design samples number is finally reported.

9.3.1 *ROM Based on Deep Learning*

Deep Learning is a type of Neural Network that imitates the way humans gain certain types of knowledge, and it got more popular over the years compared to standard models. While traditional algorithms are linear, Deep Learning models, generally Artificial Neural Networks (ANN), are stacked in a hierarchy of increasing complexity and abstraction [14].

ANN are made of layers with an input and an output dimension (Fig. 9.9). The latter is determined by the number of neurons (also called "nodes"), a computational unit that connects the weighted inputs through an activation function (which helps the neuron to switch on/off). The weights, like in most of the machine learning algorithms, are randomly initialized and optimized during the training to minimize a loss function. Generally speaking, "Deep" Learning (DL) applies when the algorithm has at least 2 hidden layers (so 4 layers in total including input and output).

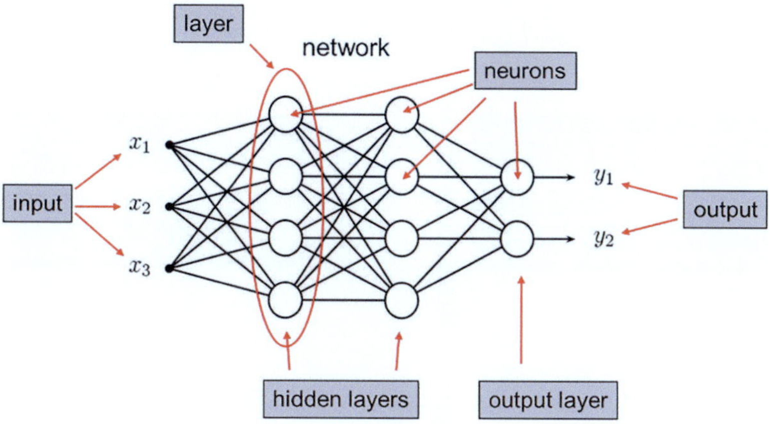

Fig. 9.9 Representation of a neural network training process

Since the DL metamodel are able to predict simultaneously several outputs by a single training process, they are particularly indicated to build a ROM (Reduced Order Model) of any output vector field of interest in a CFD simulation (for instance, the pressure field around an airfoil), which is the topic investigated in this paper.

The Keras Neural Network [15] has been implemented in the Python environment of modeFRONTIER to obtain the prediction of each pressure point in the grid of a NACA 0012 airfoil (the same LF model used for the application of previous section), in function of Mach number and angle of attack.

For this application, 4 total layers (with 2-200-2470-2470 number of neurons each, after a calibration trying different combinations, and following the recommendations in [16]) were defined, as activation function the "*relu*" (rectified linear unit), and a total of 10,000 epochs of training. Note that 2470 is the grid dimension in the region of interest around the airfoil, for which the training process has been setup.

A different number of training fields, at the variation of M and alpha parameters accordingly to the uncertainty distribution, has been evaluated. In Fig. 9.10, the field predicted by the DL related to mean (left) and standard deviation (right) of pressure is reported for a training set of 10 designs (above) and compared for a training set of 150 samples (middle); finally, the relative difference in each grid point is reported (bottom).

If the average error in the prediction of the mean pressure field by using only 10 samples is very small (0.8%), the one for the prediction of pressure field standard deviation is higher (17%) therefore an higher number of samples to get higher accuracy may be needed. Before reporting a complete curve of convergence, a different methodology to define a ROM model is introduced in the next section.

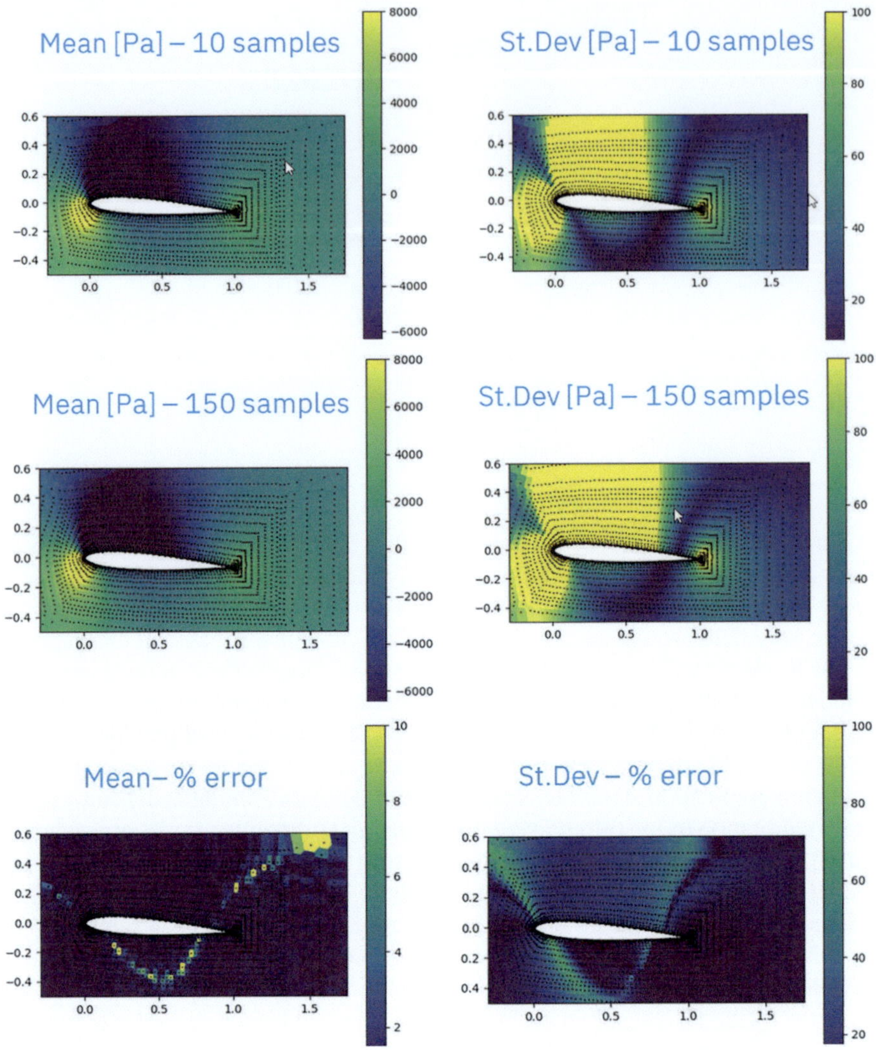

Fig. 9.10 Mean/st. dev (left/right) of pressure filed for 10 samples/150 samples/error % (top/middle/bottom) computed by DL

9.3.2 ROM Based on POD (Proper Orthogonal Decomposition)

ROM models can take advantage of the POD (Proper Orthogonal Decomposition) to represent a vector field (e.g., the velocity field around a vehicle) as a function of a reduced basis of the field eigenvectors [17].

We indicate a parameters set $\Xi = [\boldsymbol{\mu}_1, \ldots, \boldsymbol{\mu}_{\mathbf{N}}] \epsilon \mathbb{R}^{P \times N}$, where $\boldsymbol{\mu}_i \epsilon \mathbb{R}^P$ with i = 1,..N are *N* different combinations of a number P of design parameters (i.e., *N* different designs defined by a DOE). Once the simulations are completed for each design, we indicate with $\boldsymbol{U} = [\boldsymbol{y}(\boldsymbol{\mu}_1), \ldots, \boldsymbol{y}(\boldsymbol{\mu}_{\mathbf{N}})] \epsilon \mathbb{R}^{m \times N}$ the corresponding matrix of solution snapshots, where $\boldsymbol{y}_{\mathbf{i}} \epsilon \mathbb{R}^m$ with *i* = *1,..N* are the field solutions computed for each one of the *m* points of the grid.

The POD approach consists in representing the Covariance matrix $\boldsymbol{C} = \boldsymbol{U}^T \boldsymbol{U}$ as diagonal matrix $\boldsymbol{\Theta} \boldsymbol{\Lambda} \boldsymbol{\Theta}^T$ of rank M_r where $\boldsymbol{\Theta}$ is the matrix of the eigenvectors or principal modes of the field. Using the basis transformation $\boldsymbol{A} = \boldsymbol{U}\boldsymbol{\Theta}$, the reduced field solution $\boldsymbol{U}^*$ in function of a generic set of parameters μ can be expressed as:

$$\boldsymbol{U}^*(\boldsymbol{\mu}) = \sum_{i=1}^{M_r} \alpha_i(\boldsymbol{\mu}) \phi_i \quad \text{for i} = 1, \ldots, \mathrm{M}_r \tag{9.6}$$

with $\alpha_i(\boldsymbol{\mu})$ a weight coefficient that can be determined by interpolation in function of the design database.

It can be proved [18] that it is more convenient to diagonalize the inverse covariance matrix defined by $\boldsymbol{C_S} = \boldsymbol{U}\boldsymbol{U}^T$ because its dimension is *NxN* (where *N* can be in the order of 10–100) instead of *mxm* (that can be in the order of 10^5–10^8). In this case, the diagonalization of $\boldsymbol{C_S}$ obtains $\boldsymbol{A_S} \boldsymbol{\Lambda} \boldsymbol{A_S^T}$, where $\mathbf{A_S}$ (*NxN*) is defined as matrix of the temporal eigenvectors. By evaluating $\boldsymbol{U} = \boldsymbol{A_S} \phi_{\boldsymbol{S}}^{\boldsymbol{T}}$, Eq. 9.6 will then become:

$$\boldsymbol{U}^*(\boldsymbol{\mu}) = \sum_{i=1}^{M_r} (\alpha_s)_i(\boldsymbol{\mu})(\phi_s)_i \quad \text{for i} = 1, \ldots, \mathrm{M}_r \tag{9.7}$$

Several methods are available in literature to determine the weight coefficients, including POD-Galerkin, zonal-POD or POD with residual minimization [19].

In this paper, a Deep Learning metamodels is instead applied to interpolate the $(\alpha_s)_i(\boldsymbol{\mu})$ coefficients, in function of the unknown parameter combination $\boldsymbol{\mu}$, given the training dataset made by *N* combinations of $\boldsymbol{\mu}_i$, with *i* = *1,..N*. As seen in Chap. 3 the DL is particularly efficient to interpolate the responses for multiple output points, and since in this case there are *NxN* coefficients $(\alpha_s)_i$ to predict, this is absolutely an affordable task with *N* data in the order of 10^1–10^2.

In this case, as reported in Fig. 9.11, the accuracy of POD can reach very accurate results. with just 20 training designs. The error for the prediction of mean pressure is equal to 0.02% as average, while the one for the standard deviation of pressure is equal to 0.6% as average.

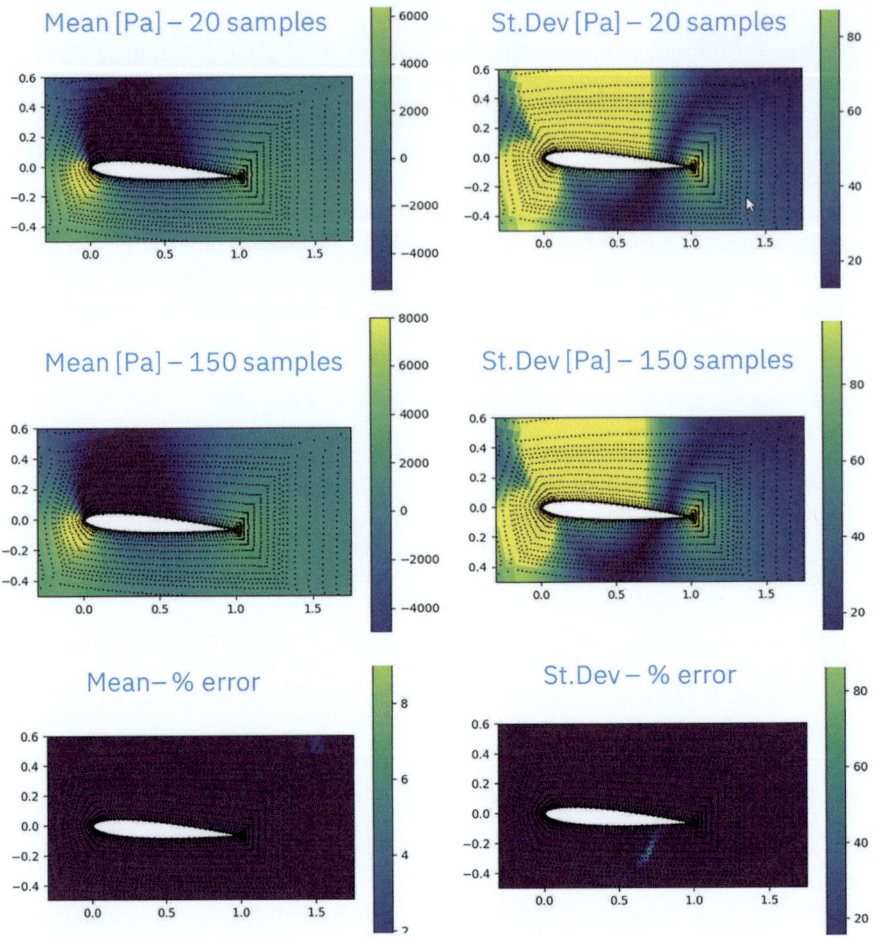

Fig. 9.11 Mean/st. dev (left/right) of pressure filed for 20 samples/150 samples/error % (above/middle/below) computed by POD

9.3.3 Effect of Samples Number for ROM Based on DL and POD

In this section, the efficiency of the two methodologies introduced in 3.1 and 3.2 is compared for the same application problem, at the variation of the number of designs in the training database (Fig. 9.12). For each training set, the average value of the pressure standard deviation is reported in the chart.

It is interesting to note that if the convergence of the DL method requires a large number of training set, the POD method converge much more rapidly, and just 20 designs are needed to obtain an error of less than 1%.

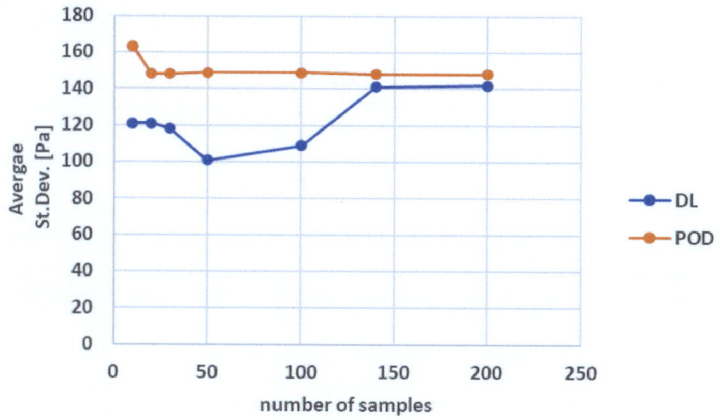

Fig. 9.12 Avg. value of pressure st. dev for DL and POD vs training set size

This result can be likely justified by the fact that, if DL is enough accurate to predict an output field for a large variation of parameters, in a UQ problem where an high level of accuracy is required, the POD method is more suitable.

9.4 Conclusions

Two different methodologies for UQ and Robust Design based on metamodels have been proposed in this paper.

Cokriging can be used to obtain an accurate RSM combining LF and HF sampling points. By using the Cokriging, a large Monte Carlo sampling can be evaluated in few seconds, to compute the distribution moments of the results. The accuracy of the results has been proven by applying the methods to an aerodynamic test case optimization (NACA0012 airfoil).

ROM (Reduced Order Model) has been introduced to build an equivalent CFD vector field in function of the uncertain parameters by the DL (Deep Learning) or POD (Proper Orthogonal Decomposition) technique. These methods can be used to evaluate also in this case a large Monte Carlo sampling in few seconds, computing the mean and standard deviation of the field of interest in each point of the grid. The POD method turned out to be more efficient than the DL for this application.

References

1. Lin G, Wan X, Su C, Karniadakis G (2007) Stochastic computational fluid mechanics. Comput Sci Eng 9(2):21–29

2. Allaire G (2015) A review of adjoint methods for sensitivity analysis, uncertainty quantification and optimization in numerical codes
3. Loeven G, Witteveen J, Bijl H (2007) Probabilistic collocation: an efficient non-intrusive approach for arbitrarily distributed parametric uncertainties. In: 45th AIAA aerospace science meeting and exhibit, Reno, Nevada
4. Parussini L, Pediroda V (2008) Investigation of multi geometric uncertainties by different polynomial chaos methodologies using a fictitious domain solver. CMES Comput Model Eng Sci 23(1):29–52
5. Clarich A, Pediroda V, Poloni C (2005) A fast and robust adaptive methodology for design under uncertainties based on DACE response surface and game theory. In: Evolutionary algorithms and intelligent tools in engineering optimization, CIMNE, Barcelona, Spain. ISBN: 1-84564-038-1
6. Vasile M, Minisci E, Quagliarella D, Guénot M, Lepot I, Sainvitu C, Coelho RF (2013) Adaptive sampling strategies for non-intrusive POD-based surrogates. Eng Comput
7. Hosder S, Walters R, Perez R (2006) A non-intrusive polynomial chaos method for uncertainty propagation in CFD simulation. In: 44th AIAA aerospace science meeting and exhibit, 9–12 January 2006, 2006–891, Reno, Nevada
8. Ernst OG, Mugler A, Starkloff HJ, Ullmann E (2012) On the convergence of generalized polynomial chaos expansion. ESAIM: M2AN 46, pp 317–339
9. Forrester AIJ, Sóbester A, Keane AJ (2007) Multi-fidelity optimization via surrogate modelling. In: Proceedings of the royal society, Mathematical, physical and engineering sciences
10. Le Gratièt L (2013) Multi-fidelity Gaussian process regression for computer experiments. Dissemination at Université Paris-Diderot-Paris VII
11. Kennedy MC, O'Hagan A (2000) Predicting the output from a complex computer code when fast approximations are available. Biometrika 87:1
12. Perron C, Rajaram D, Mavris DN (2021) Multi-fidelity non-intrusive reduced-order modelling based on manifold alignment. https://doi.org/10.1098/rspa.2021.0495
13. Ladson CL, Rill S (1987) High reynolds number transonic tests of an NACA 0012 airfoil in the Langley 0.3-meter transonic cryogenic tunnel. NASA Tech Memo 100527:89
14. Wu M-Y, Wu Y, Yuan X-Y, Chen Z-H, Wu W-T, Aubry N (2022) Fast prediction of flow field around airfoils based on deep convolutional neural network. Appl Sci 12:12075. https://doi.org/10.3390/app122312075
15. Di Pietro M. Deep learning with python: neural networks. https://towardsdatascience.com/deep-learning-with-python-neural-networks-complete-tutorial-6b53c0b06af0
16. Crichigno N (2021) Deep learning for prediction of aerodynamic simulations. Master Thesis, Politecnico di Torino
17. Genesio R, Milanese M (1976) A note on the derivation and use of reduced-order models. IEEE Trans Autom Control 21(1):118–122
18. Weiss J (2019) A Tutorial on the proper orthogonal decomposition. In: 2019 AIAA aviation forum. 17–21 June 2019, Dallas, Texas, United States
19. Ullmann S, Rotkvic M, Lang J (2016) POD-Galerkin reduced-order modeling with adaptive finite element snapshots. J Comput Phys 325: 244–258. ISSN 0021-9991

Part II
Numerical Methods, Tools and Applications

Chapter 10
Simulation of 3D Radiative Transfers with Anisotropic Scattering for Contrails

Olivier Pironneau and Pierre-Henri Tournier

Abstract In two earlier publications we have proposed a numerical algorithm to solve the RTE in 3D based on an integro-differential formulation and iterations using $\mathcal{H}$-matrices to speed-up the computation of the integrals. In this article we show how the method can be extended to handle an important class of non-isotropic scattering for the atmosphere. An application to the Earth atmosphere is given.

Keywords Radiative transfer equation (RTE) · Anisotropic scattering · Hierarchical matrices (H-matrices) · Stratified atmospheric modeling

10.1 Introduction

Airplane combustion engines generate $\mathrm{CO_2}$ (carbon dioxide), a gas which stays in the Earth atmosphere for ~500 years before being partially absorbed by the oceans.

Currently used fuels and potential future substitutes like hydrogen, synthetic or agricultural fuel combustion generate also NO_x (Nitrous oxides). Typically a nitrous oxide stays in the atmosphere ~100 years.

Carbon dioxide, nitrous oxides, methane, ozone and water vapor are "green house gases", (GHG), in the sense that atmospheric and past climates measurements indicate that they are most likely responsible for the additional heat received by Earth (global warming).

Air travel—around 100.000 flights/day—produces 3% of $\mathrm{CO_2}$ (or 12% of what is produced by all transport vehicles). It is still small, but the contribution of airplanes to global warming is expected to rise from today's 0.024 to 0.084 W/m^2 by 2050 [15].

O. Pironneau (✉) · P.-H. Tournier
Applied Mathematics, Jacques-Louis Lions Lab, Sorbonne Université, Paris, France
e-mail: olivier.pironneau@sorbonne-universite.fr

P.-H. Tournier
e-mail: Pierre-Henri.Tournier@sorbonne-universite.fr

T. Tuovinen et al. (eds.), *Challenges in Design Methods, Numerical Tools and Technologies for Sustainable Aviation, Transport and Industry*, Computational Methods in Applied Sciences 17, https://doi.org/10.1007/978-3-031-98675-8_10

A cloud similar to a cirrus, called "contrail (short for condensation trail) may appear in the airplane wake if atmospheric pressure and temperature are on the left side of a threshold curve, and further left of that pressure versus temperature curve the contrail will persist [16]. Several studies claim that these clouds have a warming effect perhaps 3 times stronger than the one caused by the combustion gases [14].

In the past few years the authors have worked to see if these claims could be validated by a numerical simulation of the fundamental equations of physics for these problems [2, 9–11]. We approach the problem from an applied mathematics point of view, with convergence error estimation and computational efficiency in mind.

The formation of a contrail has been simulated by solving the Navier-Stokes equations (NSE) with chemistry for the engine exhaust and phase change for the ice formation in the airplane wake (see [16]); it is the right approach to understand the mechanisms of the formation of contrails. Once the cloud is sufficiently developed one may study its effect on the absorption and scattering coefficients of the Radiative Transfer Equations (RTE) in the Earth atmosphere (see the data in [6]) and then solve the RTE-NSE system. The numerical simulations of RTE in one dimension is the object of intense research [5, 6], but mostly without coupling with a temperature equation or NSE.

In [8, 17] we have proposed a numerical algorithm to solve the RTE in 3D based on an integro-differential formulation and iterations using $\mathcal{H}$-matrices to speed-up the computation of the integrals. In this article we show how the method can be extended to handle an important class of non-isotropic scattering for the atmosphere.

It is known that the preferred direction of scattering in clouds is the initial direction. It is modelled by an anisotropic probability of scattering (also called phase function) from $\boldsymbol{\omega}'$ to $\boldsymbol{\omega}$, $p_\nu(\boldsymbol{\omega}, \boldsymbol{\omega}')$ (see Fig. 10.1).

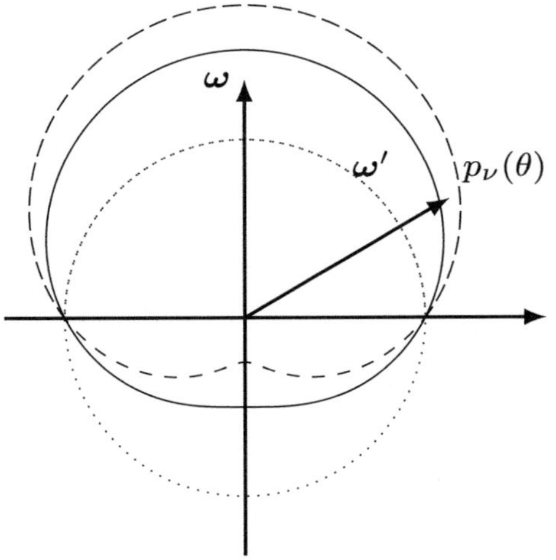

Fig. 10.1 This polar plot shows the probability $p(\theta)$ that a photon in the direction $\boldsymbol{\omega}$ scatters in the direction $\boldsymbol{\omega}'$: $p_\nu(\theta) = 1 + \frac{1}{2}\boldsymbol{\omega} \cdot \boldsymbol{\omega}' = 1 + \beta\cos(\theta)$ where θ is the angle $(\boldsymbol{\omega}, \boldsymbol{\omega}')$. The solid curve is for $\beta = \frac{1}{2}$, the dashed curve is for $\beta = 0.75$ and the dotted circle is the isotropic case is $\beta = 0$

10.2 The Radiative Transfer Equations

When molecular viscosity and wind convection are ignored, the temperature T in a medium exposed to electromagnetic waves satisfies the RTE as explained in [18]. It involves a frequency dependent radiation intensity field $I_\nu(\boldsymbol{x}, \boldsymbol{\omega})$ at position $\boldsymbol{x}$ in the physical domain Ω and in each direction $\boldsymbol{\omega}$. For all $\{\boldsymbol{x}, \boldsymbol{\omega}, \nu\} \in \Omega \times \mathbb{S}_2 \times \mathbb{R}^+$,

$$\boldsymbol{\omega} \cdot \nabla_{\boldsymbol{x}} I_\nu + \rho\bar{\kappa}_\nu a_\nu \left[I_\nu - \frac{1}{4\pi}\int_{\mathbb{S}^2} p_\nu(\boldsymbol{\omega}, \boldsymbol{\omega}') I_\nu(\boldsymbol{\omega}')\mathrm{d}\omega' \right] = \rho\bar{\kappa}_\nu(1 - a_\nu)[B_\nu(T) - I_\nu], \quad (10.1)$$

$$\int_0^\infty \int_{\mathbb{S}^2} \rho\bar{\kappa}_\nu(1 - a_\nu)[B_\nu(T) - I_\nu]\mathrm{d}\omega\mathrm{d}\nu = 0,$$

where $\mathbb{S}^2$ is the unit sphere, $B_\nu(T) = \dfrac{2\hbar\nu^3}{c^2[\mathrm{e}^{\frac{\hbar\nu}{kT}} - 1]}$ is the Planck function, $\hbar, c, k$ are the Planck constant, the speed of light in the medium and the Boltzmann constant. The absorption coefficient $\kappa_\nu := \rho\bar{\kappa}_\nu$, where ρ is the medium density, comes from nuclear physics, but for our purpose it is seen as the percentage of radiation absorbed per unit length. The scattering albedo is $a_\nu \in (0, 1)$.

With appropriate boundary conditions, existence of solution has been established by [9, 19]. In the latter, convergence of the following scheme was proved:

$$\boldsymbol{\omega} \cdot \nabla I_\nu^{n+1} + \kappa_\nu I_\nu^{n+1} = \frac{\kappa_\nu a_\nu}{4\pi}\int_{\mathbb{S}^2} p_\nu(\boldsymbol{\omega}, \boldsymbol{\omega}') I_\nu^n(\boldsymbol{\omega}')d\omega' + \kappa_\nu(1 - a_\nu)B_\nu(T^n),$$

$$\int_0^\infty \int_{\mathbb{S}^2} \kappa_\nu(1 - a_\nu)[B_\nu(T^{n+1}) - I_\nu^{n+1}]d\omega d\nu = 0. \quad (10.2)$$

By the maximum principle for the first equation and the monotony of $T \mapsto B_\nu(T)$,

$$I^n \prec I^{n-1}, T^n \prec T^{n-1} \quad \Longrightarrow \quad I^{n+1} \prec I^n, \quad \Longrightarrow T^{n+1} \prec T^n,$$

where $a \prec b$ means $a(\boldsymbol{x}) \le b(\boldsymbol{x})$, for all $\boldsymbol{x}$ but not $a(\boldsymbol{x}) = b(\boldsymbol{x})$ everywhere. Hence by choosing $T^0 = 0$ and $I^0 = 0$, the positivity of T^1 and I^1 implies that $T^1 \prec T^0$, $I^1 \prec I^0$ so that, by induction, a strictly increasing sequence is obtained. Similarly a decreasing sequence is obtained if we manage $T^1 \succ T^0, I^1 \succ I^0$. As it was observed in [11] that the increasing sequences converge faster we will focus on that one only.

10.3 Integral Formulation

Consider an anisotropic scattering density

$$p_\nu(\omega, \omega') = 1 + \frac{\beta_\nu(1-a_\nu)}{a_\nu}\omega \cdot \omega'.$$

Denote by Γ the boundary of Ω. One must find the radiation intensity $I_\nu(\boldsymbol{x}, \boldsymbol{\omega})$ at all points $\boldsymbol{x} \in \Omega$, for all directions all $\boldsymbol{\omega} \in \mathbb{S}^2$ and all radiation frequencies $\nu \in \mathbb{R}_+$, by solving the radiative transfer equations (RTE):

$$\boldsymbol{\omega} \cdot \nabla_x I_\nu + \kappa_\nu I_\nu = \kappa_\nu (1-a_\nu)\left(B_\nu(T) + \beta_\nu \boldsymbol{\omega} \cdot \boldsymbol{K}_\nu\right) + \kappa_\nu a_\nu J_\nu \,, \tag{10.3}$$

$$J_\nu(\boldsymbol{x}) := \tfrac{1}{4\pi}\int_{\mathbb{S}^2} I_\nu \mathrm{d}\omega, \quad \boldsymbol{K}_\nu(\boldsymbol{x}) := \tfrac{1}{4\pi}\int_{\mathbb{S}^2} \boldsymbol{\omega} I_\nu(\boldsymbol{x}, \boldsymbol{\omega})\mathrm{d}\omega \,, \tag{10.4}$$

$$\int_0^\infty \kappa_\nu(1-a_\nu)(J_\nu - B_\nu(T))\mathrm{d}\nu = 0 \,, \tag{10.5}$$

$$I_\nu(\boldsymbol{x}, \boldsymbol{\omega}) = R_\nu(\boldsymbol{x}, \boldsymbol{\omega}) I_\nu(\boldsymbol{x}, \boldsymbol{\omega} - 2(\boldsymbol{n} \cdot \boldsymbol{\omega})\boldsymbol{n}) + Q_\nu(\boldsymbol{x}, \boldsymbol{\omega}),$$
$$\text{on } \Sigma := \{(\boldsymbol{x}, \boldsymbol{\omega}) \in \Gamma \times \mathbb{S}^2 \,:\, \boldsymbol{\omega} \cdot \boldsymbol{n}(\boldsymbol{x}) < 0\}. \tag{10.6}$$

In (10.6), Q_ν is the radiation source at the boundary and R_ν is the portion of radiation which is reflected by the boundary; $\boldsymbol{n}(\boldsymbol{x})$ is the outer normal of Γ at $\boldsymbol{x}$. $\kappa_\nu > 0$ and $a_\nu \in [0, 1]$ are the absorption and scattering coefficients; in general they depend on ν and $\boldsymbol{x}$.

The general solution of (10.3) is

$$I(\boldsymbol{x}, \boldsymbol{\omega}) = I(\boldsymbol{x}_\Sigma(\boldsymbol{x}, \boldsymbol{\omega}), \boldsymbol{\omega})\mathrm{e}^{-\int_0^{\tau_{x,\omega}} \kappa(\boldsymbol{x}-\boldsymbol{\omega}s)\mathrm{d}s} + \int_0^{\tau_{x,\omega}} \mathrm{e}^{-\int_0^s \kappa(\boldsymbol{x}-\boldsymbol{\omega}s')\mathrm{d}s'} S_\nu(\boldsymbol{x} - \boldsymbol{\omega}s, \boldsymbol{\omega})\mathrm{d}s,$$

where $\tau_{x,\omega}$ is the length $|\boldsymbol{x} - \boldsymbol{x}_\Sigma(\boldsymbol{x}, \boldsymbol{\omega})|$, and $\mathcal{S}_\nu$ denotes its right-hand side,

$$\mathcal{S}_\nu(\boldsymbol{x}, \boldsymbol{\omega}) = \kappa_\nu(\boldsymbol{x})(1 - a_\nu(\boldsymbol{x}))\left(B_\nu(T(\boldsymbol{x})) + \beta_\nu(\boldsymbol{x})\boldsymbol{\omega} \cdot \boldsymbol{K}_\nu(\boldsymbol{x})\right) + \kappa_\nu(\boldsymbol{x}) a_\nu(\boldsymbol{x}) J_\nu(\boldsymbol{x}) \,. \tag{10.7}$$

10.3.1 Solution in Absence of Reflective Boundaries

Let us denote $\int_{[x,y]} \kappa := \int_0^{|y-x|} \kappa(\boldsymbol{x} + s(\boldsymbol{y} - \boldsymbol{x}))\mathrm{d}s$. When $R_\nu = 0$, the following holds:

Proposition 1

$$J_\nu(\boldsymbol{x}) = S_\nu^E(\boldsymbol{x}) + \mathcal{J}[\mathcal{S}_\nu](\boldsymbol{x}), \tag{10.8}$$

$$S_\nu^E(\boldsymbol{x}) = \frac{1}{4\pi}\int_\Gamma Q_\nu(\boldsymbol{y}, \tfrac{\boldsymbol{y}-\boldsymbol{x}}{|\boldsymbol{y}-\boldsymbol{x}|})\frac{[(\boldsymbol{y}-\boldsymbol{x})\cdot \boldsymbol{n}(\boldsymbol{y})]_-}{|\boldsymbol{y}-\boldsymbol{x}|^3}\mathrm{e}^{-\int_{[x,y]}\kappa} d\Gamma(\boldsymbol{y}), \tag{10.9}$$

$$\mathcal{J}[\mathcal{S}_\nu](\boldsymbol{x}) = \frac{1}{4\pi}\int_\Omega \frac{\mathrm{e}^{-\int_{[x,y]}\kappa}}{|\boldsymbol{y}-\boldsymbol{x}|^2}\mathcal{S}_\nu(\boldsymbol{y}, \frac{\boldsymbol{y}-\boldsymbol{x}}{|\boldsymbol{y}-\boldsymbol{x}|})d\boldsymbol{y}. \tag{10.10}$$

Averaging (10.7) on $\mathbb{S}^2$ after multiplication by $\boldsymbol{\omega}$ leads to

$$\begin{aligned}
\boldsymbol{K}_\nu(\boldsymbol{x}) &:= \frac{1}{4\pi}\int_{\mathbb{S}^2}\boldsymbol{\omega} I(\boldsymbol{x},\boldsymbol{\omega})\mathrm{d}\omega = \boldsymbol{K}_\nu^E(\boldsymbol{x}) + \boldsymbol{K}[\mathcal{S}_\nu](\boldsymbol{x}) \quad \text{with}\\
\boldsymbol{K}_\nu^E(\boldsymbol{x}) &:= \frac{1}{4\pi}\int_{\mathbb{S}^2}\boldsymbol{\omega} I(\boldsymbol{x}_\Sigma(\boldsymbol{x},\boldsymbol{\omega}),\boldsymbol{\omega})\mathrm{e}^{-\int_0^{\tau_{x,\omega}}\kappa(\boldsymbol{x}-\boldsymbol{\omega}s)\mathrm{d}s}\mathrm{d}\omega\\
&= \frac{1}{4\pi}\int_\Gamma (\boldsymbol{y}-\boldsymbol{x})Q_\nu(\boldsymbol{y}, \tfrac{\boldsymbol{y}-\boldsymbol{x}}{|\boldsymbol{y}-\boldsymbol{x}|})\frac{[(\boldsymbol{y}-\boldsymbol{x})\cdot \boldsymbol{n}(\boldsymbol{y})]_-}{|\boldsymbol{y}-\boldsymbol{x}|^4}\mathrm{e}^{-\int_{[x,y]}\kappa}\mathrm{d}\Gamma(\boldsymbol{y}), \quad (10.11)\\
\boldsymbol{K}[\mathcal{S}_\nu](\boldsymbol{x}) &:= \frac{1}{4\pi}\int_{\mathbb{S}^2}\int_0^{\tau_{x,\omega}}\boldsymbol{\omega}\mathrm{e}^{-\int_0^s\kappa(\boldsymbol{x}-\boldsymbol{\omega}s')\mathrm{d}s'}\mathcal{S}_\nu(\boldsymbol{x}-\boldsymbol{\omega}s,\boldsymbol{\omega})\mathrm{d}s\mathrm{d}\omega\\
&= \frac{1}{4\pi}\int_\Omega (\boldsymbol{y}-\boldsymbol{x})\mathcal{S}_\nu(\boldsymbol{y}, \tfrac{\boldsymbol{y}-\boldsymbol{x}}{|\boldsymbol{y}-\boldsymbol{x}|})\frac{\mathrm{e}^{-\int_{[x,y]}\kappa}}{|\boldsymbol{y}-\boldsymbol{x}|^3}\mathrm{d}\boldsymbol{y}.
\end{aligned}$$

10.3.2 Iterative Method

In [8] it was shown that, in absence of $\boldsymbol{K}_\nu$, i.e. $\beta_\nu = 0$, the following is monotone and convergent. Its extension to $\beta_\nu > 0$ is:

Algorithm

1. Initialize J_ν, $\boldsymbol{K}_\nu$ and T (by zero, for instance).
2. Compute $\mathcal{S}_\nu$ by (10.7).
3. Update J_ν by (10.11), (10.9), (10.10) and $\boldsymbol{K}_\nu$ by (10.11).
4. Update T by solving (10.5).

10.4 Extension to Reflective Conditions (RC)

As in [17], consider boundary condition (10.6). Proposition 1 can be extended to case with RC and in the case of non-multiple reflection, it becomes:

Proposition 2 *The same iterations are proposed with*

$$\begin{aligned}
J_\nu(\boldsymbol{x}) &= S_{\nu,1}^E(\boldsymbol{x}) + S_{\nu,2}^E(\boldsymbol{x}) + \bar{\mathcal{J}}[\mathcal{S}_\nu](\boldsymbol{x}),\\
\boldsymbol{K}_\nu(\boldsymbol{x}) &= \boldsymbol{K}_{\nu,1}^E(\boldsymbol{x}) + \boldsymbol{K}_{\nu,2}^E(\boldsymbol{x}) + \bar{\mathcal{K}}[\mathcal{S}_\nu](\boldsymbol{x}),
\end{aligned} \tag{10.12}$$

$$
\begin{aligned}
S^E_{\nu,1}(\boldsymbol{x}) &= \tfrac{1}{4\pi}\int_\Gamma Q_\nu\Big(\boldsymbol{y}, \frac{\boldsymbol{y}-\boldsymbol{x}}{|\boldsymbol{y}-\boldsymbol{x}|}\Big)\frac{[(\boldsymbol{y}-\boldsymbol{x})\cdot\boldsymbol{n}(\boldsymbol{y})]_-}{|\boldsymbol{y}-\boldsymbol{x}|^3}\mathrm{e}^{-\int_{[x,y]}\kappa}d\Gamma(\boldsymbol{y}),\\
S^E_{\nu,2}(\boldsymbol{x}) &= \sum_{n=1}^{M}\tfrac{1}{4\pi}\int_\Gamma R_\nu\big(\boldsymbol{x}'_n, \tfrac{\boldsymbol{x}-\boldsymbol{x}'_n}{|\boldsymbol{x}-\boldsymbol{x}'_n|}\big)Q_\nu\big(\boldsymbol{y}, \tfrac{\boldsymbol{x}'_n-\boldsymbol{y}}{|\boldsymbol{x}'_n-\boldsymbol{y}|}\big)\\
&\qquad\times\frac{[(\boldsymbol{x}'_n-\boldsymbol{y})\cdot\boldsymbol{n}(\boldsymbol{y})]_-\mathrm{e}^{-\int_{[x,x'_n]\cup[x'_n,y]}\kappa}}{|\boldsymbol{x}'_n-\boldsymbol{y}|\,(|\boldsymbol{x}-\boldsymbol{x}'_n|+|\boldsymbol{x}'_n-\boldsymbol{y}|)^2}d\Gamma(\boldsymbol{y}).\\
\bar{\mathcal{J}}[\mathcal{S}_\nu](\boldsymbol{x}) &= \tfrac{1}{4\pi}\int_\Omega\Bigg[\frac{\mathrm{e}^{-\int_{[x,y]}\kappa}}{|\boldsymbol{y}-\boldsymbol{x}|^2}\\
&\qquad+\sum_{n=1}^{M}\frac{\mathrm{e}^{-\int_{[x,x'_n]\cup[x'_n,y]}\kappa}}{(|\boldsymbol{x}-\boldsymbol{x}'_n|+|\boldsymbol{x}'_n-\boldsymbol{y}|)^2}R_\nu\big(\boldsymbol{x}'_n, \tfrac{\boldsymbol{x}-\boldsymbol{x}'_n}{|\boldsymbol{x}-\boldsymbol{x}'_n|}\big)\Bigg]\mathcal{S}_\nu\big(\boldsymbol{y}, \tfrac{\boldsymbol{y}-\boldsymbol{x}}{|\boldsymbol{y}-\boldsymbol{x}|}\big)d\boldsymbol{y}.
\end{aligned}
\tag{10.13}
$$

$$
\begin{aligned}
\boldsymbol{K}^E_{\nu,1}(\boldsymbol{x}) &= \tfrac{1}{4\pi}\int_\Gamma(\boldsymbol{y}-\boldsymbol{x})Q_\nu\big(\boldsymbol{y}, \tfrac{\boldsymbol{y}-\boldsymbol{x}}{|\boldsymbol{y}-\boldsymbol{x}|}\big)\frac{[(\boldsymbol{y}-\boldsymbol{x})\cdot\boldsymbol{n}(\boldsymbol{y})]_-}{|\boldsymbol{y}-\boldsymbol{x}|^3}\mathrm{e}^{-\int_{[x,y]}\kappa}d\Gamma(\boldsymbol{y}),\\
\boldsymbol{K}^E_{\nu,2}(\boldsymbol{x}) &= \sum_{n=1}^{M}\tfrac{1}{4\pi}\int_\Gamma(\boldsymbol{x}-\boldsymbol{x}'_n)R_\nu\big(\boldsymbol{x}'_n, \tfrac{\boldsymbol{x}-\boldsymbol{x}'_n}{|\boldsymbol{x}-\boldsymbol{x}'_n|}\big)Q_\nu\big(\boldsymbol{y}, \tfrac{\boldsymbol{x}'_n-\boldsymbol{y}}{|\boldsymbol{x}'_n-\boldsymbol{y}|}\big)\\
&\qquad\times\frac{[(\boldsymbol{x}'_n-\boldsymbol{y})\cdot\boldsymbol{n}(\boldsymbol{y})]_-\mathrm{e}^{-\int_{[x,x'_n]\cup[x'_n,y]}\kappa}}{|\boldsymbol{x}'_n-\boldsymbol{y}|\,(|\boldsymbol{x}-\boldsymbol{x}'_n|+|\boldsymbol{x}'_n-\boldsymbol{y}|)^2}d\Gamma(\boldsymbol{y}),\\
\bar{\boldsymbol{K}}[\mathcal{S}_\nu](\boldsymbol{x}) &= \tfrac{1}{4\pi}\int_\Omega\Bigg[\frac{\boldsymbol{y}-\boldsymbol{x}}{|\boldsymbol{y}-\boldsymbol{x}|^3}\mathrm{e}^{-\int_{[x,y]}\kappa}\\
&\qquad+\sum_{n=1}^{M}(\boldsymbol{x}-\boldsymbol{x}'_n)\frac{\mathrm{e}^{-\int_{[x,x'_n]\cup[x'_n,y]}\kappa}}{(|\boldsymbol{x}-\boldsymbol{x}'_n|+|\boldsymbol{x}'_n-\boldsymbol{y}|)^2}R_\nu\big(\boldsymbol{x}'_n, \tfrac{\boldsymbol{x}-\boldsymbol{x}'_n}{|\boldsymbol{x}-\boldsymbol{x}'_n|}\big)\Bigg]\mathcal{S}_\nu\big(\boldsymbol{y}, \tfrac{\boldsymbol{y}-\boldsymbol{x}}{|\boldsymbol{y}-\boldsymbol{x}|}\big)d\boldsymbol{y}.
\end{aligned}
\tag{10.14}
$$

where M is the number of RC boundaries and $\boldsymbol{x}'_n$ is the point of reflection on the boundary of the ray going from $\boldsymbol{x}$ to $\boldsymbol{y}$ via $\boldsymbol{x}'_n$.

All equations are discretized using a P^1 finite element framework. Some integrals have a singular kernel, so a careful quadrature should be used.

In [8, 17] a strategy is explained to compute all the integrals as a matrix vector product where the matrices are hierarchical compressed $\mathcal{H}$-matrices [3, 4, 12].

In the grey case (κ_ν and a_ν independent of ν) we need 5+M matrices. This could be taxing in computer memory, but not in computing time because the core of the method is $N \ln N$ where N is the number of finite element vertices (Fig. 10.2).

In the non grey case we need P(5+M) matrices where P is the number of values taken by the numerical approximation of κ_ν.

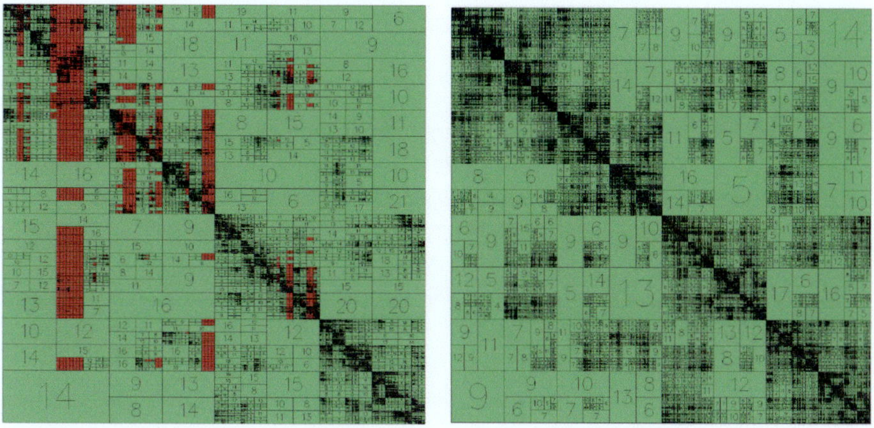

Fig. 10.2 Visual representations of the $\mathcal{H}$-matrices surface-to-volume on the left and volume-to-volume on the right, for the computation of the temperature in the Chamonix valley. The level of compression is shown by light green-to-red and the numbers correspond to the rank of the truncated SVD approximation of each block

10.5 The Stratified One Dimensional Approximation

Let us consider the Earth's atmosphere submitted to black body radiations from the sun at temperature T_S and earth at T_E. The atmosphere is thin and the Sun is far. As a first approximation the ground is locally flat and has a negligible relief, except near high mountains. Then all quantities are only function of the altitude z and independent of x and y. This, so called, stratified approximation of the RTE is (see [2])

$$(\mu\partial_\tau + \kappa_\nu)I_\nu(\tau,\mu) = \kappa_\nu(1-a_\nu)B_\nu(T(\tau)) + \tfrac{1}{2}\kappa_\nu a_\nu \int_{-1}^{1} p_\nu(\mu,\mu')I_\nu(\tau,\mu')\mathrm{d}\mu' , \tag{10.15}$$

$$I_\nu(0,\mu) = r_\nu I(0,-\mu) + \mu Q_\nu^+ , \quad I_\nu(Z,-\mu) = \mu Q_\nu^- , \qquad 0<\mu<1 , \tag{10.16}$$

$$\int_0^\infty \kappa_\nu(1-a_\nu)B_\nu(T(\tau))\mathrm{d}\nu = \int_0^\infty \kappa_\nu(1-a_\nu)J_\nu(\tau)\mathrm{d}\nu, \tag{10.17}$$

with

$$Q_\nu^+ = Q^E B_\nu(T_E), \quad Q_\nu^- = Q^S B_\nu(T_S), \quad J_\nu(\tau) := \tfrac{1}{2}\int_{-1}^{1} I_\nu(\tau,\mu)\mathrm{d}\mu, \tag{10.18}$$

where μ is the cosine of $\boldsymbol{\omega}$ with the vertical direction and τ is the optical thickness:

$$\tau = \int_0^z \rho(z')\mathrm{d}z'.$$

10.5.1 RTE with Non-isotropic and Rayleigh Phase Function

The scattering function is $p_\nu(\mu,\mu') = 1 + \beta_\nu \mu'$ in a cloud between altitude Z_m and Z_M and at high altitude $z > Z_M$ it is the Rayleigh phase function $p_\nu(\mu,\mu') = \frac{3}{16}(3 - \mu^2) + \frac{3}{16}(3\mu^2 - 1)\mu'^2$. It is combined into one formula with two given altitudes and possibly frequency dependent functions $b_\nu(\tau)$ and $\beta_\nu(\tau)$ in [0, 1],

$$p_\nu(\mu,\mu') = b_\nu + \beta_\nu \mu' + \tfrac{3}{16}(1-b_\nu)(3-\mu^2+(3\mu^2-1)\mu'^2). \qquad (10.19)$$

Observe that $p_\nu(\mu,\mu') \geq 0$ and $\frac{1}{2}\int_{-1}^{1} p_\nu(\mu,\mu')\mathrm{d}\mu' = 1$.
Keeping (10.17) as the defining equation for T, given I, the problem becomes

$$\begin{cases} (\mu\partial_\tau + \kappa_\nu) I_\nu(\tau,\mu) = \mathcal{S}_\nu(\tau) := \mathcal{R}_\nu(\tau) + \mathcal{P}_\nu(\tau)\mu^2\,, \\ I_\nu(0,\mu) = r_\nu I(0,-\mu) + \mu Q_\nu^+\,, \quad I_\nu(Z,-\mu) = \mu Q_\nu^-\,, \qquad 0 < \mu < 1\,, \end{cases} \qquad (10.20)$$

with

$$\begin{aligned} K_\nu(\tau) &:= \tfrac{1}{2}\int_{-1}^{1} \mu I_\nu(\tau,\mu)\mathrm{d}\mu\,, \qquad L_\nu(\tau) = \tfrac{1}{2}\int_{-1}^{1}\mu^2 I_\nu \mathrm{d}\mu\,, \\ \mathcal{R}_\nu(\tau) &:= \kappa_\nu a_\nu\left((\tfrac{7}{16}b + \tfrac{9}{16}) J_\nu(\tau) + \beta K_\nu(\tau) - \tfrac{3}{16}(1-b)L_\nu(\tau)\right) + \kappa_\nu(1-a_\nu)B_\nu(T(\tau))\,, \\ \mathcal{P}_\nu(\tau) &:= \tfrac{3}{16}(1-b)(-J_\nu(\tau) + 3L_\nu(\tau)) \end{aligned} \qquad (10.21)$$

and (10.17).

Algorithm

1. Initialize $J_\nu^0(\tau,\mu) = 0$, $K_\nu^0(\tau,\mu) = 0$, $L_\nu^0(\tau,\mu) = 0$ and $T^0(\tau) = 0$.
2. Compute I^{n+1} by (10.20) with $\mathcal{S}_\nu^n$.
3. Update $J_\nu^{n+1}(\tau,\mu)$, $K_\nu^{n+1}(\tau,\mu)$, $L_\nu^{n+1}(\tau,\mu)$ by (10.21), then $\mathcal{R}_\nu(\tau)$ and $\mathcal{P}_\nu(\tau)$.
4. Then, compute T^{n+1} by solving (10.17).

10.5.2 Implementation

Recall the definition of the exponential integrals

$$E_p(X) := \int_0^1 e^{-X/\mu}\mu^{p-2}\mathrm{d}\mu\,, \qquad X > 0\,. \qquad (10.22)$$

For $i = 3, 4, 5$, let

$$\begin{aligned} S_i(\nu,\tau) &:= \tfrac{1}{2}E_i(\kappa_\nu\tau)Q^+(\tau) + \tfrac{(-1)^{i-1}}{2}(E_i(\kappa_\nu(Z-\tau)) + r_\nu E_i(\kappa_\nu(Z+\tau)))Q^-(\tau),\\ F_i(\tau,t) &:= \tfrac{1}{2}E_3(\kappa_\nu|\tau-t|) + \tfrac{r_\nu}{2}E_3(\kappa_\nu|\tau+t|). \end{aligned} \tag{10.23}$$

Proposition 3 *With* (10.21) *and* (10.23) *the quantities needed by the algorithm above are given by*

$$\begin{aligned} J_\nu^{n+1}(\tau) &= S_3(\nu,\tau) + \tfrac{1}{2}\int_0^Z F_1(\tau,t)\mathcal{R}_\nu^n(\tau)dt + \tfrac{1}{2}\int_0^Z F_3(\tau,t)\mathcal{P}_\nu^n(\tau)dt\,,\\ K_\nu^{n+1}(\tau) &= S_4(\nu,\tau) + \tfrac{1}{2}\int_0^\tau F_2(\tau,t)\mathcal{R}_\nu^n(t)dt + \tfrac{1}{2}\int_0^\tau F_4(\tau,t)\mathcal{P}_\nu^n(t)dt\\ &\quad - \tfrac{1}{2}\int_\tau^Z F_2(\tau,t)\mathcal{R}_\nu^n(t)dt - \tfrac{1}{2}\int_\tau^Z F_4(\tau,t)\mathcal{P}_\nu^n(t)dt\,,\\ L_\nu^{n+1}(\tau) &= S_5(\nu,\tau) + \tfrac{1}{2}\int_0^Z F_3(\tau,t)\mathcal{R}_\nu^n(\tau)dt + \tfrac{1}{2}\int_0^Z F_5(\tau,t)\mathcal{P}_\nu^n(\tau)dt\,. \end{aligned} \tag{10.24}$$

Proof For clarity let us assume that $r_\nu = 0$. For the general case see [10]. Applying the method of characteristics shows that

$$\begin{aligned} I_\nu(\tau,\mu) &= \mu e^{-\frac{\kappa_\nu\tau}{\mu}}Q_\nu^+\mathbf{1}_{\mu>0} + |\mu|e^{-\frac{\kappa_\nu(Z-\tau)}{|\mu|}}Q_\nu^-\mathbf{1}_{\mu<0}\\ &\quad + \mathbf{1}_{\mu>0}\int_0^\tau e^{-\frac{\kappa_\nu(\tau-t)}{\mu}}\frac{\kappa_\nu}{\mu}S_\nu(t)\mathrm{d}t + \mathbf{1}_{\mu<0}\int_\tau^Z e^{-\frac{\kappa_\nu(t-\tau)}{|\mu|}}\frac{\kappa_\nu}{|\mu|}S_\nu(t)\mathrm{d}t\,. \end{aligned} \tag{10.25}$$

Therefore with $k = 0$ or $k = 2$,

$$\begin{aligned} \int_{-1}^1 \mu^k I_\nu(\tau,\mu)\mathrm{d}\mu &= \int_{-1}^1 |\mu|^{k+1}\Big(e^{-\frac{\kappa_\nu\tau}{\mu}}Q_\nu^+\mathbf{1}_{\mu>0} + e^{-\frac{\kappa_\nu(Z-\tau)}{|\mu|}}Q_\nu^-\mathbf{1}_{\mu<0}\Big)\mathrm{d}\mu\\ &+ \int_{-1}^1 |\mu|^{k-1}\Big(\mathbf{1}_{\mu>0}\int_0^\tau e^{-\frac{\kappa_\nu(\tau-t)}{\mu}}\kappa_\nu S_\nu(t)\mathrm{d}t + \mathbf{1}_{\mu<0}\int_\tau^Z e^{-\frac{\kappa_\nu(t-\tau)}{|\mu|}}\kappa_\nu S_\nu(t)\mathrm{d}t\Big)\mathrm{d}\mu\\ &= \int_0^1 |\mu|^{k+1}e^{-\frac{\kappa_\nu\tau}{\mu}}Q_\nu^+\mathrm{d}\mu + \int_{-1}^0 |\mu|^{k+1}e^{-\frac{\kappa_\nu(Z-\tau)}{|\mu|}}Q_\nu^-\mathrm{d}\mu\\ &+ \int_0^1 |\mu|^{k-1}\int_0^\tau e^{-\frac{\kappa_\nu(\tau-t)}{\mu}}\kappa_\nu S_\nu(t)\mathrm{d}t + \int_{-1}^0 |\mu|^{k-1}\int_\tau^Z e^{-\frac{\kappa_\nu(t-\tau)}{|\mu|}}\kappa_\nu S_\nu(t)\mathrm{d}t\mathrm{d}\mu \end{aligned} \tag{10.26}$$

$$
\begin{aligned}
&= \int_0^1 |\mu|^{k+1} \left(e^{-\frac{\kappa_\nu \tau}{\mu}} Q_\nu^+ + e^{-\frac{\kappa_\nu (Z-\tau)}{|\mu|}} Q_\nu^- \right) \mathrm{d}\mu \\
&+ \int_0^\tau \int_0^1 |\mu|^{k-1} e^{-\frac{\kappa_\nu(\tau - t)}{\mu}} \kappa_\nu (\mathcal{R}_\nu(t) + \mathcal{P}_\nu(t)\mu^2) \mathrm{d}\mu \mathrm{d}t \\
&+ \int_\tau^Z \int_0^1 |\mu|^{k-1} e^{-\frac{\kappa_\nu(t-\tau)}{|\mu|}} \kappa_\nu (\mathcal{R}_\nu(t) + \mathcal{P}_\nu(t)\mu^2) \mathrm{d}\mu \mathrm{d}t \\
&= E_{k+3}(\kappa_\nu \tau) Q_\nu^+ + E_{k+3}(\kappa_\nu (Z-\tau)) Q^- \\
&+ \int_0^\tau E_{k+1}(\kappa_\nu(\tau - t)) R_\nu(t) \mathrm{d}t + \int_\tau^Z E_{k+1}(\kappa_\nu(t-\tau)) R_\nu(t) \mathrm{d}t \\
&+ \int_0^\tau E_{k+3}(\kappa_\nu(\tau - t)) P_\nu(t) \mathrm{d}t + \int_\tau^Z E_{k+3}(\kappa_\nu(t-\tau)) P_\nu(t) \mathrm{d}t.
\end{aligned}
\tag{10.27}
$$

For $k = 1$, the same computation gives

$$
\begin{aligned}
&\int_{-1}^1 \mu I_\nu(\tau, \mu) \mathrm{d}\mu = \int_{-1}^1 \mu^2 \Big(e^{-\frac{\kappa_\nu \tau}{\mu}} Q_\nu^+ \mathbf{1}_{\mu>0} - e^{-\frac{\kappa_\nu (Z-\tau)}{|\mu|}} Q_\nu^- \mathbf{1}_{\mu<0} \Big) \mathrm{d}\mu \\
&+ \int_{-1}^1 \Big(\mathbf{1}_{\mu>0} \int_0^\tau e^{-\frac{\kappa_\nu(\tau-t)}{\mu}} \kappa_\nu \mathcal{S}_\nu(t) \mathrm{d}t - \mathbf{1}_{\mu<0} \int_\tau^Z e^{-\frac{\kappa_\nu(t-\tau)}{|\mu|}} \kappa_\nu \mathcal{S}_\nu(t) \mathrm{d}t \Big) \mathrm{d}\mu \\
&= \int_0^1 \mu^2 e^{-\frac{\kappa_\nu \tau}{\mu}} Q_\nu^+ \mathrm{d}\mu - \int_{-1}^0 \mu^2 e^{-\frac{\kappa_\nu (Z-\tau)}{|\mu|}} Q_\nu^- \mathrm{d}\mu \\
&+ \int_0^1 \int_0^\tau e^{-\frac{\kappa_\nu(\tau-t)}{\mu}} \kappa_\nu \mathcal{S}_\nu(t) \mathrm{d}t - \int_{-1}^0 \int_\tau^Z e^{-\frac{\kappa_\nu(t-\tau)}{|\mu|}} \kappa_\nu \mathcal{S}_\nu(t) \mathrm{d}t \mathrm{d}\mu
\end{aligned}
\tag{10.28}
$$

$$
\begin{aligned}
&= \int_0^1 \mu^2 \left(e^{-\frac{\kappa_\nu \tau}{\mu}} Q_\nu^+ - e^{-\frac{\kappa_\nu (Z-\tau)}{|\mu|}} Q_\nu^- \right) \mathrm{d}\mu \\
&+ \int_0^\tau \int_0^1 e^{-\frac{\kappa_\nu(\tau - t)}{\mu}} \kappa_\nu (\mathcal{R}_\nu(t) + \mathcal{P}_\nu(t)\mu^2) \mathrm{d}\mu \mathrm{d}t \\
&- \int_\tau^Z \int_0^1 e^{-\frac{\kappa_\nu(t-\tau)}{|\mu|}} \kappa_\nu (\mathcal{R}_\nu(t) + \mathcal{P}_\nu(t)\mu^2) \mathrm{d}\mu \mathrm{d}t \\
&= E_4(\kappa_\nu \tau) Q_\nu^+ - E_4(\kappa_\nu (Z-\tau)) Q^- \\
&+ \int_0^\tau E_2(\kappa_\nu(\tau - t)) \mathcal{R}_\nu(t) \mathrm{d}t - \int_\tau^Z E_2(\kappa_\nu(t-\tau)) \mathcal{R}_\nu(t) \mathrm{d}t \\
&+ \int_0^\tau E_4(\kappa_\nu(\tau - t)) \mathcal{P}_\nu(t) \mathrm{d}t - \int_\tau^Z E_4(\kappa_\nu(t-\tau)) \mathcal{P}_\nu(t) \mathrm{d}t.
\end{aligned}
\tag{10.29}
$$

□

10.5.3 Numerical Simulations

It is estimated [7] that at the top of the atmosphere the solar radiation intensity is 340 W/m^2 but the atmosphere reflects 30% of it and our RTE does not handle volumic reflection. At 10 or 12km the sunlight is diffused, so we apply an intensity proportional to ω_3. Paris being at latitude 45°, the radiation is divided by $\sqrt{2}$. Hence an effective number is $Q_{sun} = 162$ W/m^2. And yet a good portion is spent for the evaporation of water and into convection. Arbitrarily we keep only half 80 W/m^2.

Recall the scalings defined in [11]. The frequencies and the temperatures are scaled as follows (primes label scaled variables),

$$\nu' = 10^{-14}\nu, \quad T' = 10^{-14}\frac{k}{h}T = 10^{-14}\frac{1.381 \cdot 10^{-23}}{6.626 \cdot 10^{-34}}T = \frac{T}{4798}.$$

Planck's function is written as

$$B_0\frac{\nu'^3}{e^{\frac{\nu'}{T'}} - 1}, \quad \text{with } B_0 = \frac{2h}{c^2}10^{42} = \frac{2 \times 6.626 \cdot 10^{-34}}{2.998^2 \cdot 10^{16}}10^{42} = 1.4744 \cdot 10^{-8}.$$

Consequently the computed intensity, I'_ν, is the physical intensity divided by B_0. The scaled Sun temperature is $T'_S = 5800/4798 = 1.209$. According to (10.18),

$$\int_0^\infty Q_\nu^+ = 80 = Q^S \int_0^\infty B_0\frac{\nu'^3}{e^{\frac{\nu'}{T'_S}} - 1}(10^{14}\mathrm{d}\nu') = Q^S 1.4744 \cdot 10^6\frac{\pi^4}{15}1.209^4.$$

This leads to $Q^S = 4 \cdot 10^{-6}$.

Similarly Earth is at $T'_E = 288/4798 = 0.06$ and emits about 350 W/m^2, but some is used for the evaporation of water so we have kept only 300 W/m^2:

$$300 = Q^E \int_0^\infty B_0\frac{\nu'^3}{e^{\frac{\nu'}{T'_E}} - 1}(10^{14}\mathrm{d}\nu') = Q^E 1.4744 \cdot 10^6\frac{\pi^4}{15}0.06^4 = Q^E 124.3.$$

This leads to $Q^E = 2.41$.

The air density is $1 - \frac{3}{4}z$. The earth reflective albedo is set at $r_\nu = 0.3$. Isotropic scattering is set at $a_\nu = 0.3$.

κ_ν is given by the Gemini experimental program. To measure the temperature perturbation due to an increase of CO_2 (resp NO_x) in the atmosphere, we set $\kappa_\nu = 1$ in the range $(3/16, 3/14)$ (resp. $(3/7, 3/5)$).

10.5.3.1 Isotropic Scattering

Two tests were performed with the stratified 1D approximation, isotropic scattering, no cloud and Gemini κ_ν + Rayleigh scattering equal to 0.3 for $\nu > 3$ and $\tau > 0.7$. One computation is with an infrared source Q^E at $\tau = 0$ only (similar to the situation at night) and another with a sunlight source Q^S at $\tau = Z$ only.

Results are on Figs. 10.3 and 10.4. Computations are done with 3 different atmospheres. One is conform to the Gemini measurements, the other two imitate an addition of CO_2.

One can observe a small additional greenhouse effect due to addition of CO_2 but a cooling effect at high altitude. The opposite is observed for addition of NO_x.

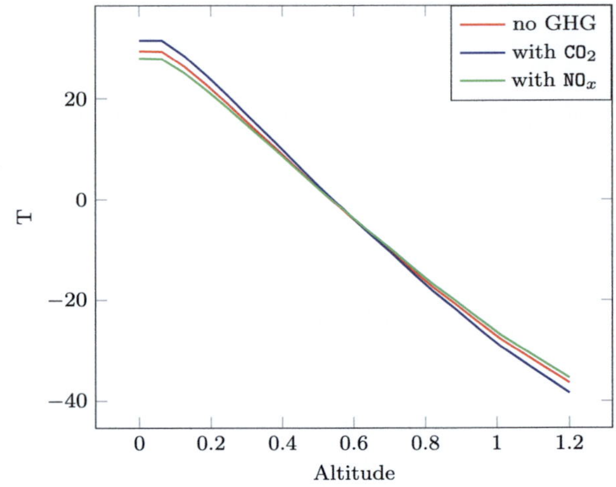

Fig. 10.3 Temperature T versus altitude z for 3 values of $\nu \mapsto \kappa_\nu$. Only the earth radiation is taken as a source (night time)

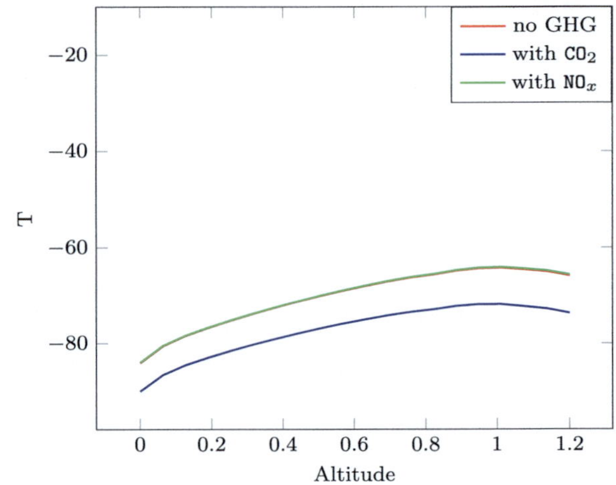

Fig. 10.4 T vs z for 3 values of $\nu \mapsto \kappa_\nu$. Only the Sun radiation is taken as a source

10.5.3.2 Anisotropic Scattering in a Cloud

Next, still with the stratified approximation, a cloud is added between altitude $Z_m = 7000$ m and $Z_M = 9000$ m with anisotropic scattering coefficient $\beta = \frac{a_\nu}{2}, b = 1,\ a_\nu = 0.3\frac{4}{(Z_M - Z_m)^2}(z - Z_m)^+(Z_M - z)^+$. Above Z_M Rayleigh scattering $0.3 1_{\nu>3}$, replaces the anisotropic scattering.

Results are shown on Figs. 10.5, 10.6 and 10.7. The strong effect of the cloud is shown on Fig. 10.6. Figure 10.7 shows how important it is to take all the details of κ_ν into account.

Finally, in Table 10.1 total radiative intensities $\int_0^\infty J_\nu(\tau)$ are shown at $\tau = 0$ and $\tau = Z$ in a variety of situations to see the effect of clouds, scattering, earth albedo.

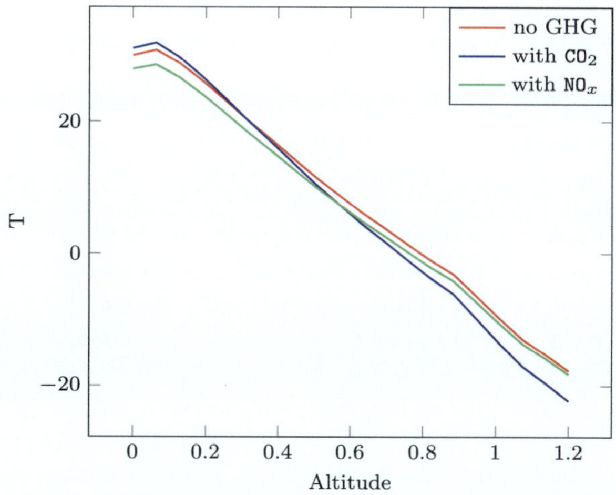

Fig. 10.5 T versus z computed with a cloud

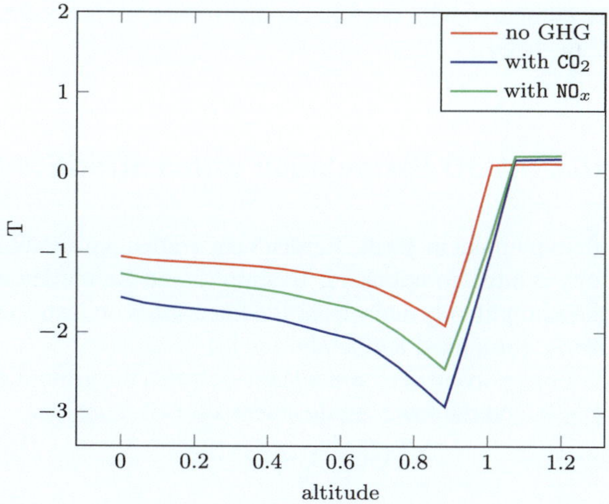

Fig. 10.6 Difference $T_{nocloud} - T_{cloud}$

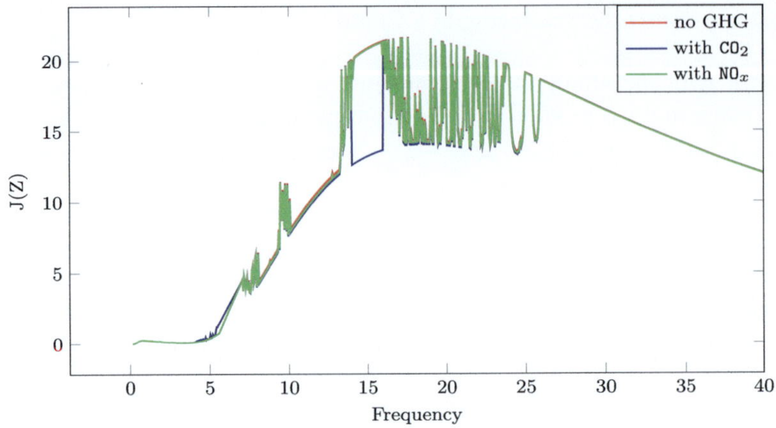

Fig. 10.7 Total radiative intensity $J_\nu(Z)$ versus ν for the 3 $\nu \mapsto \kappa_\nu$ cases

Table 10.1 Radiative intensities for the Earth radiations + Sunlight (first 2 rows) and the Sunlight only (last 2 rows) when all terms are present and when one of the terms is removed

	Everything	No cloud	No scattering	No earth albedo	No absorption
$\tau = 0$	9.52495	9.39105	9.52441	7.34548	7.21413
$\tau = Z$	5.98738	5.99558	5.98622	4.93935	7.14319
$\tau = 0$	1.44484	1.43589	1.4431	1.51029	1.51029
$\tau = Z$	1.91376	1.90286	1.91093	1.73976	1.51995

In the presence of CO_2 the total radiative energy at ground level (resp. $\tau = Z$), is 9.91316 (resp. 5.72292). With NO_x it is 9.55564 (resp. 5.8434). These numbers must be compared with the first number on the left in the first (resp 2nd) row in Table 10.1 (Fig. 10.8).

10.6 A 3D Formulation with a Stratified Part

To implement in 3D the Earth albedo (reflective condition by the ground) in all generality is quite complicated. To simplify the computation we consider the reflection of the sunlight only and we use the linearity, with respect to the sources, of the equation for I_ν, for a fixed temperature field.

For a portion Ω of atmosphere between the ground $\boldsymbol{g} = (\boldsymbol{x}_1, \boldsymbol{x}_2, \boldsymbol{x}_3 = g(\boldsymbol{x}_1, \boldsymbol{x}_2))$, $g \geq 0$, and the lower stratosphere $\boldsymbol{x}_3 = Z$, consider

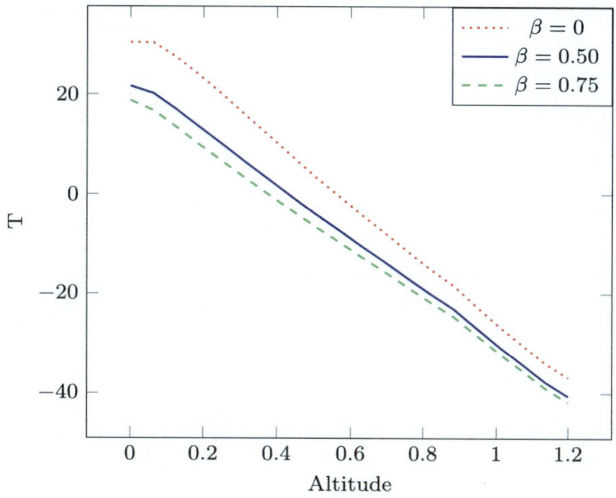

Fig. 10.8 T versus z computed with a cloud with anisotropic scattering of coefficient β

$$\omega \cdot \nabla_x I_\nu + \rho\bar{\kappa}_\nu a_\nu \left[I_\nu - \frac{1}{4\pi}\int_{\mathbb{S}^2} p_\nu(\omega, \omega') I_\nu(\omega') d\omega' \right] = \rho\bar{\kappa}_\nu (1 - a_\nu)[B_\nu(T) - I_\nu], \tag{10.30}$$

$$I_\nu|_{g,\omega\cdot n<0} = Q^E (\omega \cdot n)^- B_\nu(T_E), \quad I_\nu|_{Z,\omega_3<0} = Q^S \omega_3^- B_\nu(T_S).$$

The solution is $I_\nu = I_\nu^E + I_\nu^S$ with I_ν^E computed with the condition $I_\nu|_g$ only and the source term in T, and I_ν^S computed with $I_\nu|_Z$ only and $T = 0$:

$$\begin{cases} \omega \cdot \nabla_x I_\nu^S + \rho\bar{\kappa}_\nu a_\nu \left[I_\nu^S - \frac{1}{4\pi}\int_{\mathbb{S}^2} p_\nu(\omega, \omega') I_\nu^S(\omega') d\omega' \right] + \rho\bar{\kappa}_\nu (1 - a_\nu) I_\nu^S = 0, \\ I_\nu^S|_{Z,\omega_3<0} = Q^S \omega_3^- B_\nu(T_S), \qquad I_\nu^S|_{0,\omega_3>0} = 0, \end{cases}$$
$$\begin{cases} \omega \cdot \nabla_x I_\nu^E + \rho\bar{\kappa}_\nu a_\nu \left[I_\nu^E - \frac{1}{4\pi}\int_{\mathbb{S}^2} p_\nu(\omega, \omega') I_\nu^E(\omega') d\omega' \right] = \rho\bar{\kappa}_\nu (1 - a_\nu)[B_\nu(T) - I_\nu^E], \\ I_\nu^E|_{g,\omega\cdot n<0} = Q^E (\omega \cdot n)^- B_\nu(T_E) - I_\nu^S|_{g,\omega}, \quad I_\nu^E|_{Z,\omega_3<0} = 0. \end{cases} \tag{10.31}$$

For I_ν^S the stratified approximation is justified. Furthermore, it can be precomputed beforehand.

To account for the fact that a fraction $r_\nu(x)$ of the sunlight is reflected by the Earth surface, we consider a partial reflective condition

$$I_\nu^E|_{g,\omega\cdot n<0} = -I_\nu^S|_{g,\omega} + Q^E (\omega \cdot n)^- B_\nu(T_E) + r_\nu(g) I_\nu^S|_{g,\omega-2(\omega\cdot n)n},$$

but $I_\nu^S|_{g,\omega} \approx 0$ when $\omega \cdot n < 0$, so, neglecting scattering, $a_\nu = 0$ (see also [1])

$$I_\nu^S(x, \omega) = \omega_3^- Q^S B_\nu(T_S) e^{-\kappa_\nu \frac{Z-\tau(x_3)}{\omega_3^-}}.$$

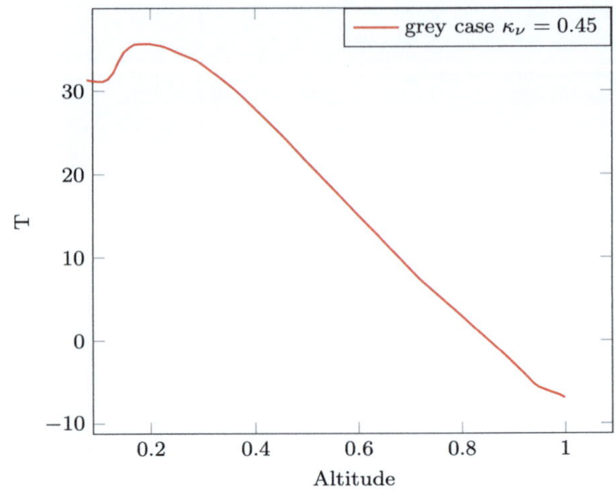

Fig. 10.9 Temperature above the city of Chamonix versus altitude in a grey case. The growth of T from the ground up is partially due to quadrature errors, partially to the parabolic effect of mountains

Therefore, with $\boldsymbol{\omega} \cdot \boldsymbol{n} < 0$ and $\boldsymbol{x}$ on the ground,

$$\begin{aligned} -I_\nu^S|_{\boldsymbol{g},\boldsymbol{\omega}} + r_\nu(\boldsymbol{x}) I_\nu^S|_{\boldsymbol{g},\boldsymbol{\omega}-2(\boldsymbol{\omega}\cdot\boldsymbol{n})\boldsymbol{n}} \\ = (\omega_3 - 2(\boldsymbol{\omega}\cdot\boldsymbol{n})n_3)^- Q^S B_\nu(T_S) r_\nu(\boldsymbol{g}) e^{-\frac{\kappa_\nu(Z-\tau(\boldsymbol{g}))}{(\omega_3-2(\boldsymbol{\omega}\cdot\boldsymbol{n})n_3)^-}}. \end{aligned} \tag{10.32}$$

For a flat ground, it is equal to $\omega_3^+ Q^S B_\nu(T_S) r_\nu(\boldsymbol{g}) e^{-\frac{\kappa_\nu Z}{\omega_3^+}}$ and its angle average on $\mathbb{S}^2$ is $\frac{1}{2} Q^S B_\nu(T_S) r_\nu(\boldsymbol{g}) E_3(\kappa_\nu Z)$ so that it makes sense to solve the problem with the following conditions,

$$\begin{aligned} I_\nu|_{\boldsymbol{g},\boldsymbol{\omega}\cdot\boldsymbol{n}<0} &= (\boldsymbol{\omega}\cdot\boldsymbol{n})^- \left(Q^E B_\nu(T_E) + Q^S B_\nu(T_S) 2 r_\nu(\boldsymbol{g}) E_3(\kappa_\nu Z)\right), \\ I_\nu|_{Z,\omega_3<0} &= (\boldsymbol{\omega}\cdot\boldsymbol{n})^- Q^S B_\nu(T_S). \end{aligned} \tag{10.33}$$

The valley of Chamonix is considered in the same physical conditions as above, i.e. with a cloud between 7000 and 9000 m. However we have not implemented the anisotropic scattering yet, so $\beta = 0$.

All 4 vertical boundaries which limit the domain Ω are reflective. Results are shown on Figs. 10.9 and 10.10.

10.7 Conclusion

The integro-differential formulation of the RTE and its solution by iterations on the source has been extended here to handle anisotropic scattering.

The iterative part of the method is $O(N \ln N)$, thanks to an efficient use of $\mathcal{H}$-matrices.

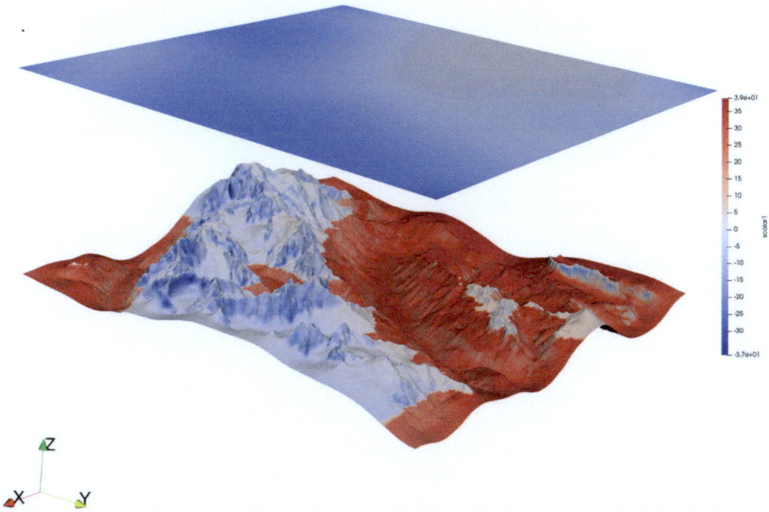

Fig. 10.10 Ground and high altitude temperature in an atmosphere receiving sunlight and Earth infrared radiation. In addition 0.3% of the sunlight is reflected in the normal direction of the ground and 0.7% if the ground is covered by snow

The precision is good enough to evaluate the effect of sensitive parameters for the study of contrails. Most of the time the stratified 1D approximation should suffice, but in complex cases with high relief the 3D formulation is needed.

Acknowledgements The 3D code is written using `FreeFEM` [13] and `htool`.[1]
`FreeFEM` is a popular open-source software package for solving PDE systems by the finite element method (FEM). Our gratitude to the team leader Frédéric Hecht for his valuable help.
`htool` is a parallel C++ toolbox implementing $\mathcal{H}$-Matrices, used in particular for the boundary element method in electromagnetism.

References

1. Bardos C, Golse F, Pironneau O (2023) Remarks on the radiative transfer equations for climatology. Impact of scientific computing on science and society of series computational methods in applied sciences. Springer Nature
2. Bardos C, Pironneau O (2021) Radiative transfer for the greenhouse effect. SeMA J (Springer) 79:489–525. Special Issue ICIAM 2019
3. Bebendorf M (2000) Approximation of boundary element matrices 86:565–589
4. Bebendorf M (2008) Hierarchical matrices. Lecture notes in science and engineering. Springer, Heidelberg
5. Cahalan R, Oreopoulos L, Marshak A, Evans K, Davis A et al (2005) Bringing together the most advanced radiative transfer tools for cloudy atmospheres. Am Meteorol Soc 86:1275–1294

[1] https://www.github.com/htool-ddm/htool.

6. Emde C, Buras-Schnell R, Kylling A, Mayer B, Gasteiger J, Hamann U, Kylling J, Richter B, Pause C, Dowling T, Bugliaro L (2016) The libradtran software package for radiative transfer calculations. Geosci Model Dev 9:1647–1672
7. Fowler A (2011) Mathematical geoscience. Springer, Berlin
8. Golse F, Hecht F, Pironneau O, Smetz D, Tournier P-H (2023) Radiative transfer for variable 3D atmospheres. J Comp Phys 475:1–19
9. Golse F, Pironneau O (2022) Radiative transfer in a fluid, RACSAM, Springer, Volume dedicated to I. Diaz. 10.1007/s13398-022-01362-x
10. Golse F, Pironneau O (2022) Stratified radiative transfer for multidimensional fluids. Comptes Rendus Mécanique 350:1–15. https://doi.org/10.5802/crmeca.136
11. Golse F, Pironneau O (2022) Stratified radiative transfer in a fluid and numerical applications to earth science. SIAM J Numer Anal 60:2963–3000. https://doi.org/10.1137/21M1459009
12. Hackbusch W (1999) A sparse matrix arithmetic based on h-matrices. Part I: Introduction to h-matrices. Computing 62:89–108
13. Hecht F (2012) New developments in freefem++. J Numer Math 20:251–265
14. Lee DS, Fahey DW, Skowron A, Allen MR, Burkhardt U, Chen Q (2021) The contribution of global aviation to anthropogenic climate forcing for 2000 to 2018. Atmos Environ 244:117834
15. Marquart S, Ponater M, Mager F, Sausen R (2003) Future development of contrail cover, optical depth, and radiative forcing: impacts of increasing air traffic and climate change. J Clim 16:2890–2904
16. Paoli R, Shariff K (2016) Contrail modeling and simulation. Ann Rev Fluid Mech 48:393–427
17. Pironneau O, Tournier P-H (2023) Reflective conditions for radiative transfer in integral form with h-matrices. Submitted to JCP
18. Pomraning G (1973) The equations of radiation hydrodynamics. Pergamon Press, NY
19. Porzio M, Lopez-Pouso O (2004) Application of accretive operators theory to evolutive combined conduction, convection and radiation. Rev Mat Iberoamericana 20:257–275

Chapter 11
Discrete Exterior Calculus and Controlled Time Integration for Time-Harmonic Acoustics

Mikael Myyrä

Abstract We present an efficient numerical method for time-harmonic acoustics problems, consisting of a discrete exterior calculus (DEC) discretization of the time-dependent wave equation and a controllability-based optimization procedure. For improved performance in the harmonic problem domain, the method is modified with an exact timestepping scheme and a Hodge star operator optimized for harmonic waves. The presented method's effectiveness is investigated experimentally by solving a scattering problem with varying degrees of geometric complexity. The controllability method shows most benefits in highly nonconvex geometries, where a simple asymptotic iteration does not reliably converge to a solution. The harmonic Hodge star operator is found to increase accuracy significantly without imposing a runtime performance cost. Quality of the computation mesh is identified as a key issue affecting the method's accuracy and reliability.

Keywords Discrete exterior calculus · Acoustics · Wave equation · Time-harmonic waves · Controllability

11.1 Introduction

We present an approach to time-harmonic acoustic wave problems in $\mathbb{R}^2$ based on solving a mixed formulation of the time-dependent wave equation and searching for a time-periodic state. We're particularly interested in scattering problems with nonconvex geometry, which aren't easily solved by simple integration forward in time. The wave equation is expressed in terms of differential forms, discretized using the framework of discrete exterior calculus (DEC), and advanced forward in time with leapfrog integration.

DEC was first developed by [4] and can be seen as a generalization of the finite difference time domain (FDTD) method of [26]. A notable advantage of the method

M. Myyrä (✉)
University of Jyväskylä,Jyväskylä, Finland
e-mail: mikael.b.myyra@jyu.fi

T. Tuovinen et al. (eds.), *Challenges in Design Methods, Numerical Tools and Technologies for Sustainable Aviation, Transport and Industry*, Computational Methods in Applied Sciences 17, https://doi.org/10.1007/978-3-031-98675-8_11

is that the diagonal Hodge star approximation lends itself to an efficient explicit time discretization which does not require the solution of linear systems. DEC has a wide range of applications in computational physics, including electromagnetics [15, 24], quantum mechanics [10, 23], and fluid dynamics [14, 19]. Räbinä et al. [22] have developed a generalized theory of wave propagation problems with DEC as the basis, of which our acoustics model is a special case.

We augment the generic framework of DEC with an exact timestepping procedure and a Hodge star operator optimized for harmonic problems, the latter originally introduced by [21] in the context of three-dimensional electromagnetics, where the Hodge star operates on values on a tetrahedral mesh. We derive the equivalent operators for 2D triangle meshes. The efficacy of these optimizations is verified by comparison of numerical results with a known analytic solution.

Time-harmonic problems are approached using the idea of exact controllability developed by [11], formulating the time-harmonic condition as an optimization problem which is solved using the conjugate gradient (CG) algorithm. Controllability problems were first applied to the computation of time-periodic solutions by [2], and the method was later adapted to the mixed formulation we employ here by [9]. An advantage of the latter formulation is that the related controllability problem does not require preconditioning. We note that the control method is independent of the choice of discretization for the time-dependent problem; we have chosen DEC for its advantages in computational efficiency.

This article is structured as follows. Continuous equations and their differential form representations are presented in Sect. 11.2. The next Sect. 11.3 details our numerical methods, beginning with a DEC discretization of the continuous equations, followed by a derivation of the harmonically optimized operators and an overview of the control method. Finally, our experiments and results are discussed in Sect. 11.4.

11.2 Model

We consider acoustic waves described by the differential equation

$$\frac{\partial^2 \phi}{\partial t^2} - c^2 \nabla^2 \phi = 0 \tag{11.1}$$

where ϕ is a velocity potential, t is time, and c is the speed of sound. This equation can be expressed as a first-order system

$$\begin{cases} \frac{\partial p}{\partial t} - c^2 \nabla \cdot \mathbf{v} = 0 \\ \frac{\partial \mathbf{v}}{\partial t} - \nabla p = 0 \end{cases} \tag{11.2}$$

where $p = \frac{\partial \phi}{\partial t}$ is the scalar-valued acoustic pressure and $\mathbf{v} = \nabla \phi$ is the vector-valued velocity of particles perturbed by the wave.

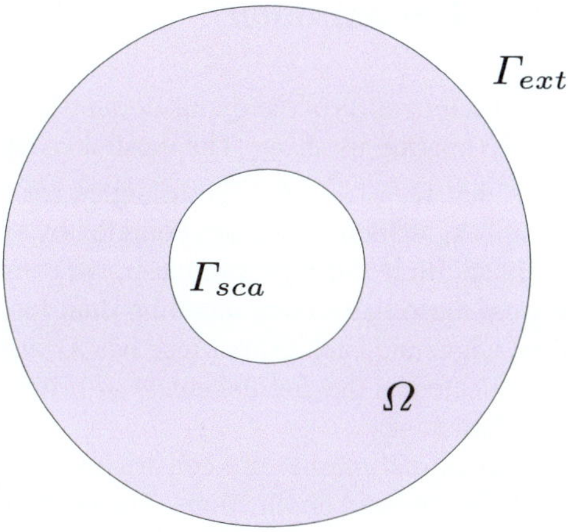

Fig. 11.1 The domain Ω and boundaries Γ_{ext} and Γ_{sca} of a scattering model

Represented in terms of the differential 0-form pressure $\tilde{p} = p$ and 1-form flux $\tilde{q} = -v_y dx + v_x dy$, the system (11.2) is

$$\begin{cases} \frac{\partial \tilde{p}}{\partial t} - c^2 \star d\tilde{q} = 0 \\ \frac{\partial \tilde{q}}{\partial t} - \star d\tilde{p} = 0. \end{cases} \tag{11.3}$$

The selection of the flux $\tilde{q} = \star\tilde{v}$ as our second variable is motivated by the need for flux-based boundary conditions, which we elaborate on next.

Our primary test case (Sect. 11.4.2) models the scattering of a plane wave by a sound-soft obstacle. The surface of the obstacle is represented by a boundary curve Γ_{sca} of our computational domain Ω, illustrated in Fig. 11.1, where the incident plane wave's velocity potential ϕ_{inc} is applied as a Dirichlet boundary condition.

To model empty space around the obstacle while truncating the computational domain to a finite size, we apply a first-order absorbing Engqvist-Majda boundary condition [6]

$$\frac{1}{c}\frac{\partial \phi}{\partial t} + \mathbf{n} \cdot \nabla \phi = 0 \tag{11.4}$$

on the outer boundary Γ_{ext}.

In terms of the differential form variables $\tilde{p}$ and $\tilde{q}$, Eq. (11.4) translates to

$$\frac{1}{c}\tilde{p} + \tilde{q} = 0. \tag{11.5}$$

11.3 Discretization

Our discretization of the spatial domain Ω is based on an unstructured simplicial mesh approximating its shape. The mesh consists of N_f triangular faces (2-simplices) f_j such that $\Omega \approx \cup_{j=1}^{N_f} f_j$. The triangles are bounded by edges (1-simplices) e_j, $j = 1, \ldots, N_e$, which in turn are bounded by vertices (0-simplices) v_j, $j = 1, \ldots, N_v$.

In addition to this *primal* mesh, we construct its *dual* such that each vertex of the primal mesh has a corresponding dual face, each primal edge has a corresponding dual edge, and each primal face has a corresponding dual vertex. The dual vertices are selected as the circumcenters of primal faces, resulting in perpendicular primal and dual edges.

Our primal meshes are constructed using `gmsh` [7], a mesh generator designed for finite element applications. The circumcentric dual is generated by PyDEC [1].

11.3.1 Discrete Exterior Calculus

The discrete counterparts of the differential forms $\tilde{p}$ and $\tilde{q}$ are *cochains* represented as column vectors $P = (P_1, \ldots, P_{N_v})^T$, $Q = (Q_1, \ldots, Q_{N_e})^T$ with P associated with vertices v_i^* of the dual mesh and Q with edges e_i of the primal mesh such that

$$P_i = \tilde{p}(v_i^*), \quad Q_i = \int_{e_i} (\mathbf{v} \cdot \mathbf{n})\, ds \tag{11.6}$$

(see [5]) where $\mathbf{n}$ is the normal vector of an edge and $\mathbf{v} \cdot \mathbf{n}$ computes the flux of velocity $\mathbf{v}$ over that edge.

The discrete analogues for the operators d and $\star$ are then matrices which operate on the column vector cochains by multiplication. The discrete k-dimensional exterior derivative $\mathbf{d}_k$ is defined as the coboundary operator mapping primal k-cochains to $(k+1)$-cochains such that the Stokes theorem

$$\int_\sigma \mathbf{d}\alpha = \int_{\partial\sigma} \alpha \tag{11.7}$$

is satisfied for every k-simplex σ and k-form α. This results in a matrix where the jth column on the ith row is 0 if the k-simplex σ_j is not on the boundary of the $(k+1)$-simplex σ_i, 1 if the orientation of σ_j matches that of σ_i, and –1 otherwise.

The dual mesh is constructed such that each coboundary on the primal mesh is a boundary on the dual mesh and vice versa, and thus a dual exterior derivative $\mathbf{d}^*_{n-k-1}$ can also be defined in terms of the corresponding primal operator as $\mathbf{d}_k^T$.

The discrete k-dimensional Hodge star $\star_k$ is constructed as a diagonal matrix such that

$$(\star_k)_{ii} = \frac{|\sigma_i^*|}{|\sigma_i|}. \tag{11.8}$$

$\star_k$ takes a k-cochain on the primal mesh and produces a $(n-k)$-cochain on the dual mesh.

A dual $\star^*_{n-k}$ operating on dual $(n-k)$-cochains and producing primal k-cochains is obtained by taking the inverse of the previously defined $\star_k$ while accounting for the sign change defined by $\star\star\alpha = (-1)^{k(n-k)}\alpha$. Here the circumcentric duality of the meshes is important to ensure the diagonality of the Hodge star matrix, an essential factor in the efficiency of our time-stepping procedure.

The spatially discretized form of Eq. (11.3) expressed in terms of the cochains and operators defined here is

$$\begin{cases} \frac{\partial p}{\partial t} - c^2 \star_2 \mathbf{d}_1 Q = 0 \\ \frac{\partial q}{\partial t} - \star_1^{-1} \mathbf{d}_1^T P = 0. \end{cases} \tag{11.9}$$

11.3.2 Time Discretization

To discretize the time derivatives in Eq. (11.9) we employ the leapfrog method, a second-order finite difference scheme commonly used in FDTD [26] and related methods.

Values of P and Q are placed at intervals of Δt seconds in time, offset from each other by $\frac{\Delta t}{2}$. Denoting $t^k = k\Delta t$ the time after k timesteps, P^k the value of P at this time, and $Q^{k+\frac{1}{2}}$ the value of Q at time $t^{k+\frac{1}{2}} = (k+\frac{1}{2})\Delta t$, we approximate the time derivatives of P and Q as

$$\frac{\partial P}{\partial t}(t^{k+\frac{1}{2}}) = \frac{P^{k+1} - P^k}{\Delta t} \tag{11.10}$$

$$\frac{\partial Q}{\partial t}(t^k) = \frac{Q^{k+\frac{1}{2}} - Q^{k-\frac{1}{2}}}{\Delta t} \tag{11.11}$$

Substituting Eqs. (11.10) and (11.11) into Eq. (11.9) yields the explicit timestep formulas

$$P^{k+1} = P^k + \Delta t c^2 \star_2 \mathbf{d}_1 Q^{k+\frac{1}{2}} \tag{11.12}$$

$$Q^{k+\frac{3}{2}} = Q^{k+\frac{1}{2}} + \Delta t \star_1^{-1} \mathbf{d}_1^T P^{k+1}. \tag{11.13}$$

11.3.3 Harmonic Optimizations

The time derivative and Hodge star discretizations presented in Sects. 11.3.2 and 11.3.1 are based on linear and locally constant approximations. When simulating harmonic waves, we can improve the accuracy of these operators by instead assuming a time-harmonic solution.

A similar approximation of the time derivative has been previously applied in the context of electromagnetics by [13, 20, 21]. The harmonic Hodge star was derived for three-dimensional electromagnetics by [21]. In this section we apply the same method of approximation to obtain optimized Hodge star operators for our two-dimensional acoustics equations.

Exact Timestepping We begin by assuming that the state of our system at a point can be expressed as

$$P(t) = \mathrm{Re}(\hat{P}_0 \exp(-i\omega t)) \tag{11.14}$$

$$Q(t + \frac{\Delta t}{2}) = \mathrm{Re}(\hat{Q}_0 \exp(-i\omega(t + \frac{\Delta t}{2}))) \tag{11.15}$$

where $\hat{P}_0$ and $\hat{Q}_0$ are complex initial values for P and Q, ω is the angular velocity of the wave, and i is the imaginary unit.

Substituting Eq. (11.14) into the numerator of Eq. (11.10) with $t^k = t + \frac{\Delta t}{2}$ we get

$$\begin{aligned} &P\left(t^k + \frac{\Delta t}{2}\right) - P\left(t_k - \frac{\Delta t}{2}\right) \\ &= \mathrm{Re}\Big[\hat{P}_0 \exp\left(-i\omega\left(t^k + \frac{\Delta t}{2}\right)\right)\Big] - \mathrm{Re}\Big[\hat{P}_0 \exp\left(-i\omega\left(t^k - \frac{\Delta t}{2}\right)\right)\Big] \\ &= \frac{\partial P}{\partial t}(t^k)\Big(\frac{2}{\omega}\sin\frac{\omega\Delta t}{2}\Big). \end{aligned}$$

The same consideration for the flux Q produces the same coefficient, leading to the time derivative approximations

$$\frac{\partial P}{\partial t} = \frac{P^{n+1} - P^n}{\frac{2}{\omega}\sin\frac{\omega\Delta t}{2}} \tag{11.16}$$

$$\frac{\partial Q}{\partial t} = \frac{Q^{n+\frac{3}{2}} - Q^{n+\frac{1}{2}}}{\frac{2}{\omega}\sin\frac{\omega\Delta t}{2}}. \tag{11.17}$$

Substituting these into the timestep formulas Eqs. (11.12), (11.13) gives

$$P^{k+1} = P^k + \frac{2}{\omega}\sin\frac{\omega\Delta t}{2}(c^2 \star_2 \mathbf{d}_1 Q^{k+\frac{1}{2}}) \tag{11.18}$$

$$Q^{k+\frac{3}{2}} = Q^{k+\frac{1}{2}} + \frac{2}{\omega}\sin\frac{\omega\Delta t}{2}(\star_1^{-1}\mathbf{d}_1^T P^{k+1}). \tag{11.19}$$

Harmonic Hodge Star To derive an optimized Hodge star operator we begin by assuming that the solution can be locally approximated as a complex plane wave

$$\hat{u}(\mathbf{x}) := \hat{u}_0 \exp(i\kappa \mathbf{d} \cdot \mathbf{x}) \tag{11.20}$$

where $\hat{u}_0$ is a complex initial value, κ is the wavenumber, and $\mathbf{d} = (\cos\theta, \sin\theta)$ is the propagating direction of the wave.

Our optimization procedure depends on the input and output dimensions of the Hodge star operator and thus needs to be done separately for each unique $\star_k$ present in the equations.

We begin with $\star_1$, which maps a 1-cochain representing flux over a primal edge into a 1-cochain representing total velocity over a dual edge. Given a primal edge e_i and corresponding dual edge e_i^*, we seek to minimize the error

$$||\int_{e_i} \hat{u} \cdot \mathbf{n}\, ds - (\star_1^{-1})_{ii} \int_{e_i^*} \hat{u} \cdot \mathbf{t}\, ds||^2 \tag{11.21}$$

where $\mathbf{n}$ and $\mathbf{t}$ are the unit vectors normal to e_i and tangent to e_i^*, respectively.

Denoting $\hat{u}_{e_i} = \int_{e_i} \hat{u} \cdot \mathbf{n}\, ds$ and $\hat{u}_{e_i^*} = \int_{e_i^*} \hat{u} \cdot \mathbf{t}\, ds$ the error (11.21) is minimized when

$$(\star_1)_{ii} = \frac{\hat{u}_{e_i^*}^2}{\hat{u}_{e_i} \hat{u}_{e_i^*}}.$$

This solution depends on the wavevector $\kappa\mathbf{d}$, which we generally do not know in advance. To obtain a more useful approximation we integrate over the unit circle,

$$(\star_1)_{ii} = \frac{\int_0^{2\pi} \hat{u}_{e_i^*}^2\, d\theta}{\int_0^{2\pi} \hat{u}_{e_i} \hat{u}_{e_i^*}\, d\theta}. \tag{11.22}$$

For the following computation we assume that the edges are orthogonal, axis-aligned with e_i along the x-axis and e_i^* along the y-axis, and centered at the origin. A Taylor series expansion of $\hat{u}_{e_i}$ then yields

$$\begin{aligned}
\hat{u}_{e_i} &= (\hat{u}_0)_y \int_{-\frac{1}{2}|e_i|}^{\frac{1}{2}|e_i|} \exp(i\kappa x \cos\theta)\, dx \\
&= (\hat{u}_0)_y |e_i| \sum_{n=0}^{\infty} \frac{(\alpha|e_i|)^{2n}}{2^{2n}(2n+1)!} \\
&= (\hat{u}_0)_y |e_i| \Big(1 + \frac{(\alpha|e_i|)^2}{24} + \frac{(\alpha|e_i|)^4}{1920} + \dots\Big)
\end{aligned} \tag{11.23}$$

where $\alpha = i\kappa\cos\theta$, $(\hat{u}_0)_y = |\hat{u}_0|\sin\theta$ is the y coordinate of $\hat{u}_0$, and $|\cdot|$ denotes the length of an edge.

Likewise for $\hat{u}_{e_i^*}$, we get

$$\hat{u}_{e_i^*} = (\hat{u}_0)_y |e_i^*| \sum_{n=0}^{\infty} \frac{(\beta |e_i^*|)^{2n}}{2^{2n}(2n+1)!} \tag{11.24}$$

with $\beta = i\kappa \sin\theta$.

Finally, we use the series expressions (11.23) and (11.24) to compute the integrals $\int_0^{2\pi} \hat{u}_{e_i^*}^2 \, d\theta$ and $\int_0^{2\pi} \hat{u}_{e_i} \hat{u}_{e_i^*} \, d\theta$, truncate the resulting series to second order, and substitute the results into Eq. (11.22) to obtain

$$(\star_1)_{ii} \approx \frac{|e_i^*|}{|e_i|} \left(\frac{1 - \frac{\kappa^2 |e_i^*|^2}{16}}{1 - \frac{\kappa^2 |e_i|^2}{96} - \frac{\kappa^2 |e_i^*|^2}{32}} \right). \tag{11.25}$$

To obtain a similar approximation for $\star_2$ which associates a primal face f_i with a dual vertex v_i^*, we first assume that the face is an origin-centered circle with radius r and the dual vertex is at the origin. We will generalize this result to polygon faces afterwards.

Minimizing an error analogous to Eq. (11.21) for $\star_2$ gives the diagonal element

$$(\star_2)_{ii} = \frac{\int_0^{2\pi} \hat{u}(v_i^*)^2 \, d\theta}{\int_0^{2\pi} \hat{u}_{f_i} \hat{u}(v_i^*) \, d\theta} \tag{11.26}$$

where $\hat{u}_{f_i}$ is the plane wave (11.20) integrated over the face and $\hat{u}(v_i^*) = \exp(0) = 1$ is the plane wave evaluated at the dual vertex.

Computing $\hat{u}_{f_i} = \int_{f_i} \hat{u}$ integrating over the unit circle, truncating the resulting series to second order as before, and substituting into Eq. (11.26), we obtain

$$(\star_2)_{ii} \approx \frac{1}{|f_i|} \left(\frac{1}{1 - \frac{\kappa^2 r^2}{8}} \right). \tag{11.27}$$

To generalize this result to triangular faces, we approximate r^2 in Eq. (11.27) by the maximum internal sphere radius r and minimum external sphere radius R. In the case of a regular triangle, r is the distance from the circumcenter c to each edge midpoint, and R is the distance from c to each vertex. Computing the integrals of Eq. (11.26) over a regular triangle results in a series whose first three terms match those obtained by Eq. (11.27) with $r^2 \approx \frac{1}{3}(2r^2 + R^2)$.

In the general case of irregular triangles, each vertex remains equidistant from the circumcenter, but the perpendicular distance to each edge varies, as illustrated in Fig. 11.2. We approximate r^2 for each triangle face as

$$r^2 \approx \frac{1}{9} \sum_{k=1}^{3} (2r_k^2 + R^2). \tag{11.28}$$

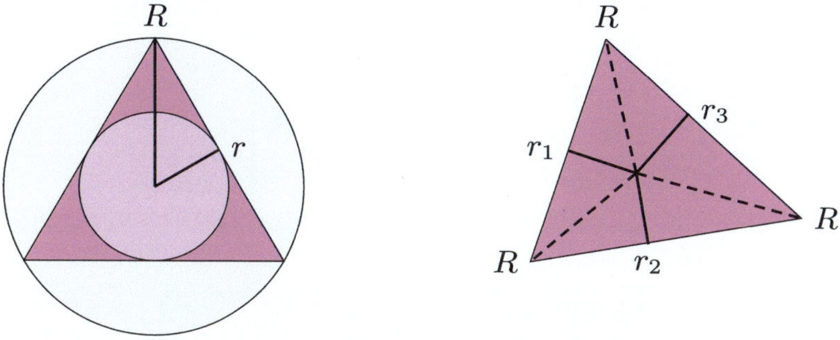

Fig. 11.2 Internal and external radii of a regular and irregular triangle

11.3.4 Controlled Time Integration

Our DEC solver detailed in the preceding sections simulates a system forward in time. In order to find a time-periodic solution, a simple approach is to run this solver until the system converges to a time-periodic state. We call this approach the *asymptotic iteration method*. However, convergence of this method may be slow or unreliable, particularly in the presence of nonconvex geometry blocking and reflecting waves, as we find in the experiments of Sect. 11.4. To accelerate convergence, we employ the concept of exact controllability, as discussed in the introduction. We define a least-squares optimization problem minimizing the difference between states before and after a time period T, solved using the conjugate gradient (CG) algorithm.

Optimization Problem Let $U^n = (P^n, Q^{n+1})^T$ be the state of the simulation at timestep n, $\mathbf{e} = U^0$ an unknown initial state, and N the number of timesteps needed to simulate the time period T. The solution is T-periodic if $U^N - \mathbf{e} = 0$. To guide the simulation into a T-periodic state we minimize the quadratic *control energy*

$$J(\mathbf{e}) = \frac{1}{2}(U^N - \mathbf{e})^T (U^N - \mathbf{e}). \tag{11.29}$$

Solving this minimization problem with the CG algorithm requires the gradient $\nabla J(\mathbf{e})$, which we will compute next.

A single timestep using the formulas (11.12) and (11.13) can be expressed as a block matrix equation

$$\begin{bmatrix} I & \Delta t c^2 \star_2 \mathbf{d}_1 & -I & \\ & I & \Delta t \star_1^{-1} \mathbf{d}_1^T & -I \end{bmatrix} \begin{bmatrix} P^n \\ Q^{n+\frac{1}{2}} \\ P^{n+1} \\ Q^{n+\frac{3}{2}} \end{bmatrix} = 0. \tag{11.30}$$

A repeated application of Eq. (11.30) produces the *state equation*

$$S(\mathbf{e}) = \begin{bmatrix} I & & & & & \\ C & \mathcal{D} & & & & \\ & C & \mathcal{D} & & & \\ & & \ddots & & & \\ & & & C & \mathcal{D} & \\ & & & & C & \mathcal{D} \end{bmatrix} \begin{bmatrix} \mathbf{e} \\ U^1 \\ \vdots \\ U^{N-1} \\ U^N \end{bmatrix} - \begin{bmatrix} \mathbf{e} \\ 0 \\ \vdots \\ 0 \\ 0 \end{bmatrix} + \begin{bmatrix} 0 \\ \mathcal{F}^0 \\ \vdots \\ \mathcal{F}^{N-2} \\ \mathcal{F}^{N-1} \end{bmatrix} = 0 \tag{11.31}$$

where

$$C = \begin{bmatrix} I & \Delta t c^2 \star_2 \mathbf{d}_1 \\ 0 & I \end{bmatrix}, \quad \mathcal{D} = \begin{bmatrix} -I & 0 \\ \Delta t \star_1^{-1} \mathbf{d}_1^T & -I \end{bmatrix},$$

and the vectors $\mathcal{F}$ have been added to account for source terms.

It should be noted that the state Eq. (11.31) as stated here does not address any boundary conditions, which would manifest as individual matrix rows deviating from the overall block structure. This section's analysis is carried out without consideration for boundary conditions, leaving treatment of the boundary to Sect. 11.3.5.

A single timestep can be expressed in this matrix notation in explicit form as

$$U^{n+1} = -\mathcal{D}^{-1}(CU^k + \mathcal{F}^k). \tag{11.32}$$

Denoting $Q = -\mathcal{D}^{-1}C$, the state after N timesteps is

$$U^N = Q^N\mathbf{e} - \sum_{n=0}^{N-1} Q^n \mathcal{D}^{-1} \mathcal{F}^{N-1-n} \tag{11.33}$$

where Q^N denotes the Nth power of the square matrix Q.

We can now express the control energy (11.29) as

$$\begin{aligned} J(\mathbf{e}) &= \frac{1}{2}(U^N - \mathbf{e})^T(U^N - \mathbf{e}) \\ &= \frac{1}{2}(\mathbf{e})^T(Q^N - I)^T(Q^N - I)\mathbf{e} - G^T(Q^N - I)\mathbf{e} + \frac{1}{2}G^T G \\ &=: \frac{1}{2}(\mathbf{e})^T A\mathbf{e} - \mathbf{b}^T\mathbf{e} + c \end{aligned}$$

where $G = \sum_{n=0}^{N-1} Q^n \mathcal{D}^{-1} \mathcal{F}^{N-1-n}$, resulting in an expression for the gradient

$$\nabla J(\mathbf{e}) = A\mathbf{e} - \mathbf{b} \tag{11.34}$$

with

$$A = (Q^N - I)^T(Q^N - I), \quad \mathbf{b} = G^T(Q^N - I).$$

Adjoint Method For efficient computation of the gradient $\nabla J(\mathbf{e})$ we apply the *adjoint method* (see e.g. [8]), solving the *adjoint equation*

$$\left(\frac{\partial S}{\partial \mathbf{u}}\right)^T \mathbf{z} = \left(\frac{\partial J}{\partial \mathbf{u}}\right)^T \tag{11.35}$$

where S is the state Eq. (11.31) and $\mathbf{z} = (Z^0, \ldots, Z^N)^T$ has the same dimension as the unknown variable $\mathbf{u} = (U^0, \ldots, U^N)^T$ of the state equation.

Using the adjoint solution $\mathbf{z}$, the gradient $\nabla J(\mathbf{e})$ can be expressed as

$$\nabla J(\mathbf{e}) = \frac{\partial J}{\partial \mathbf{e}} - \mathbf{z}^T \frac{\partial S}{\partial \mathbf{e}}. \tag{11.36}$$

Substituting $\frac{\partial S}{\partial \mathbf{e}} = (-I, 0, 0, \ldots, 0)^T$ and $\frac{\partial J}{\partial \mathbf{e}} = -(U^N - \mathbf{e})$ yields

$$\nabla J(\mathbf{e}) = -(U^N - \mathbf{e}) + Z^0. \tag{11.37}$$

To solve for Z^0 in Eq. (11.37) we turn to the adjoint Eq. (11.35). Noting that $\frac{\partial J}{\partial \mathbf{u}} = (0, 0, \ldots, 0, U^N - \mathbf{e})$, we can express Eq. (11.35) in matrix form as

$$\begin{bmatrix} I & C^T & & & & \\ & \mathcal{D}^T & C^T & & & \\ & & \mathcal{D}^T & C^T & & \\ & & & \ddots & \ddots & \\ & & & & \mathcal{D}^T & C^T \\ & & & & & \mathcal{D}^T \end{bmatrix} \begin{bmatrix} Z^0 \\ Z^1 \\ \vdots \\ Z^{N-1} \\ Z^N \end{bmatrix} = \begin{bmatrix} 0 \\ 0 \\ \vdots \\ 0 \\ U^N - \mathbf{e} \end{bmatrix} \tag{11.38}$$

where C and $\mathcal{D}$ are the block matrices defined in the state Eq. (11.31).

Equation (11.38) can be solved with a backward procedure starting with

$$Z^N = (\mathcal{D}^T)^{-1}(U^N - \mathbf{e}), \tag{11.39}$$

followed by repeating

$$Z^n = -(\mathcal{D}^T)^{-1} C^T Z^{n+1} \tag{11.40}$$

for $k = N - 1, \ldots, 1$, and finally

$$Z^0 = -C^T Z^1. \tag{11.41}$$

The intermediate steps according to Eq. (11.40) are solved using an explicit timestepping procedure analogous to that of the forward problem (Eqs. 11.12 and 11.13) with the formulas

$$Z_1^{n+\frac{1}{2}} = Z_1^{n+\frac{3}{2}} + \Delta t c^2 (\star_2 \mathbf{d}_1)^T Z_0^{n+1} \tag{11.42}$$

$$Z_0^n = Z_0^{n+1} + \Delta t (\star_1^{-1} \mathbf{d}_1^T)^T Z_1^{n+\frac{1}{2}} \tag{11.43}$$

where $Z^n = (Z_0^n, Z_1^{n+\frac{1}{2}})^T$ is the adjoint variable at the nth timestep decomposed into vectors matching the dimensions of $U^n = (P^n, Q^{n+\frac{1}{2}})^T$ in the forward problem.

With this procedure and Eq. (11.37) we can obtain $\nabla J(\mathbf{e})$ everywhere in the domain Ω by advancing the system forward in time N steps according to the state Eq. (11.31) and then backwards N steps according to the adjoint Eq. (11.35). Equipped with this gradient, we can apply the conjugate gradient algorithm to search for a value of $\mathbf{e}$ that satisfies the time-harmonic condition.

Conjugate Gradient Algorithm The conjugate gradient (CG) algorithm (see e.g. [25]) solves a symmetric positive definite linear system $A\mathbf{e} = \mathbf{b}$ iteratively by searching along an A-conjugate sequence of directions

$$\{\mathbf{p}_0, \ldots, \mathbf{p}_{n-1} : \mathbf{p}_i^T A \mathbf{p}_j = 0 \text{ for all } i \neq j\}. \tag{11.44}$$

Given an initial guess $\mathbf{e}_0$, the exact solution $\mathbf{e}_*$ can be expressed in terms of a sequence of approximations (e_j) as

$$\mathbf{e}_* = \mathbf{e}_0 + \sum_{i=0}^{n-1} \alpha_i \mathbf{p}_i = \mathbf{e}_j + \sum_{i=j}^{n-1} \alpha_i \mathbf{p}_i \tag{11.45}$$

where

$$\mathbf{e}_j = \mathbf{e}_0 + \sum_{i=0}^{j-1} \alpha_i \mathbf{p}_i = \mathbf{e}_{j-1} + \alpha_{j-1} \mathbf{p}_{j-1} \tag{11.46}$$

is the estimated solution at step j and α_i is a real coefficient chosen such that the residuals $\mathbf{r}_i = \mathbf{b} - A\mathbf{e}_i$ are mutually orthogonal. To find the value of α_i we observe that

$$\mathbf{r}_{j+1} = \mathbf{r}_j - \alpha_j A \mathbf{p}_j, \tag{11.47}$$

implying we must impose

$$\langle \mathbf{r}_{j+1}, \mathbf{r}_j \rangle = \langle \mathbf{r}_j - \alpha_j A \mathbf{p}_j, \mathbf{r}_j \rangle = 0 \tag{11.48}$$

where $\langle \cdot, \cdot \rangle$ denotes the dot product. Solving Eq. (11.48) gives

$$\alpha_j = \frac{\langle \mathbf{r}_j, \mathbf{r}_j \rangle}{\langle A\mathbf{p}_j, \mathbf{r}_j \rangle}. \tag{11.49}$$

We can then apply α_j to Eqs. (11.46) and (11.47) to obtain $\mathbf{e}_{j+1}$ and $\mathbf{r}_{j+1}$, which can in turn be used to construct the next search direction as

$$\mathbf{p}_{j+1} = \mathbf{r}_{j+1} + \beta_j \mathbf{p} \tag{11.50}$$

where β_j is obtained by requiring $\mathbf{p}_{j+1}$ and $\mathbf{p}_j$ to be A-conjugate, implying

$$\beta_j = -\frac{\langle \mathbf{r}_{j+1}, A\mathbf{p}_j \rangle}{\langle \mathbf{p}_j, A\mathbf{p}j \rangle}. \tag{11.51}$$

Applying the relations $\langle A\mathbf{p}_j, \mathbf{r}_j \rangle = \langle A\mathbf{p}_j, \mathbf{p}_j \rangle$, $A\mathbf{p}_j = -\frac{1}{\alpha_j}(\mathbf{r}_{j+1} - \mathbf{r}_j)$, $\langle \mathbf{r}_{j+1}, A\mathbf{p}_j \rangle = -\frac{1}{\alpha_j}\langle \mathbf{r}_{j+1}, \mathbf{r}_{j+1} \rangle$, and $\langle A\mathbf{p}_j, \mathbf{p}_j \rangle = \frac{1}{\alpha_j}\langle \mathbf{r}_j, \mathbf{r}_j \rangle$, we obtain more convenient expressions for the coefficients α_j and β_j,

$$\alpha_j = \frac{\langle \mathbf{r}_j, \mathbf{r}_j \rangle}{\langle A\mathbf{p}_j, \mathbf{p}_j \rangle} \tag{11.52}$$

$$\beta_j = \frac{\langle \mathbf{r}_{j+1}, \mathbf{r}_{j+1} \rangle}{\langle \mathbf{r}_j, \mathbf{r}_j \rangle}. \tag{11.53}$$

Applying these expressions in an iterative sequence produces the following algorithm:

1: $\mathbf{r}_0 \leftarrow \mathbf{b} - A\mathbf{e}_0$ ▷ Initial residual
2: $\mathbf{p}_0 \leftarrow \mathbf{r}_0$ ▷ Initial search direction
3: **for** $j = 0, 1, \ldots$ until convergence **do**
4: $\alpha_j \leftarrow \frac{\langle \mathbf{r}_j, \mathbf{r}_j \rangle}{\langle A\mathbf{p}_j, \mathbf{p}_j \rangle}$ ▷ Enforce orthogonality of residuals
5: $\mathbf{e}_{j+1} \leftarrow \mathbf{e}_j + \alpha_j \mathbf{p}_j$ ▷ Next approximate solution
6: $\mathbf{r}_{j+1} \leftarrow \mathbf{r}_j - \alpha_j A\mathbf{p}_j$ ▷ Next residual
7: $\beta_j \leftarrow \frac{\langle \mathbf{r}_{j+1}, \mathbf{r}_{j+1} \rangle}{\langle \mathbf{r}_j, \mathbf{r}_j \rangle}$ ▷ Enforce A-conjugacy of search directions
8: $\mathbf{p}_{j+1} \leftarrow \mathbf{r}_{j+1} + \beta_j \mathbf{p}_j$ ▷ Next search direction
9: **end for**

Recalling that $\nabla J(\mathbf{e}) = A\mathbf{e} - \mathbf{b}$ (Eq. 11.34) where $\mathbf{b}$ depends on source terms and A does not, we can compute $\mathbf{r}_0 = \mathbf{b} - A\mathbf{e} = -\nabla J(\mathbf{e}_0)$ using the adjoint method as detailed in the previous section. In subsequent steps, $A\mathbf{p}_j = \nabla J(\mathbf{p}_j) - \mathbf{b}$ is computed using the adjoint method with source terms set to zero to eliminate $\mathbf{b}$. With these procedures, each step of the CG algorithm only requires us to compute two simulations over the harmonic time period, once forward and once backward.

11.3.5 Boundary Conditions

Discretization of Dirichlet boundary conditions is straightforward, applying the prescribed values of $\tilde{q}$ on boundary edges as $Q_i = \int_{e_i} \tilde{q}$. The absorbing boundary (11.5), on the other hand, requires some care. Specifically, pressure values P_i stored on dual vertices must be related to flux values Q_i stored on primal edges. We approximate the pressure as constant per triangle, yielding the discrete boundary condition

$$Q_i = -\frac{1}{c}|e_i|P_{e_i} \tag{11.54}$$

where $|e_i|$ denotes the length of the boundary edge with index i and P_{e_i} the pressure value stored on the adjacent dual vertex.

In the adjoint problem, values of Dirichlet boundaries are treated as source terms and therefore set to zero. The absorbing boundary (11.54), on the other hand, swaps a sign:

$$Q_i = \frac{1}{c}|e_i|P_{e_i}, \tag{11.55}$$

absorbing incoming energy instead of outgoing energy, which accounts for the time-reversed nature of the problem [8].

11.4 Experiments

We performed two types of numerical experiments to investigate the effectiveness of our methods. To test the accuracy of the harmonically optimized operators and to approximate the CFL condition for this problem, a simple plane wave with a known analytical solution was simulated. The controlled time integration method was tested by simulating sound-soft scattering obstacles with geometries of varying complexity.

The experiments were implemented in Python using DEC operators provided by the PyDEC library [1]. The source code of the experiments is available on GitHub [18].

11.4.1 Plane Wave

We consider a plane wave

$$\phi(\mathbf{x}, t) = -\cos(\omega t - \mathbf{k} \cdot \mathbf{x}) \tag{11.56}$$

with angular velocity $\omega = 2$ and wavevector $\mathbf{k} = (0, 2)$ in a square domain $\Omega = [0, \pi]^2$.

We apply the plane wave (11.56) on the boundary Γ as a Dirichlet condition on flux

$$\tilde{q} = \star(\nabla\phi)^\flat \quad \text{in } \Gamma \times [0, T] \tag{11.57}$$

with initial conditions

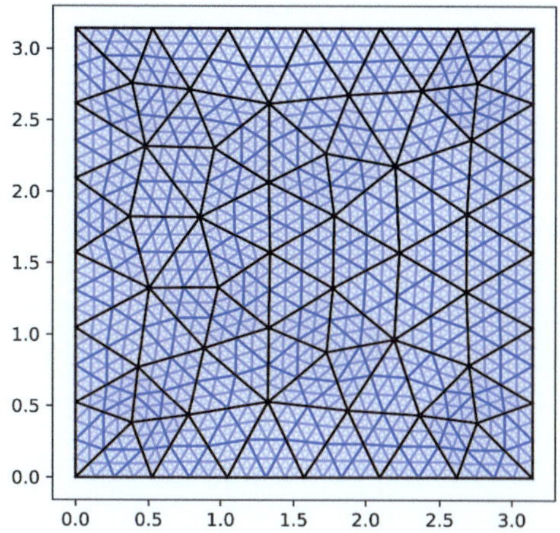

Fig. 11.3 Sequence of meshes used in the plane wave experiments

$$\tilde{p}_0 = \left.\frac{\partial \phi}{\partial t}\right|_{t=0} \tag{11.58}$$

$$\tilde{q}_0 = \left.\star(\nabla\phi)^\flat\right|_{t=0} \tag{11.59}$$

discretized according to Eq. (11.6). In the interior of the domain we simulate the wave according to Sects. 11.3.2 and 11.3.3.

The simulation is repeated on a sequence of progressively finer meshes, constructed from an initial unstructured mesh by cutting each edge in half and connecting the midpoints, as illustrated in Fig. 11.3. This refinement strategy produces simular triangles on each step, minimizing the influence of varying angles on the results.

The first result we're interested in is an approximation of a Courant-Friedrichs-Lewy (CFL) condition [3] for the problem, giving a stable range for the term $\frac{\Delta t}{h}$, where $h = \min |e_i|$ is the length of the shortest edge in the mesh. If this range is known, the timestep Δt can be automatically selected to maximize efficiency of the simulation. To obtain a numerical approximation for the maximum stable timestep Δt as a function of h, we simulated the plane wave (11.56) on each mesh while varying Δt. A binary search was performed to find the highest Δt where the simulation variables remain bounded. The results are plotted in Fig. 11.4.

The measurements suggest that $\Delta t = 0.3h$ may be a safe choice of timestep for our subsequent experiments. However, in practice this coefficient was found to vary significantly depending on specific mesh geometry, and a much lower value of $\Delta t = 0.1h$ was used in Sect. 11.4.2.

Next, we examine the accuracy of the harmonically optimized operators detailed in Sect. 11.3.3, relative to the Hodge star defined by Eq. (11.8) and the timestep coefficient Δt, which we refer to here as the "generic" operators. Error against the exact solution ϕ is measured after a time period of 2π by comparing the simulated values P and Q elementwise to the cochains P_* and Q_* obtained by integrating $\tilde{p} = \frac{\partial \phi}{\partial t}$ and $\tilde{q} = \star(\nabla\phi)^\flat$ over the corresponding mesh elements. The experiment is

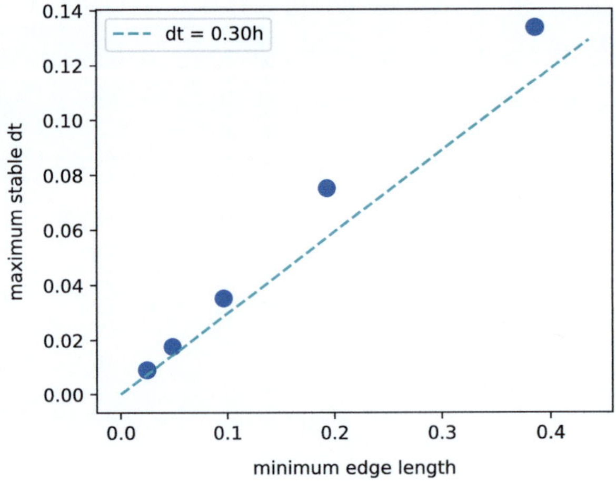

Fig. 11.4 Maximum stable value of Δt plotted against the shortest edge length

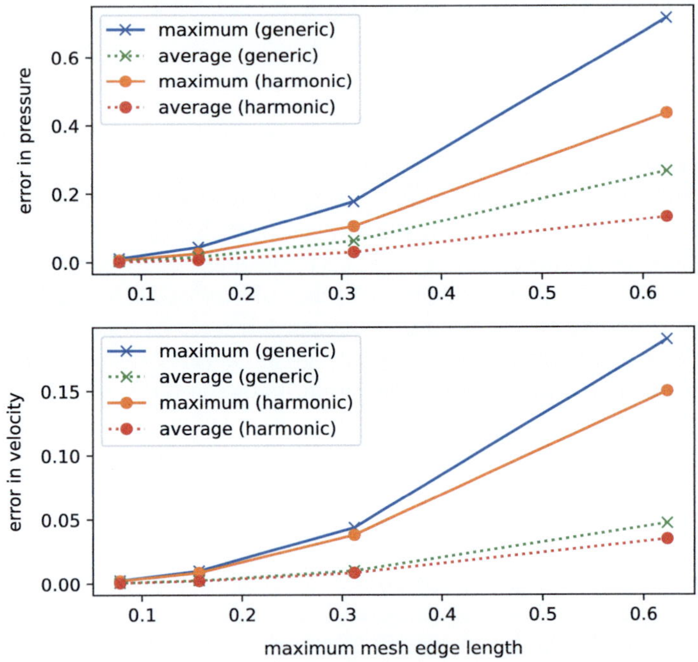

Fig. 11.5 Error of the simulated plane wave relative to maximum edge length of the mesh

then repeated with the harmonically optimized Hodge stars (11.25), (11.27) and the exact timestep Eqs. (11.18), and (11.19). Results are plotted in Fig. 11.5.

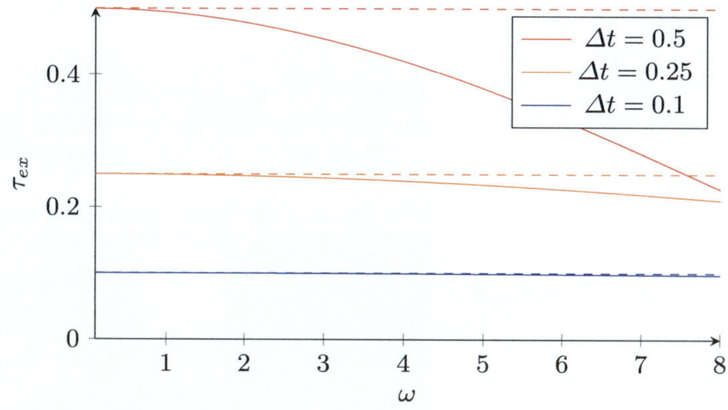

Fig. 11.6 Effect of angular velocity on the exact timestep coefficient τ_{ex}

We find a decrease in error of approximately 40% in pressure and 8–10% in velocity when the harmonic optimizations are applied. Since the only cost of these optimizations is a small amount of preprocessing, leaving the cost of the timestepping procedure unaffected, this is a notable advantage.

To examine the proportional effects of the exact timestep and the harmonic Hodge star, we note that Eqs. (11.12) and (11.18) only differ by a scalar coefficient. Δt in Eq. (11.12) is replaced by $\frac{2}{\omega}\sin\frac{\omega\Delta t}{2} =: \tau_{ex}$ in Eq. (11.18). Figure 11.6 plots τ_{ex} as a function of angular velocity ω with a few selected values of Δt.

The difference between Δt and τ_{ex} only becomes significant at high values of both ω and Δt. In our experiments, Δt was limited by stability demands to values less than 0.1, where the difference is negligible. We can thus conclude that the measured difference in these experiments is predominantly due to the optimized Hodge star operators. However, when stability permits large values of Δt, the exact timestep may provide an advantage for an insignificant computational cost.

11.4.2 Scatterer

In this experiment we investigate the controlled time integration method (Sect. 11.3.4) in the computation of time-harmonic scattering patterns. We decompose the total wave solution into an incident and scattered component, $\phi = \phi_{inc} + \phi_{sca}$, where ϕ_{inc} is a plane wave $\phi_{inc} = -\cos(\omega t - \mathbf{k}\cdot\mathbf{x})$. We apply ϕ_{inc} on the scattering obstacle's surface Γ_{sca} as a Dirichlet boundary condition and solve for the scattered wave ϕ_{sca}. On the domain's outer boundary Γ_{ext} an absorbing boundary condition is applied according to Sect. 11.3.5.

To obtain smooth initial conditions satisfying both boundary conditions, we apply an easing procedure proposed by [17]. We begin the simulation with pressure, flux,

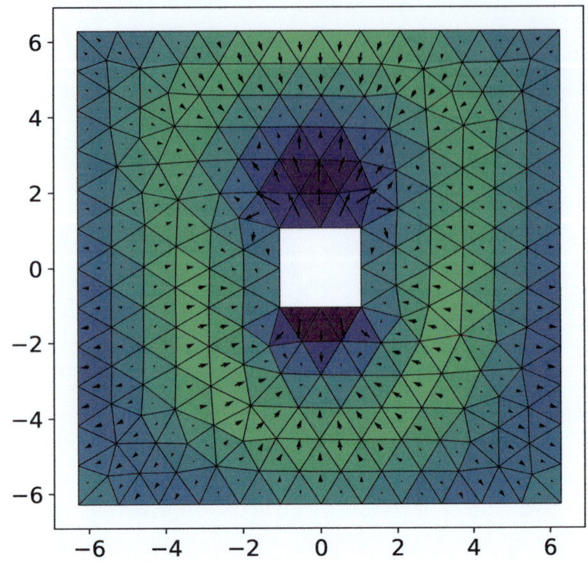

Fig. 11.7 Scatterer geometry, case 1: square

and source terms all set to zero. Source terms are then eased in over a time period t_{tr} by multiplying them by

$$f_{tr}(t) = \left(2 - \sin\left(\frac{\pi}{2}\frac{t}{t_{tr}}\right)\right)\sin\left(\frac{\pi}{2}\frac{t}{t_{tr}}\right). \tag{11.60}$$

The state at the end of the easing time period t_{tr} is taken as the initial guess $\mathbf{e}_0$ of the CG algorithm.

We measure the rate of convergence by computing the control energy (11.29) after each iteration of the CG algorithm. This is compared against the asymptotic iteration method, where an "iteration" consists of simulating forward in time for one harmonic time period T. We repeat the experiment on four different meshes of increasing geometric complexity, using both the generic DEC operators and our harmonically optimized ones. The resulting scattered waves ϕ_{sca} and convergence rates are visualized in Figs. 11.7, 11.8, 11.9, 11.10, 11.11, 11.12, 11.13 and 11.14, where the twelve-pointed star case (Fig. 11.11) has wavenumber $\kappa = 2$, while $\kappa = 1$ in the other cases.

Pressure is indicated by the colors of triangles, and arrows denote velocity obtained from neighboring edge fluxes via Whitney interpolation.

In the square case we find that the asymptotic iteration method outperforms the control method, and in the eight-pointed star case neither has a strong advantage. In these cases there are few obstacles preventing the scattered wave from reaching the absorbing boundary, which results in fast convergence for the asymptotic iteration method. The other two cases feature more consequentially nonconvex geometry,

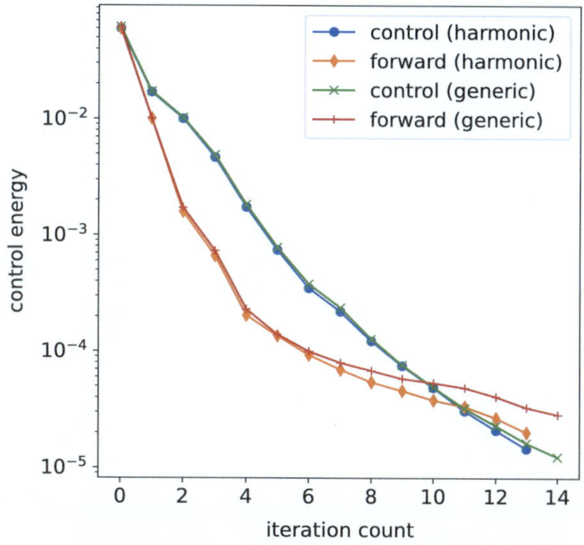

Fig. 11.8 Energy convergence of the square scatterer

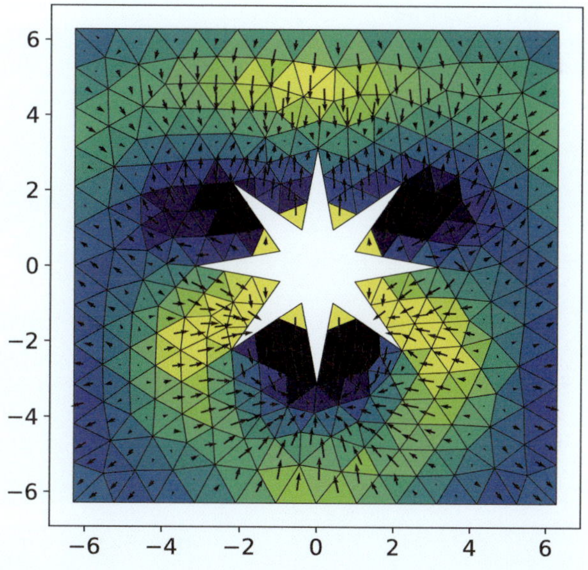

Fig. 11.9 Scatterer geometry, case 2: eight-pointed star

trapping waves and causing difficulties for asymptotic iteration. In these cases, the control method still gives reliable results.

The harmonic operators are not found to significantly impact the rate of convergence. We note, however, that the control energy is also computed using these operators, meaning the solution obtained is different. Referring to the results of our plane wave experiment (Sect. 11.4.1), we therefore expect that the solution obtained using the harmonic operators is more accurate.

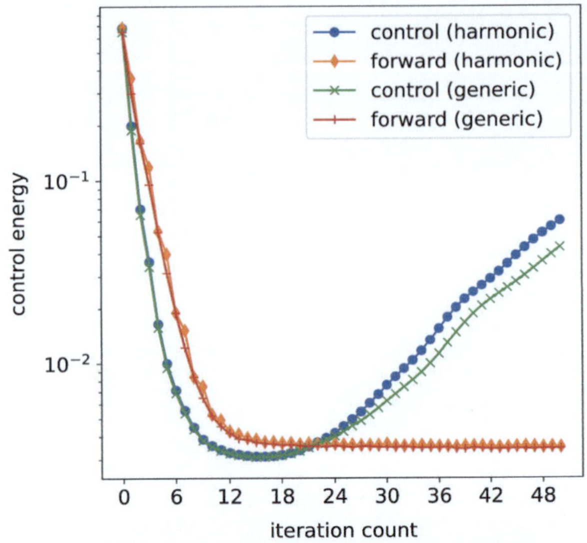

Fig. 11.10 Energy convergence of the eight-pointed star scatterer

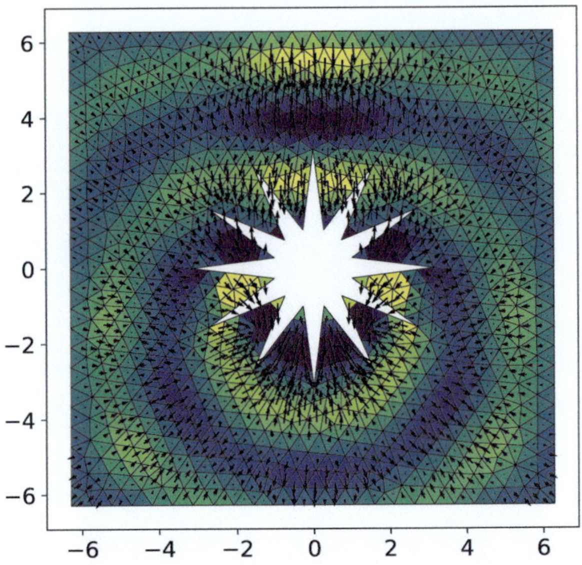

Fig. 11.11 Scatterer geometry, case 3: twelve-pointed star

We also note that in the eight-pointed star case, the control method fails to converge to the desired level and energy begins to increase. Similar failures were observed in other test cases, particularly in ones featuring extremely short edges. We believe this is due to suboptimal quality of the computation mesh, which produces a high amount of error in the adjoint solution when the gradient is close to zero, causing the CG algorithm's invariants of orthogonal residuals and A-conjugate search directions to

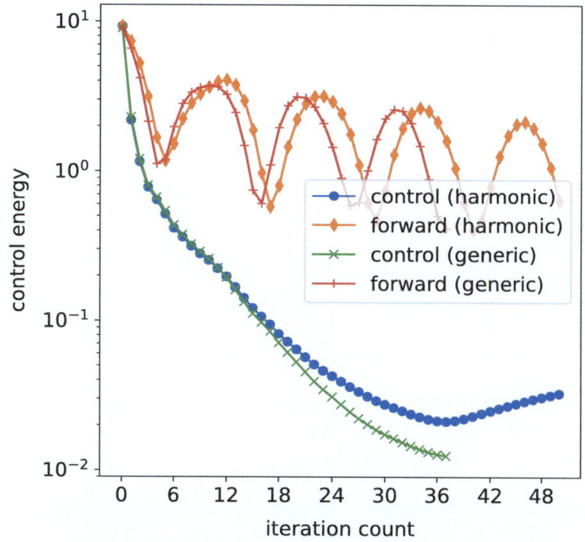

Fig. 11.12 Energy convergence of the twelve-pointed star scatterer

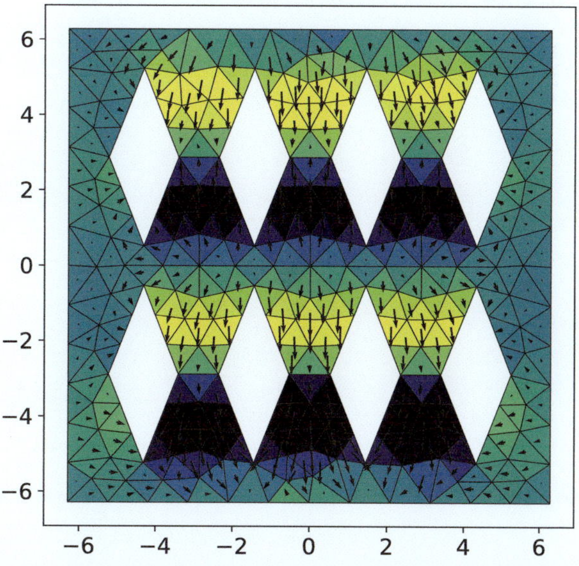

Fig. 11.13 Scatterer geometry, case 4: diamonds

be violated. Measuring and improving the quality of the mesh is thus an important research direction to increase the reliability of convergence.

For a cursory investigation into the method's computational efficiency, we measured computation time per iteration of the CG algorithm as mesh size increases. Meshes were expanded in space while maximum triangle size was kept constant, so that the number of timesteps required for a CG iteration does not vary significantly. Results are plotted for each of our scatterer shapes in Fig. 11.15, with mesh size

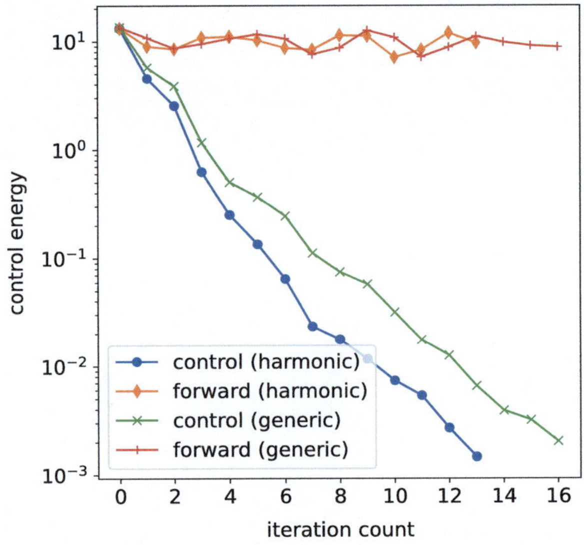

Fig. 11.14 Energy convergence of the diamond scatterer

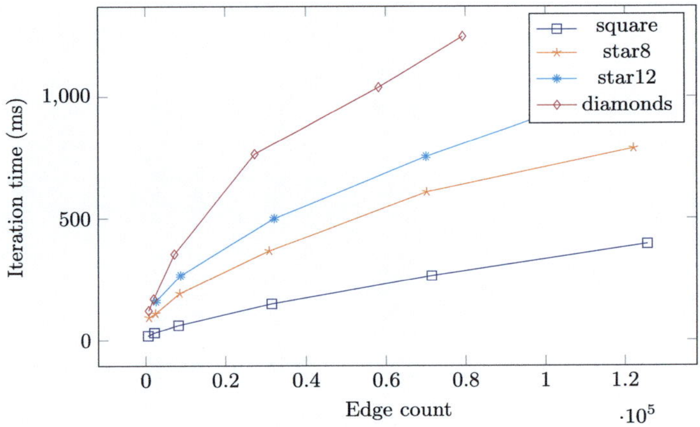

Fig. 11.15 Computation times relative to number of edges in the mesh

represented by the number of edges in the mesh and time per CG iteration in units of milliseconds. The measurements were performed on a consumer PC with an AMD Ryzen 7 5800X CPU.

The results display sublinear scaling with respect to edge count. The relative difference between scatterer shapes correlates with the number of boundary edges in the mesh, likely explained by our implementation of boundary conditions as Python loops, whereas timesteps within the domain are matrix operations accelerated by the `numpy` library's native C routines. This is also the likely cause of sublinear scaling, as the fraction of edges on the boundary drops with mesh area. If all edges were treated identically, one would expect linear scaling, since each edge adds one row with predictable sparsity to the timestep matrices.

11.5 Conclusion

We've presented a numerical method for time-harmonic acoustic wave problems based on discrete exterior calculus (DEC) and a controlled optimization process in the time domain. An alternative approximation of the Hodge star and an exact timestepping scheme accounting for the harmonic nature of the problem were also investigated. Numerical experiments were performed, testing the accuracy of the harmonic operators against a known analytical solution and the convergence of the control method in a scattering problem.

The accuracy benefits of the harmonic operators in our tests simulating a plane wave were significant, dominated in our cases by the Hodge star. The exact timestep's effect is maximized at high wavenumbers and long timesteps, the ratio of which is constrained by the CFL condition. Neither of these operators changes the matrix sparsity pattern relative to their generic counterparts, and therefore they only incur a small preprocessing cost, making them an attractive choice for harmonic problems.

In scattering problems with predominantly convex geometry, where waves can exit the domain unimpeded, the control method was outperformed by a simple asymptotic iteration. However, with complicated nonconvex geometry the asymptotic iteration method is not applicable, whereas the control method still converges to a solution. It is thus useful in a much wider range of problems.

The control method was found to be more sensitive to mesh quality than the asymptotic iteration, owing to the fact that state variables in the control method are very close to zero when approaching a solution, and the relative effect of numerical errors becomes large. This can lead to a failure to converge to the desired level of accuracy. It is thus an important direction for future research to measure and maximize the quality of the mesh. Some relevant metrics are the minimum and maximum values of the Hodge star matrices and well-centeredness of mesh elements. The regular crystal meshes of [24] and the Hodge-optimized triangulations of [16] are promising ideas towards higher-quality meshes.

We have presented only two-dimensional test cases, but one of the benefits of DEC is straightforward generalization into higher dimensions. We nonetheless leave 3D experiments to future work. Another interesting future direction is the higher-order approximations recently developed by [12].

Acknowledgements The author wishes to thank Sanna Mönkölä, Tuomo Rossi, and Tytti Saksa for their longstanding guidance and encouragement. This work has been supported by the Finnish Ministry of Education and Culture's Pilot for Doctoral Programmes (Pilot project Mathematics of Sensing, Imaging and Modelling).

References

1. Bell N, Hirani AN (2012) PyDEC: software and algorithms for discretization of exterior calculus. ACM Trans Math Softw 39(1)(November):3:1–3:41. https://doi.org/10.1145/2382585.2382588
2. Bristeau MO, Glowinski R, Periaux J (1993) Numerical simulation of high frequency scattering waves using exact controllability methods. In: Donato A, Oliveri F (eds) Nonlinear hyperbolic problems: theoretical, applied, and computational aspects: proceedings of the fourth international conference on hyperbolic problems, Taormina, Italy, April 3 to 8, 1992. Vieweg+Teubner Verlag, Wiesbaden, pp 86–108. https://doi.org/10.1007/978-3-322-87871-712
3. Courant R, Friedrichs K, Lewy H (1967) On the partial difference equations of mathematical physics. IBM J Res Dev 11(2)(March):215–234. https://doi.org/10.1147/rd.112.0215
4. Desbrun M, Hirani AN, Leok M, Marsden JE (2005) Discrete exterior calculus. https://doi.org/10.48550/arXiv.math/0508341
5. Desbrun M, Kanso E, Tong Y (2006) Discrete differential forms for computational modeling. In: Discrete differential geometry
6. Engquist B, Majda A (1977) Absorbing boundary conditions for the numerical simulation of waves. Math Comput 31(139):629–651. https://doi.org/10.1090/S0025-5718-1977-0436612-4
7. Geuzaine C, Remacle J-F (2009) Gmsh: a three-dimensional finite element mesh generator with built-in pre- and post-processing facilities. Int J Numer Methods Eng 79:1309–1331
8. Givoli D (2021) A tutorial on the adjoint method for inverse problems. Comput Methods Appl Mech Eng 380(July):113810. https://doi.org/10.1016/j.cma.2021.113810
9. Glowinski R, Rossi T (2006) A mixed formulation and exact controllability approach for the computation of the periodic solutions of the scalar wave equation. (I): controllability problem formulation and related iterative solution. Comptes Rendus Math 343(7):493–498. https://doi.org/10.1016/j.crma.2006.08.002
10. Kivioja M, Zamora-Zamora R, Blinova A, Mönkölä S, Rossi T, Möttönen M (2023) Evolution and decay of an Alice ring in a spinor Bose-Einstein condensate. Phys Rev Res 5(2)(May):023104. American Physical Society. https://doi.org/10.1103/PhysRevResearch.5.023104
11. Lions JL (1988) Exact controllability, stabilization and perturbations for distributed systems. SIAM Rev 30(1)(March):1–68. https://doi.org/10.1137/1030001
12. Lohi J (2022) Systematic implementation of higher order Whitney forms in methods based on discrete exterior calculus. Numer Algorithms 91(3)(November):1261–1285. https://doi.org/10.1007/s11075-022-01301-2
13. Ma C, Chen Z (2005) Stability and numerical dispersion analysis of CE-FDTD method. IEEE Trans Antennas Propag 53(1)(January):332–338. https://doi.org/10.1109/TAP.2004.838760
14. Mohamed MS, Hirani AN, Samtaney R (2016) Discrete exterior calculus discretization of incompressible Navier-Stokes equations over surface simplicial meshes. J Comput Phys 312(May):175–191. https://doi.org/10.1016/j.jcp.2016.02.028
15. Mönkölä S, Räty J (2022) Discrete exterior calculus for photonic crystal waveguides. Numer Meth Eng (November):7144. https://doi.org/10.1002/nme.7144
16. Mullen P, Memari P, de Goes F, Desbrun M (2011) HOT: Hodge-optimized triangulations. In: ACM SIGGRAPH 2011 papers, 1–12. SIGGRAPH '11. Association for Computing Machinery. https://doi.org/10.1145/1964921.1964998
17. Mur G (1993) The finite-element modeling of three-dimensional electromagnetic fields using edge and nodal elements. IEEE Trans Antennas Propag 41(7)(July):948–953. https://doi.org/10.1109/8.237627
18. Myyrä M (2025) Wavepool. https://github.com/m0lentum/wavepool/tree/f4d69e9c3132a483cbe4d8907a56dbc127828193
19. Nitschke I, Reuther S, Voigt A (2017) Discrete exterior calculus (DEC) for the surface Navier-Stokes equation, pp 177–197. https://doi.org/10.1007/978-3-319-56602-37

20. Peng Z, Appelö D (2022) EM-WaveHoltz: a flexible frequency domain method built from time-domain solvers. IEEE Trans Antennas Propag 70(7)(July):5659–5671. https://doi.org/10.1109/TAP.2022.3161448
21. Räbinä J (2014) On a numerical solution of the Maxwell equations by discrete exterior calculus. Doctoral dissertation, University of Jyväskylä. http://urn.fi/URN:ISBN:978-951-39-5951-7
22. Räbinä J, Kettunen L, Mönkälä S, Rossi T (2018) Generalized wave propagation problems and discrete exterior calculus. ESAIM: M2AN 52(3)(May):1195–1218. https://doi.org/10.1051/m2an/2018017
23. Räbinä J, Kuopanportti P, Kivioja MI, Möttönen M, Rossi T (2018) Three-dimensional splitting dynamics of giant vortices in Bose-Einstein condensates. Phys Rev A 98(2)(August):023624. https://doi.org/10.1103/PhysRevA.98.023624
24. Räbinä J, Mönkölä S, Rossi T (2015) Efficient time integration of maxwell's equations with generalized finite differences. SIAM J Sci Comput 37(6)(January):B834–B854. ISSN: 1064-8275. https://doi.org/10.1137/140988759
25. Saad Y (2003) Iterative methods for sparse linear systems. Other titles in Applied mathematics. Society for Industrial/Applied Mathematics. https://doi.org/10.1137/1.9780898718003
26. Yee K (1966) Numerical solution of initial boundary value problems involving Maxwell's equations in isotropic media. IEEE Trans Antennas Propag 14(3)(May):302–307. https://doi.org/10.1109/TAP.1966.1138693

Chapter 12
Convergence Analysis of the Explicit Nonstandard Runge-Kutta Methods

Istvan Faragó and Rahele Mosleh

Abstract This paper explores the properties of explicit nonstandard Runge–Kutta (ENRK) methods for solving dynamical systems. It focuses on their application to higher-order methods that use distinct step-size functions for the main and internal stages. Standard explicit Runge–Kutta (SRK) methods often suffer from sensitivity to the discretization step size, Δt, which can lead to a loss of dynamical consistency, particularly in long-term simulations where stability and dynamical consistency are crucial. To overcome these challenges, we apply nonstandard finite difference (NSFD) techniques to improve numerical consistency. The study investigates the impact of different step-size functions on the preservation of essential numerical properties, such as the method's order, consistency, and convergence. Specifically, we identify key properties of the step-size functions that ensure the ENRK method's order is comparable to that of the corresponding SRK methods. Based on the findings, under certain conditions on the step-size functions, the order of consistency and convergence of ENRK methods matches that of their corresponding SRK methods.

Keywords Explicit standard Runge–Kutta methods · Explicit nonstandard Runge–Kutta methods · Elementary stability · Step-size functions

---

I. Faragó (✉)
Department of Differential Equations, Budapest University of Technology and Economics, Budapest, Hungary
e-mail: faragois@cs.elte.hu

Department of Applied Analysis and Computational Mathematics, Eötvös Loránd University, Budapest, Hungary

Large Networks Research Group, ELKH, Budapest, Hungary

R. Mosleh
Laboratory of Mathematics for Public Health (MfPH), Department of Mathematics and Statistics, York University, Toronto, ON, Canada
e-mail: rmosleh@yorku.ca

T. Tuovinen et al. (eds.), *Challenges in Design Methods, Numerical Tools and Technologies for Sustainable Aviation, Transport and Industry*, Computational Methods in Applied Sciences 17, https://doi.org/10.1007/978-3-031-98675-8_12

12.1 Introduction and Motivation

In numerical analysis, developing effective methods for solving differential equations is crucial for various scientific and engineering applications. While traditional finite difference methods (FDMs) are widely used, nonstandard finite difference (NSFD) methods offer a more robust alternative for preserving the qualitative properties of dynamical systems. Numerical methods are particularly valuable for solving nonlinear dynamical systems, where analytical solutions are often impractical or impossible to obtain. Besides ensuring convergence, numerical solutions should maintain dynamical consistency, meaning they preserve essential properties such as stability, positivity, and boundedness. Standard FDMs achieve this only under specific conditions, often requiring adjustments that introduce computational or theoretical challenges. To address these limitations, NSFD methods were developed to systematically maintain the qualitative behavior of differential equations. Various NSFD approaches have been introduced, including those by Mickens [11], Erdogan and Ozis [8], Anguelov and Lubuma [2], Dimitrov and Kojouharov [4], and others [8]. Among these, Mickens' method stands out as a pioneering framework due to its versatility and practical advantages. Unlike traditional schemes, Mickens' method employs nonstandard discretization rules, ensuring better stability when solving nonlinear and complex differential systems. This approach has been particularly effective in applications such as mathematical modeling in biology, epidemiology [11–13], and physics [10]. For numerical solutions, explicit standard Runge–Kutta (ESRK) methods are widely used. However, their dynamical consistency depends heavily on the choice of time step Δt [6, 7]. While decreasing Δt improves stability, it also increases computational cost, especially for long-term simulations. Implicit methods offer better stability but require solving nonlinear equations at each time step, making them computationally expensive. To overcome these challenges, explicit nonstandard Runge–Kutta (ENRK) methods introduce carefully designed step-size functions that replace the conventional time step Δt. These modifications help preserve key properties of the original continuous system while maintaining computational efficiency. As a result, ENRK methods provide a more reliable framework for long-term simulations compared to their standard counterparts. The primary goal of NSFD methods is to enhance the stability and dynamical consistency of classical FDMs, particularly for nonlinear and complex systems. By incorporating nonstandard differential operators in time and space discretization, NSFD methods align more closely with the physical characteristics of the underlying models. In this paper, we consider an autonomous system of ordinary differential equations (ODEs)

$$\begin{cases} \dot{U} = f(U), & t \in [0, T) \\ U(0) = U_0, \end{cases} \tag{12.1}$$

where $U = (u_1, u_2, ..., u_N)^\top$ and $f : \mathbb{R}^N \to \mathbb{R}^N$ is a given sufficiently smooth function. We define $t_n = t_0 + n\Delta t$ over the time interval $[0, t^*]$, where $t^* \in [0, T)$, Δt

is a positive step size defined with $\Delta t = \frac{t^*}{N}$ for n=1, …, N and the numerical solution $Y^{(n)} = (y_1^{(n)}, y_2^{(n)}, \ldots, y_N^{(n)})$ serves an approximation of the analytical solution $U(t_n)$ for system (12.1) at time t_n. Then descritized version of system (12.1), using the classical FDMs, is given by

$$D_{\Delta t} Y^{(n)} = F_n(t_n, f, Y^{(n)}) \tag{12.2}$$

where $D_{\Delta t} Y^{(n)}$ represents the discretized version of $\dot{U}$ at time t_n and $F_n(t_n, f, Y^{(n)})$ approximates $f(t_n, U(t_n))$ at time t_n. NSFD methods can be considered a generalized form of classical FDMs. In the NSFD method, the approximation of $\dot{u}_i(t_n)$ is more general than the standard finite different schemes and takes the form

$$\dot{u}_i(t_n) \approx \frac{y_i^{(n)} - y_i^{(n-1)}}{\Phi(\Delta t)} \tag{12.3}$$

where $\Phi(\Delta t)$ is a given function with the following properties [6]

1. It is defined for non-negative numbers,
2. It is a sufficiently smooth function,
3. It is positive for all positive arguments.

Beyond step-size functions, NSFD methods employ nonlocal approximations, where nonlinear terms are often represented in nonlocal forms. This approach ensures that values at future time steps are influenced by both current and previous steps, allowing for a simplified representation of the dynamical system. Unlike standard numerical methods, which prioritize accuracy and convergence rates, NSFD methods are designed to maintain the fundamental structure and qualitative behavior of the underlying dynamical system, making them particularly valuable in solving complex differential equations. NSFD methods offer significant advantages in real-world applications where standard finite difference methods often struggle to accurately capture complex system dynamics. A key area of application is biological modeling, particularly in population dynamics and epidemiological models [14], where maintaining positivity and boundedness is crucial for obtaining meaningful results. Traditional numerical methods can introduce artificial oscillations or yield negative values that are biologically implausible, while NSFD techniques ensure realistic solutions by preserving these essential properties. Beyond biological modeling, NSFD methods are also widely used in fluid dynamics, heat transfer modeling, and various engineering fields, where stability and long-term accuracy are vital [11, 13]. These schemes effectively prevent numerical instabilities that could otherwise compromise the reliability of simulations. In this framework, we mainly focus on step-size functions and investigate the consistency and convergence of the explicit nonstandard Runge-Kutta (ENRK) method when the step-size function differs between the main stage and the internal stages, following classical numerical theory. As an extension of previous works, Anguelov and Lubuma [2] and Dimitrov and Kojouharov [4, 5] studied ENRK methods with identical step-size functions for both the main and internal stages. In this paper, Sect. 12.2 investigates the Lipschitz property of ENRK methods, emphasizing the impact of step-size functions on their performance. Section 12.3 examines the consistency and order of ENRK methods, with support from step-size functions. Section 12.4 analyzes the convergence order of ENRK methods, providing concrete examples with suitably chosen step-size functions. Finally, Sect. 12.5 summarizes the findings and explores potential directions for future research.

12.2 Lipschitz Property in the NSFD Methods Combined with Explicit RK Methods

In this section, for the Cauchy problem (12.1) we introduce NSFD methods combined with explicit RK methods (ENRK methods) and analyze the inheritance of the Lipschitz property in (12.1) to the discretization.

We assume that the function $f : \mathbb{R}^N \to \mathbb{R}^N$ is satisfying the global Lipschitz condition

$$|| f(V_1) - f(V_2) || \leq L || V_1 - V_2 || \tag{12.4}$$

for any $V_1, V_2 \in \mathcal{D}(f) \subset \mathbb{R}^N$ with some constant $L > 0$. Under this assumption, the Cauchy problem (12.1) has a unique solution.

In what follows, $Y^{(n)}$ denotes the approximation to the exact solution $U(t)$ at time $t = t_n$ on the fixed equidistant mesh

$$\Omega_{\Delta t} = \{t_n = n\Delta t, n = 0, 1, 2, ..., N\} \tag{12.5}$$

with the step-size $\Delta t > 0$. The basic question is how we define the approximation $Y^{(n)}$? In the case of the classical explicit Runge–Kutta method the approximations are defined by the formulas [1, 9]

$$Y^{(n)} = Y^{(n-1)} + \Delta t \sum_{i=1}^{s} \sigma_i K_i(Y^{(n-1)}), \tag{12.6}$$

where

$$K_1(Y^{(n-1)}) = f(Y^{(n-1)}), \tag{12.7}$$

$$K_i(Y^{(n-1)}) = f(Y^{(n-1)} + \Delta t \sum_{j=1}^{i-1} a_{ij} K_j(Y^{(n-1)})), \quad i = 2, 3, \ldots, s. \tag{12.8}$$

with $Y^{(0)} = U_0$. However, for the ENRK methods we use the formulas

$$Y^{(n)} = Y^{(n-1)} + \Phi(\Delta t) \sum_{i=1}^{s} \sigma_i K_i(Y^{(n-1)}), \tag{12.9}$$

with the newly introduced functions $\Phi(\Delta t)$ and $\Phi(\Delta t)$ where

$$K_1(Y^{(n-1)}) = f(Y^{(n-1)}),$$

$$K_i(Y^{(n-1)}) = f(Y^{(n-1)} + \Phi(\Delta t) \sum_{j=1}^{i-1} a_{ij} K_j(Y^{(n-1)}), \quad i = 2, , 3, \ldots, s \tag{12.10}$$

the step-size functions $\Phi(\Delta t)$ and $\Phi(\Delta t)$ are positive. Obviously, the ENRK methods are defined by the choice of the functions $\Phi(\Delta t)$ and $\Phi(\Delta t)$ that possess certain properties [6]. When both are the identity functions then ENRK yields the classical explicit Runge–Kutta method (12.6) and (12.8). This method can be expressed as a one-step method

$$Y^{(n)} = Y^{(n-1)} + \Phi(\Delta t) G(Y^{(n-1)}), \tag{12.11}$$

where the increment function $G : \mathbb{R}^N \to \mathbb{R}^N$ is defined as

$$G(V) = \sum_{i=1}^{s} \sigma_i K_i(V). \tag{12.12}$$

In the following, we examine the Lipschitz property of G.

First, we demonstrate that the Lipschitz property of the function f is inherited by the functions K_i.

Lemma 1 *Under the condition Lipschitz-condition* (12.4) *for any* $i = 1, 2, \ldots s$ *the estimation*

$$|| \, K_i(V_1) - K_i(V_2) \, || \leq L_i \, || \, V_1 - V_2 \, || \tag{12.13}$$

holds for any $V_1, V_2 \in \mathcal{D}(f) \subset \mathbb{R}^N$ *with*

$$L_1 = L; \tag{12.14}$$

$$L_i = L\big[1 + \Phi(\Delta t) A \sum_{j=1}^{i-1} L_j\big] \quad i = 2, 3, \ldots, s, \tag{12.15}$$

where the notation

$$A = \max_{i,j} |a_{ij}| \tag{12.16}$$

denotes the maximum absolute value of the coefficients a_{ij}.

Proof The statement (12.14) is straightforward. The statement (12.15) will be proven by induction. For $i = 2$, using (12.4) and (12.8), we have

$$\begin{aligned}\| K_2(V_1) - K_2(V_2) \| =& \| f(V_1 + \Phi(\Delta t)a_{21}K_1(V_1)) - f(V_2 + \Phi(\Delta t)a_{21}K_1(V_2)) \| \le \\ &\le L \| V_1 - V_2 + \Phi(\Delta t)a_{21}(K_1(V_1) - K_1(V_2)) \| \le \\ &\le L\, [\| V_1 - V_2 \| + \Phi(\Delta t)|a_{21}| \, \| K_1(V_1) - K_1(V_2) \|] \le \\ &\le L \| V_1 - V_2 \| \, [1 + \Phi(\Delta t)AL_1] . \end{aligned} \tag{12.17}$$

Hence,

$$L_2 = L\,[1 + \Phi(\Delta t)AL_1] ,$$

which confirms the validity of (12.15) for $i = 2$. Next, we use the induction, i.e., if (12.15) valid for $j = 1, 2, \ldots i$ then it is also valid for $j = i + 1$. Using (12.4) and (12.8), we have

$$\begin{aligned}\| K_{i+1}(V_1) - K_{i+1}(V_2) \| =& \| f(V_1 + \Phi(\Delta t)\sum_{j=1}^{i} a_{ij}K_j(V_1)) - f(V_2 + \Phi(\Delta t)\sum_{j=1}^{i} a_{ij}K_j(V_2)) \| \le \\ &\le L\left[\| V_1 - V_2 \| + \Phi(\Delta t)\sum_{j=1}^{i} |a_{ij}| \, \| K_j(V_1) - K_j(V_2) \| \right] \le \\ &\le L \| V_1 - V_2 \| \left[1 + \Phi(\Delta t)A\sum_{j=1}^{i} L_j \right] . \end{aligned} \tag{12.18}$$

This confirms the validity of (12.15) for any $i = 1, 2, \ldots s$.

In the next step, for $i = 2, 3, \ldots s$ we define L_i as a function of the Lipschitz constant L of the function f.

Lemma 2 *Under the condition* (12.4) *for any* $i = 1, 2, \ldots s$ *the relation*

$$L_i = (1 + LA\Phi(\Delta t))^{i-1}L \tag{12.19}$$

holds.

Proof The statement for $i = 1$ is trivial. For $i \ge 2$, based on (12.15) we have

$$\begin{aligned} L_i =& L\big[1 + \Phi(\Delta t)A\sum_{j=1}^{i-1} L_j\big] = L\big[1 + \Phi(\Delta t)A\sum_{j=1}^{i-2} L_j\big] + LA\Phi(\Delta t)L_{i-1} = \\ =& L_{i-1} + LA\Phi(\Delta t)L_{i-1} = (1 + LA\Phi(\Delta t))L_{i-1}. \end{aligned} \tag{12.20}$$

Therefore

$$L_{i-1} = (1 + LA\Phi(\Delta t))L_{i-2}, \tag{12.21}$$

and hence

$$L_i = (1 + LA\Phi(\Delta t))^2 L_{i-2} = \cdots = (1 + LA\Phi(\Delta t))^{i-1}L_1, \tag{12.22}$$

which proves the statement.

Now we can prove the following statement.

Theorem 1 *Under the conditions*

$$\sup_{[0,T)} \Phi(\tau) = C < \infty. \tag{12.23}$$

, the increment function G has the Lipschitz property

$$\| G(V_1) - G(V_2) \| \leq L^\star \| V_1 - V_2 \| \tag{12.24}$$

with

$$L^\star = L^\star(\Delta t) = \sigma L s \exp(\Phi(\Delta t) L A s) \tag{12.25}$$

and

$$\sigma = \max_i | \sigma_i |, \quad i = 1, 2, \ldots s. \tag{12.26}$$

Proof Based on (12.12) we have

$$\| G(V_1) - G(V_2) \| = \| \sum_{i=1}^{s} \sigma_i \left[K_i(V_1) - K_i(V_2) \right] \| \leq \sum_{i=1}^{s} |\sigma_i| \, \| (K_i(V_1) - K_i(V_2)) \| . \tag{12.27}$$

Hence, applying the notation (12.26) and Lemma 1, we obtain

$$\| G(V_1) - G(V_2) \| \leq \sigma \sum_{i=1}^{s} \| (K_i(V_1) - K_i(V_2)) \| \leq \sigma \| V_1 - V_2 \| \sum_{i=1}^{s} L_i. \tag{12.28}$$

Now, using the well-known inequality $1 + \tau \leq \exp(\tau)$, holds for any $\tau \geq 0$, and Lemma 2 we have

$$\sum_{i=1}^{s} L_i = \sum_{i=1}^{s} (1 + L A \Phi(\Delta t))^{i-1} L \leq L s (1 + L A \Phi(\Delta t))^s \leq L s \exp(s L A \Phi(\Delta t)). \tag{12.29}$$

Combination of (12.28) and (12.29) proves the statement.

Remark 1 As we have seen, the Lipschitz constant $L^\star$ depends on the discretization step-size Δt. However, taking into account the assumption (12.23) we obtain a uniform estimation

$$L^\star \leq \sigma L s \exp(C L A s). \tag{12.30}$$

Moreover, when

$$\lim_{\Delta t \to 0} \Phi(\tau) = 0 \tag{12.31}$$

then

$$\lim_{\Delta t \to 0} L^\star(\Delta t) = \sigma s L. \tag{12.32}$$

12.3 Consistency and Order Analysis of NSFD Methods Combined with Explicit RK Methods

The local truncation error (LTE) of some method is the error incurred in obtaining the numerical solution at the new mesh-point $t = t_n$ when input data in the formulas at $t = t_{n-1}$ is known exactly. (It is local in the sense that it does not include the propagation errors, i.e. the errors created at previous steps.) When its order is $\mathcal{O}(\Delta t^{p+1})$ then the method is of order p. This means that for the ENRK the LTE is defined as

$$LTE(\Delta t) = U(t_n) - \tilde{Y}^{(n)}, \tag{12.33}$$

where $t_n \in \Omega_{\Delta t}$ is an arbitrary fixed mesh-point, and $\tilde{Y}^{(n)}$ denotes the numerical solution obtained by

$$\tilde{Y}^{(n)} = U^{(n-1)} + \Phi(\Delta t) \sum_{i=1}^{s} \sigma_i K_i(U^{(n-1)}), \tag{12.34}$$

where

$$K_1(U^{(n-1)}) = f(U^{(n-1)}),$$
$$K_i(U^{(n-1)}) = f(U^{(n-1)} + \Phi(\Delta t) \sum_{j=1}^{i-1} a_{ij} K_j(U^{(n-1)}), \quad i = 2, , 3, \ldots, s. \tag{12.35}$$

Remark 2 Based on (12.33) and (12.34) we have

$$U(t_n) - \left(U^{(n-1)} + \Phi(\Delta t) \sum_{i=1}^{s} \sigma_i K_i(U^{(n-1)}) \right) = LTE(\Delta t). \tag{12.36}$$

At the same time, for the NSFD method the numerical solution satisfies the relation

$$Y^{(n)} - \left(Y^{(n-1)} + \Phi(\Delta t) \sum_{i=1}^{s} \sigma_i K_i(Y^{(n-1)}) \right) = 0. \tag{12.37}$$

Therefore, LTE shows how exactly the exact analytical solution satisfies the formula of the numerical method at the grid points.

The following theorem analyzes the order of the ENRK method based on the choice of input functions Φ and Φ that define the method.

Theorem 2 *Assume that the basic ERK method has the accuracy order p, and the functions $\Phi(\Delta t)$ and $\Phi(\Delta t)$ have the property*

$$\Phi(\Delta t) = \Delta t + O(\Delta t^{p+1}) \quad \textit{and} \quad \Phi(\Delta t) = \Delta t + O(\Delta t^{p}). \tag{12.38}$$

Then the local truncation error of NSFDM (12.9) *and* (12.10) *is* $O(\Delta t^{p+1})$, *i.e., the order of the method is* p.

Proof To prove the above theorem, we examine the order of the ENRK method for stage numbers $s = 1$ and $s = 2$, and then generalize the results to higher orders.

12.3.1 Case s = 1

According to (12.9) and (12.10), in this case the numerical solution $Y^{(n)}$ is defined as

$$Y^{(n)} = Y^{(n-1)} + \Phi(\Delta t)\sigma_1 f(Y^{(n-1)}). \tag{12.39}$$

Therefore, using the notation $f_n = f(U^{(n)})$, $\tilde{Y}^{(n)}$ is defined as

$$\tilde{Y}^n = U^{(n-1)} + \Phi(\Delta t)\sigma_1 f_{n-1}. \tag{12.40}$$

On the other hand, for the exact solution we have

$$U(t_n) = U(t_{n-1} + \Delta t) = U(t_{n-1}) + \Delta t U'(t_{n-1}) + \mathcal{O}(\Delta t^2). \tag{12.41}$$

Since $U'(t_{n-1}) = f_{n-1}$, for the truncation error we have

$$LTE(\Delta t) = (\Delta t - \sigma_1 \Phi(\Delta t)) f_{n-1} + \mathcal{O}(\Delta t^2). \tag{12.42}$$

Therefore, the method has the first-order consistency when the condition

$$\Delta t - \sigma_1 \Phi(\Delta t) = \mathcal{O}(\Delta t^2),$$

e.g., the condition

$$\Phi(\Delta t) = \frac{1}{\sigma_1}\Delta t + \mathcal{O}(\Delta t^2) \tag{12.43}$$

holds.

Remark 3 When the function $\Phi(\Delta t)$ is chosen according to the condition (12.43), then the ENRK method (12.39) takes the form

$$Y^{(n)} = Y^{(n-1)} + \Delta t f(Y^{(n-1)}) + \sigma_1 f(Y^{(n-1)}) g(\Delta t^2), \tag{12.44}$$

where g is a sufficiently small function. I.e., taking g as the identity function, we get the scheme

$$Y^{(n)} = Y^{(n-1)} + \Delta t f(Y^{(n-1)}) + \sigma_1 f(Y^{(n-1)}) \Delta t^2. \tag{12.45}$$

Remark 4 As it was mentioned in the Introduction, the NSFD method is the generalization of the classical finite difference method. Therefore, the function $\Phi(\Delta t)$

should be chosen in such a way that $U'(t_n)$ was approximated by the classical FDM having the form $\dfrac{Y_n - Y_{n-1}}{\Delta t}$. This yields that

$$\lim_{\Delta t \to 0} \frac{Y_n - Y_{n-1}}{\Delta t} = U'(t_n).$$

Since, for the NSFD, we have

$$U'(t_n) \approx \frac{Y_n - Y_{n-1}}{\Phi(\Delta t)} = \frac{Y_n - Y_{n-1}}{\Delta t} \frac{\Delta t}{\Phi(\Delta t)}$$

taking limes as $\Delta t \to 0$, we have $\lim\limits_{\Delta t \to 0} \dfrac{\Delta t}{\Phi(\Delta t)} = 1$. Consequently, in (12.43) we have $\sigma_1 = 1$, and hence

$$\Phi(\Delta t) = \Delta t + \mathcal{O}(\Delta t^2) \tag{12.46}$$

for $n = 1, 2, \ldots, N$.

12.3.2 Case $s = 2$

In the following we consider the case $s = 2$. Then the ENRK formulas are the following:

$$Y^{(n)} = Y^{(n-1)} + \Phi(\Delta t)\left(\sigma_1 K_1(Y^{(n-1)}) + \sigma_2 K_2(Y^{(n-1)})\right) \tag{12.47}$$

where

$$K_1(Y^{(n-1)}) = f(Y^{(n-1)}), \quad K_2(Y^{(n-1)}) = f(Y^{(n-1)} + a_{21}\Phi(\Delta t) K_1(Y^{(n-1)})). \tag{12.48}$$

Therefore, $\tilde{Y}^{(n)}$ is defined as

$$\tilde{Y}^n = U^{(n-1)} + \Phi(\Delta t)\left(\sigma_1 K_1(U^{(n-1)}) + \sigma_2 K_2(U^{(n-1)})\right). \tag{12.49}$$

Clearly, $K_1(U^{(n-1)}) = f(U^{(n-1)})$. Moreover, by Taylor expansion we have

$$K_2(U^{(n-1)}) = f(U^{(n-1)}) + a_{21}\Phi(\Delta t) f(U^{(n-1)}) f'(U^{(n-1)}) + \mathcal{O}(\Phi^2(\Delta t)), \tag{12.50}$$

where $f'(U^{(n-1)})$ denotes the Jacobian matrix of f at the point $U^{(n-1)}$.

Now, we consider the Taylor expansion of the exact solution $U(t)$ at $t = t_n$ around the point $t = t_{n-1}$. This is

$$U(t_n) = U(t_{n-1} + \Delta t) = U^{(n-1)} + \Delta t U'(t_{n-1}) + 0.5(\Delta t)^2 U''(t_{n-1}) + \mathcal{O}((\Delta t)^3). \quad (12.51)$$

Since

$$U'(t_{n-1}) = f(U^{(n-1)}); \quad U''(t_{n-1}) = f'(U^{(n-1)})U'(t_{n-1}) = f'(U^{(n-1)})f(U^{(n-1)}), \quad (12.52)$$

for the local truncation error we have

$$\begin{aligned}
LTE(\Delta t) &= U(t_n) - U(t_{n-1}) - \Phi(\Delta t)\sigma_1 \underbrace{U'(t_{n-1})}_{K_1} - \\
&- \Phi(\Delta t)\sigma_2 \underbrace{\left(f(U^{(n-1)}) + a_{21}\Phi(\Delta t) f(U^{(n-1)}) f'(U^{(n-1)}) + \mathcal{O}(\Phi^2(\Delta t))\right)}_{K_2} = \\
&= \underbrace{U^{(n-1)} + \Delta t U'(t_{n-1}) + 0.5(\Delta t)^2 f'(U^{(n-1)})U'(t_{n-1}) + \mathcal{O}((\Delta t)^3)}_{U(t_n)} - \\
&- U^{(n-1)} - \Phi(\Delta t)\sigma_1 U'(t_{n-1}) - \\
&- \Phi(\Delta t)\sigma_2 \left(f(U^{(n-1)}) + a_{21}\Phi(\Delta t) f(U^{(n-1)}) f'(U^{(n-1)}) + \mathcal{O}(\Phi^2(\Delta t))\right) = \\
&= U'(t_{n-1}) [\Delta t - \Phi(\Delta t)\sigma_1 - \Phi(\Delta t)\sigma_2] + \\
&+ f(U^{(n-1)}) f'(U^{(n-1)}) \left[0.5(\Delta t)^2 - \Phi(\Delta t)\sigma_2 a_{21}\Phi(\Delta t)\right] + \\
&+ \mathcal{O}(\Phi(\Delta t)\Phi^2(\Delta t)) + \mathcal{O}(\Delta t)^3. \quad (12.53)
\end{aligned}$$

Hence, the method (12.47) and (12.48) has second order when the conditions

$$\Delta t - \Phi(\Delta t)\sigma_1 - \Phi(\Delta t)\sigma_2 = \mathcal{O}(\Delta t^3), \quad (12.54)$$

$$0.5(\Delta t)^2 - \Phi(\Delta t)\sigma_2 a_{21}\Phi(\Delta t) = \mathcal{O}(\Delta t^3), \quad (12.55)$$

$$\mathcal{O}(\Phi(\Delta t)\Phi^2(\Delta t)) + \mathcal{O}(\Delta t)^3 = \mathcal{O}(\Delta t^3), \quad (12.56)$$

are satisfied. Taking into the account the Remark 4, the condition (12.54) implies that

$$\sigma_1 + \sigma_2 = 1, \quad (12.57)$$

and therefore

$$\Phi(\Delta t) = \Delta t + \mathcal{O}(\Delta t)^3. \quad (12.58)$$

Furthermore, using (12.58), the conditions (12.55) and (12.56) are satisfied when

$$\sigma_2 a_{21} = 0.5 \quad (12.59)$$

and

$$\Phi(\Delta t) = \Delta t + \mathcal{O}(\Delta t)^2. \quad (12.60)$$

Remark 5 We note that the conditions (12.57) and (12.59) are the well-known conditions for the maximal ($p = 2$) order of the two-stages ($s = 2$) classical RK method.

Remark 6 As we have seen, the choice of the step-size functions $\Phi(\Delta t)$ and $\Phi(\Delta t)$ affect the order of the NSFDM (12.9) and (12.10). We can generalize the aforementioned results to higher-order ENRK methods. Theorem 2 is proven in [3] with more restrictive case when $\Phi(\Delta t) = \Phi(\Delta t) = \Delta t + O(\Delta t^{p+1})$.

Remark 7 Considering the Taylor expansion of the functions $\Phi(\Delta t)$ and $\Phi(\Delta t)$ around $\Delta t = 0$, we can define the conditions of consistency order of the NSFDM. Namely, we have the following.

- When

$$\Phi(0) = 0 \quad \text{and} \quad \Phi'(0) = 1, \tag{12.61}$$

  then the requirement (12.43) is satisfied. This means that under the condition (12.61) the method (12.39) has first order consistency.
- When

$$\Phi(0) = 0, \quad \Phi'(0) = 1 \quad \text{and} \quad \Phi''(0) = 0, \tag{12.62}$$

  and

$$\Phi(0) = 0 \quad \text{and} \quad \Phi'(0) = 1, \tag{12.63}$$

  then the requirements (12.58) and (12.60) are satisfied. This means that under the conditions (12.64) and (12.63) the method (12.47) and (12.48) has second order consistency.
- For the general p-th order case, based on (12.38) the conditions are

$$\Phi(0) = 0, \quad \Phi'(0) = 1 \quad \text{and} \quad \Phi^{(i)}(0) = 0 \quad \text{for} \quad i = 2, 3, \dots p \tag{12.64}$$

  and

$$\Phi(0) = 0, \quad \Phi'(0) = 1 \quad \text{and} \quad \Phi^{(i)}(0) = 0 \quad \text{for} \quad i = 2, 3, \dots p - 1. \tag{12.65}$$

Remark 8 Since the case $p = 1$ yields the consistency of a numerical method, therefore the condition (12.61) can be considered as *the necessary condition of the consistency* of an arbitrary NSFD+ERK method.

12.4 Convergence

In this section, we investigate the convergence of the ENRK methods (12.9) and (12.10) with the order p, highlighting in the following theorem.

Theorem 3 *Assume that the consistency order of the NSFDM* (12.9) *and* (12.10) *is* p, *and for the initial guess* $\| e^{(0)} \| = \mathcal{O}(\Delta t^p)$. *Then, by the choice of the functions* Φ *by* (12.38) *the method is convergent in order* p.

Proof First, we consider the global error

$$e^{(n-1)} = Y^{(n-1)} - U^{(n-1)}. \tag{12.66}$$

of the one-step method (12.9) and (12.10) on the refined uniform mesh (12.5) at time t_{n-1} which is rewritten as

$$Y^{(n-1)} = e^{(n-1)} + U^{(n-1)}. \tag{12.67}$$

By employing (12.67) in one-step method (12.11) we obtain

$$e^{(n)} - e^{(n-1)} = -(U^{(n)} - U^{(n-1)}) + \Phi(\Delta t)G(U^{(n-1)} + e^{(n-1)}). \tag{12.68}$$

Alternatively,

$$\begin{aligned} e^{(n)} - e^{(n-1)} = -(U^{(n)} - U^{(n-1)} - \Phi(\Delta t)G(U^{(n-1)})) + \Phi(\Delta t)(G(U^{(n-1)} + e^{(n-1)}) \\ -\Phi(\Delta t)G(U^{(n-1)})). \end{aligned} \tag{12.69}$$

In the above equation, we name

$$g^{(n-1)} = -U^{(n)} + U^{(n-1)} + \Phi(\Delta t)G(U^{n-1}), \tag{12.70}$$

which corresponds to the local truncation error (12.36), and

$$k^{(n-1)} = G(U^{(n-1)} + e^{(n-1)}) - G(U^{(n-1)}). \tag{12.71}$$

Hence we get the relation

$$e^{(n)} - e^{(n-1)} = g^{(n-1)} + \Phi(\Delta t)k^{(n-1)} \tag{12.72}$$

From Theorem 1 the following relation

$$\| k^{(n-1)} \| \leq L^* \| e^{(n-1)} \| \tag{12.73}$$

holds for the Eq. (12.71). Then from (12.72) we attain

$$\| e^{(n)} \| \leq \| e^{(n-1)} \| (1 + L^*\Phi(\Delta t)) + \| g^{(n-1)} \|, \tag{12.74}$$

implying

$$\| e^{(n)} \| \leq e^{L^*\Phi(\Delta t)} \| e^{(n-1)} \| + \| g^{(n-1)} \| \tag{12.75}$$

for any $n = 1, \ldots, N$ and we can write the following estimation for the global error $e^{(N)}$ as follows

$$\begin{aligned}\| e^{(N)} \| &\le e^{L^*\Phi(\Delta t)} \| e^{(N-1)} \| + \| g^{(N-1)} \| \le e^{L^*\Phi(\Delta t)} (e^{L^*\Phi(\Delta t)} \| e^{(N-2)} \| + \| g^{(N-2)} \|) \\ &+ \| g^{(N-1)} \| = e^{2L^*\Phi(\Delta t)} \| e^{(N-2)} \| + e^{L^*\Phi(\Delta t)} \| g^{(N-2)} \| + \| g^{(N-1)} \| \\ &\le \cdots \le e^{NL^*\Phi(\Delta t)} \| e^{(0)} \| + \Sigma_{i=0}^{N-1} e^{iL^*\Phi(\Delta t)} \| g^{(N-1-i)} \| \\ &\le e^{NL^*\Phi(\Delta t)} (\| e^{(0)} \| + \Sigma_{i=0}^{N-1} \| g^{(N-i-1)} \|).\end{aligned} \tag{12.76}$$

Therefore

$$\| e^{(N)} \| \le e^{NL^*\Phi(\Delta t)} (| e^{(0)} | + \Sigma_{i=0}^{N-1} | g^{(N-i-1)} |). \tag{12.77}$$

We assume that the NSFD method has p-th order consistency, i.e. for the local error we have

$$\| g^{(i)} \| = \mathcal{O}(\Delta t^{p+1}). \tag{12.78}$$

Then, (12.77) results

$$\begin{aligned}\| e^{(N)} \| &\le e^{NL^*\Phi(\Delta t)} (\| e^{(0)} \| + N\mathcal{O}(\Delta t^{p+1})) = \\ &= e^{NL^*\Phi(\Delta t)} (\| e^{(0)} \| + N\Delta t\mathcal{O}(\Delta t^{p})) = \\ &= e^{NL^*\Phi(\Delta t)} (\| e^{(0)} \| + T\mathcal{O}(\Delta t^{p})).\end{aligned} \tag{12.79}$$

We assume that

$$\sup_{\Delta t} \frac{\Phi(\Delta t)}{\Delta t} \le C^\star < \infty. \tag{12.80}$$

Then, based on (12.79), we have

$$\| e^{(N)} \| \le e^{N\Delta t L^* \frac{\Phi(\Delta t)}{\Delta t}} (\| e^{(0)} \| + t^*\mathcal{O}(\Delta t^p)) \le= e^{t^*L^*C^*} (\| e^{(0)} \| + \mathcal{O}(\Delta t^p)) \tag{12.81}$$

Remark 9 When the function $\Phi(\Delta t)$ is chosen according to the condition (12.38), then we have

$$\frac{\Phi(\Delta t)}{\Delta t} = 1 + \mathcal{O}(\Delta t^p)$$

therefore the condition (12.80) is always satisfied.

Remark 10 Let us present examples of step-size functions that satisfy the consistency and convergence properties for $s = 1$ and $s = 2$. *Case* $s = 1$. For this case, we consider the following step-size function

$$\Phi(\Delta t) = \frac{1 - e^{-\rho\Delta t}}{\rho}, \quad \rho \in \mathbb{R}^+ \tag{12.82}$$

which satisfies $\Phi(\Delta t) = \Delta t + \mathcal{O}(\Delta t)^2$ with the property $\sup \Phi(\Delta t) = \Delta t$.

Case $s = 2$. In this case, we study two following functions

$$\Phi(\Delta t) = \sin \Delta t \tag{12.83}$$

as the main step-size function which satisfies $\Phi(\Delta t) = \Delta t + \mathcal{O}(\Delta t)^3$ and $\sup \Phi(\Delta t) = \Delta t$. For the internal step-size function $\Phi(\Delta t)$, we can consider a function similar to the one used in the case of $s = 1$ in the form (12.82).

12.5 Conclusion

In this study, we investigated explicit nonstandard Runge–Kutta (ENRK) methods for solving dynamical systems, focusing on their application to higher-order Runge–Kutta methods with distinct step-size functions for the main and internal stages. While standard explicit Runge–Kutta (SERK) methods are widely used due to their simplicity and efficiency, they often suffer from sensitivity to the discretization step size Δt. This sensitivity can lead to a loss of dynamical consistency, particularly in long-term simulations where stability, positivity, and qualitative accuracy are essential. To overcome these challenges, we applied nonstandard finite difference (NSFD) techniques to modify the step-size selection process, allowing for better preservation of dynamical properties. Our findings highlight that the introduction of carefully chosen, problem-dependent step-size functions significantly improves the numerical behavior of ENRK methods. These modifications ensure that the numerical solutions retain key properties of the underlying continuous system, such as stability, boundedness, and positivity, without the constraints typically associated with SERK methods. Moreover, we analyzed the impact of different step-size functions on the order, consistency, and convergence of ENRK methods. Our results demonstrate that, under appropriate conditions on step-size functions, ENRK methods can achieve an order of accuracy and convergence comparable to their SERK counterparts while maintaining superior qualitative fidelity. This is particularly advantageous in applications where preserving the structure of the dynamical system is more important than achieving high numerical accuracy. The significance of this work lies in the broader applicability of ENRK methods to a wide range of problems in mathematical biology, epidemiology, and engineering, where conventional numerical methods often struggle to maintain physical realism over extended simulations. By demonstrating the effectiveness of ENRK methods, this study provides a foundation for their further development and application in solving complex dynamical systems while ensuring that essential qualitative properties are preserved.

Acknowledgements This research has been supported by the National Research, Development and Innovation Office—NKFIH, grant no. K137699.

References

1. Ascher U, Petzold L (1998) Computer methods for ordinary differential equations and differential-algebraic equations. Society for Industrial and Applied Mathematics, Philadelphia
2. Anguelov R, Lubuma JM-S (2001) Contributions to the mathematics of the non-standard finite difference methods and applications. Numer Methods PartL Differ Equ 17:518–543
3. Dang QA, Hoang MT (2020) Positive and elementary stable explicit nonstandard Runge-Kutta methods for a class of autonomous dynamical systems. Int J Comput Math 97(10):2036–2054
4. Dimitrov DT, Kojouharov HV (2005) Nonstandard finite-difference schemes for general two-dimensional autonomous dynamical systems. Appl Math Lett 18:769–774
5. Dimitrov DT, Kojouharov HV (2005) Stability-preserving finite-difference methods for general multi-dimensional autonomous dynamical systems. Int J Numer Anal Model 4(2):280–290
6. Faragó I, Mosleh R (2023) Some qualitative properties of the discrete models for malaria propagation. Appl Math Comput 439:127628
7. Faragó I, Mincsovics ME, Mosleh R (2018) Reliable numerical modelling of malaria propagation. Appl Math (Springer) 63(3):259–271
8. Gurski KF (2013) A simple construction of nonstandard finite-difference schemes for small nonlinear systems applied to SIR models. Comput Math Appl 66:2165–2177
9. Hairer E, Nørsett SP, Wanner G (1993) Solving ordinary differential equations I: nonstiff problems. Springer
10. Markus AS, Mickens RE (1999) Suppression of numerically induced chaos with nonstandard finite difference schemes. J Comput Appl Math Elsevier 106:317–324
11. Mickens RE (1994) Nonstandard finite difference of differential equations. World Scientific Publishing Co. Pte. Ltd
12. Mickens RE (2006) Calculations of denominator functions for nonstandard finite difference schemes for differential equations satisfying a positivity condition. Wiley InterScience, pp 672–691
13. Mickens RE (2005) Advances in the applications of nonstandard finite difference schemes. World Scientific Publishing Co. Pte. Ltd
14. Keeling MJ, Rohani P (2008) Modeling infectious diseases in humans and animals. Princeton University Press

Chapter 13
Causal Inference for Power Grid Dynamics on a University Campus

Yuka Fujiki, Takashi Yanase, Hiroyasu Ando, Yuta Fujishige, Yuiko Ohta, Hiroyuki Higuchi, and Hiroshi Suito

Abstract This study is a research effort to extract causal relations from time series data of electrical consumption obtained from a power grid serving a university campus. Electrical consumption exhibits different characteristics and dynamics depending on building types and diverse external conditions such as seasons, temperatures, and school calendars. Causal relations among such factors and electrical consumption from power grid systems are expected to provide prediction and control strategies to reduce campus electricity costs. Causal graphs, which should be Directed Acyclic Graphs (DAGs) constructed from data, are complicated. Certain strategies must be developed to give them sufficient explainability. The mathematical basis for causality inference must be strengthened using mathematical logic and discrete geometry, as well as high-performance machine learning techniques and cognitive sciences to enhance their explainability. Moreover, network controllability should be evaluated and analyzed using appropriate network centralities. This research was undertaken as a part of Discovery Intelligence Laboratory co-established by Fujitsu Ltd. and

Y. Fujiki (✉)
Frontier Research Institute for Interdisciplinary Sciences (FRIS), Tohoku University, Sendai, Japan
e-mail: yfujiki@tohoku.ac.jp

T. Yanase · Y. Fujishige · Y. Ohta · H. Higuchi
Artificial Intelligence Laboratory, Fujitsu Limited, Kawasaki, Japan
e-mail: tyana@fujitsu.com

Y. Fujishige
e-mail: fujishige.yuta@fujitsu.com

Y. Ohta
e-mail: yuiko@fujitsu.com

H. Higuchi
e-mail: h-higuchi@fujitsu.com

H. Ando · H. Suito (✉)
Advanced Institute for Materials Research (WPI-AIMR), Tohoku University, Sendai, Japan
e-mail: hiroshi.suito@tohoku.ac.jp

H. Ando
e-mail: hiroyasu.ando.d1@tohoku.ac.jp

T. Tuovinen et al. (eds.), *Challenges in Design Methods, Numerical Tools and Technologies for Sustainable Aviation, Transport and Industry*, Computational Methods in Applied Sciences 17, https://doi.org/10.1007/978-3-031-98675-8_13

Tohoku University, which is intended to extend application areas to address widely diverse social issues and to construct robust mathematical bases for them.

Keywords Causal discovery in energy systems · Time-series analysis of power consumption · Condition-specific subpopulation inference · Causal graph and network centrality · Explainable AI for energy optimization

13.1 Introduction

13.1.1 Electricity Concerns at University Campuses

Because of current global tensions in natural oil and gas producing countries, escalating energy costs are manifested in electricity prices. Furthermore, reduction of carbon dioxide emissions to mitigate global warming has emerged as a crucially important social and economic concern. In response to these circumstances, energy-related issues are being considered in various countries, with specific emphasis on using renewable energy as the primary power source. This new energy usage constitutes an urgent concern for university campuses, which must accommodate large numbers of students and faculty members.

Tohoku University, which is located in a cool and green city, Sendai, Japan, has a dispersed campus. Electricity is fundamentally necessary for educational and research activities, particularly for air conditioning equipment and experimental devices. Therefore, increasing electricity prices pose important challenges for university management. Sendai city and Tohoku University are aligning themselves with global trends of decarbonization efforts using renewable energy as distributed energy resources.

However, simply decreasing electricity usage and replacing power resources with renewable energy to mitigate electricity consumption and carbon dioxide emissions might compromise educational and research quality. For constraining energy consumption while maintaining support for these activities, mathematical analysis can be employed within the infrastructure network, leading to effective strategies for the power network of the university. Although reducing electricity consumption at the individual building level is a necessary achievement, a university campus must operate as a network of interconnected buildings with complex interactions. For instance, to decrease power consumption in building A at time t, it becomes necessary to reduce power usage in building B apart from building A at time $t-1$ and to adjust schedules, such as timetables, to control people moving to building A. These intricacies are attributable to the complex interactions among diverse facilities, which are much more complicated than simple on-off switches. Nevertheless, no report has described a true mechanism for controlling the entire network.

As described in this paper, we use time series data of electricity consumption as building blocks of the electricity network comprising Tohoku University's campuses.

Using mathematical methods, we aim at discerning the causal relations underlying the data and undertake an initial investigation to propose measures aimed at reducing electricity consumption across the entire campus. These analyses will inform predictions of future power consumption and will guide the development of efficient strategies for power consumption reduction. The Facilities Department is gathering electricity consumption data, which are recorded hourly throughout the year, across the entire campus. Using those data, energy conservation proposals that leverage these data can be explored.

Although various time-series data analysis methods exist, ranging from statistical techniques [1] to machine learning [2], our emphasis is placed on "causal discovery" [3] of electricity consumption by building blocks. Understanding the causes and effects of demand enables the implementation of clear measures [4, 5]. Rather than relying solely on correlation, our objective is to establish clear measures for power forecasting and consumption reduction, with the aim of enhancing the accuracy and efficiency of time-series prediction. We would like to introduce our current efforts undertaken in this direction.

13.1.2 Campus Power Grid

Tohoku University encompasses four campuses: Katahira, Aobayama, Kawauchi, and Seiryo. The Aobayama campus is subdivided further into the East Campus, the North Campus, and the New Campus. The primary focus of student activities in terms of education and research is at the Kawauchi and Aobayama campuses. The Seiryo campus is occupied mainly by the University Hospital and the Faculty of Medicine, whereas the Katahira campus is home to the university's headquarters and research institutes. Given that these activities occur with less relation to student activities which affect variations in power usage, these analyses specifically address the Aobayama and Kawauchi campuses.

Regarding the power network, the campus grid comprises various sets of buildings organized as nodes, linked by physical connections from extra high-voltage substations within the Tohoku Electric Power Company service area. This network is presented in Fig. 13.1, where "F102" and similar labels denote building blocks, constituting 43 nodes in all. It is noteworthy that the number of buildings in each node varies. Hourly records of electricity consumption [kWh] are recorded at each node, ranging from April 1, 2021, to March 31, 2022, including school semesters and holidays. An example of a time series depicting electricity consumption is shown in Fig. 13.2. Using these data, we analyze the causal relations between building blocks.

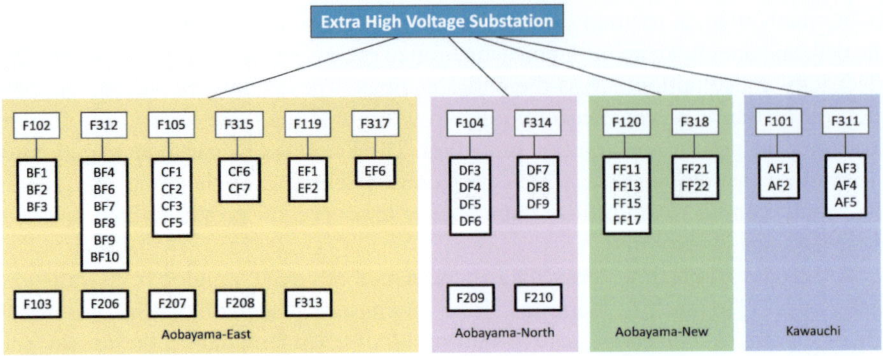

Fig. 13.1 Physical connections in the campus power grid

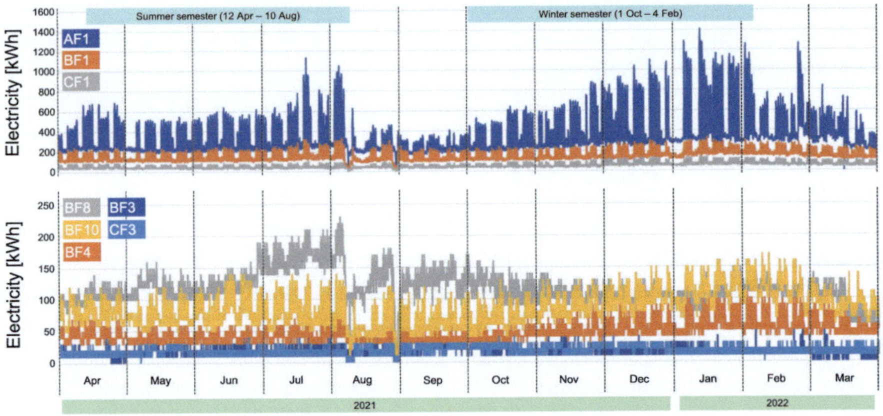

Fig. 13.2 Time series of electricity consumption data for each building block on the campus during the year

13.1.3 Organization of the Paper

The consecutive section introduces a detailed methodology of causal discovery. Section 13.3 presents examples of causal discovery, illustrating how the methodology is applied in practical scenarios. Section 13.4 analyzes the found causal relations from complex network aspects. Section 13.5 summarizes this work and discusses the future direction of this research line.

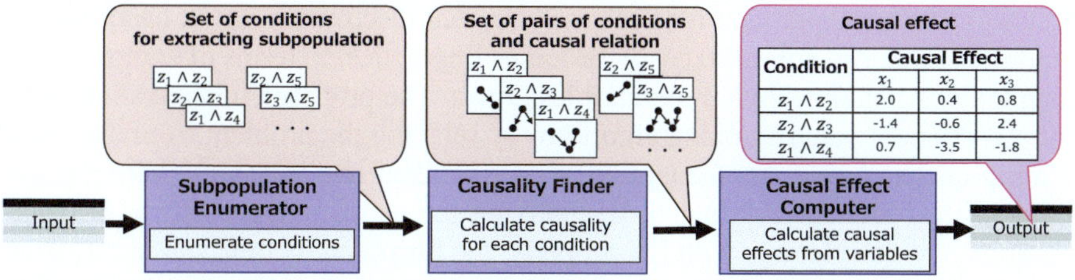

Fig. 13.3 Flowchart of the causal discovery platform

13.2 Causal Discovery Platform

To identify the factors responsible for electricity consumption from time series of electricity consumption data of buildings on campus, one must seek variables that are causally related to the amount of electricity consumption, rather than simply looking at correlation among the variables. Furthermore, even if a causal relation to electricity consumption is found for a factor, the causal relation might not be true at any time of the year. It might even vary depending on the time of day, season, etc. Therefore, it is important to consider causal relations that hold only under specific conditions, rather than causal relations that are apparent from the data overall. Therefore, we built a causal discovery platform to seek and discover causal relations specific to a specific dataset.

13.2.1 Module Configuration of Causal Discovery Platform

The causal discovery platform comprises three modules. Figure 13.3 depicts the processing flow of the causal discovery platform. For a given sample set, the Subpopulation Enumerator computes all conditions defined as product terms that consist of explanatory variables with high correlation to the target variable: subgroups of samples are then constructed according to these conditions. Thereafter, the Causality Finder computes the causal relations among variables under each condition by applying causal discovery methods for each subgroup. Finally, the Causal Effect Computer calculates and outputs causal effects from explanatory variables to the target variable for each condition.

13.2.2 Subpopulation Enumerator

The Subpopulation Enumerator enumerates all conditions representing subpopulations, i.e. subsets of data samples that potentially have characteristic causal relations.

A straightforward approach requires many causal discoveries for all candidates of conditions. Such an approach easily becomes impractical in terms of computational complexity when the number of variables grows. The proposed method alleviates it with an efficient pattern mining algorithm by relaxing the problem from the causal relation to the correlation. Generally speaking, when two variables share a causal relation, they are also correlated. By relaxing the enumeration problem of causal relations into correlations with the property, the proposed method can use efficient and practical algorithms for fast pattern discovery. For the current implementation, we employ Wide Learning [6, 7] to enumerate emerging patterns [8] efficiently.

13.2.3 Causal Finder

The Causal Finder applies a causal discovery algorithm for each subpopulation corresponding to each condition enumerated by the Subpopulation Enumerator. Subsequently, it outputs pairs of the condition and the causal relation. For the current implementation, we employ DirectLiNGAM [9] as the causal discovery algorithm. In the real data, local causality might be known beforehand from the domain knowledge, such as "there is no causality between variable x_1 and variable x_2." DirectLiNGAM has a function that is able to reflect this prior knowledge in causal search. The causal discovery platform also performs causal search using prior knowledge.

13.2.4 Causal Effect Computer

When a causal relation exists between variable x as a cause and variable y as a result, the average change in the value of y when the value of x is changed from 0 to 1 is designated as the causal effect from x to y. If the causal effect is positive, then y increases as x increases. However, if the causal effect is negative, then y decreases as x increases. After the causal effect computer summarizes causal relation outputs by the causal finder for each condition, it calculates causal effects from each variable to the target variable. Consequently, it is possible to ascertain differences in causal effects of different conditions from the same explanatory variable.

13.3 Causal Discovery of Electricity Consumption

This section presents examples of causal discovery by the causal discovery platform using campus electricity consumption data. We conducted two experiments to discover causality by particularly addressing causality of the following types.

- Causality from temperature to electricity consumption.
- Causality from past electricity consumption in the same building block.

Table 13.1 Causality effects of temperature on electricity consumption

Conditions	Causal effects of temperature
All	−3.40
School days, daytime, temperature $<15\,^{\circ}\mathrm{C}$	−15.96
School days, daytime, temperature $\geq 15\,^{\circ}\mathrm{C}$	10.21

13.3.1 Causality from Temperature to Electricity Consumption

We conducted causal discovery with the electricity consumption of one building block (AF1) in Kawauchi campus of Tohoku University as the target variable. Explanatory variables include the electricity consumption of other building blocks, as well as the date, hour and temperature. In addition, categorical variables generated by determining whether a class is held there from the date and determining the daytime, peak, and nighttime hours from the hour were added as variables for generating conditions.

Specifically examining the causal relation from temperature to electricity consumption, we calculated causal effects from temperature to electricity consumption. Table 13.1 shows the causal effects for all data and for a subset of samples corresponding to some of the enumerated conditions. According to this result, when the temperature is low, there are negative causal effects. When the temperature is high, there are positive causal effects. The causal effects from temperature to electricity consumption indicate that lower temperatures are associated with greater electricity consumption when negative, and higher temperatures are associated with greater electricity consumption when positive. In other words, cooling and heating effects on electricity consumption are regarded as indicated.

13.3.2 Causality from Past Electricity Consumption

Movements of people among campus buildings might contribute to electricity consumption. However, campus electricity consumption data do not include measurement data of human flows. Therefore, we analyzed causal relations that arise when the electricity consumption of the same building block several hours prior was used as an explanatory variable.

As in Sect. 13.3.1, the power consumption of the AF1 building block at Kawauchi Campus of Tohoku University was used as the target variable. Also, the date, hour, temperature, and electricity consumption of the same building block 1, 2, and 3 h prior were used as explanatory variables. The causal search provides prior knowledge such that no causal relation can be found going back in time, such as a causal relation from electricity consumption one hour prior to electricity consumption two hours prior.

Table 13.2 Causality from past electricity consumption

Conditions	Electricity consumption One hour prior	Electricity consumption Two hours prior	Electricity consumption Three hours prior
School days	1.48	1.66	0.80
No-school days	1.33	1.39	0.81
Daytime	1.38	1.39	0.77
Nighttime	1.11	1.03	0.69
School days, daytime	1.37	1.36	0.76
School days, nighttime	1.09	1.20	0.60
No-school days, daytime	1.27	1.27	0.77

Table 13.2 presents causal effects from electricity consumption 1, 2, and 3 h prior for some conditions. According to the result, the causal effects from power consumption 1 h and 2 h prior are equivalent under every condition, but the causal effect from power consumption 3 h prior is considerably smaller. This trend is more pronounced on days with school attendance and in the daytime, when compared by conditions. This result might derive from the fact that classes are 1.5 h long and that movement of many people on campus occurs about every 1.5 h.

13.4 Network Visualization and Analysis

The causal discovery platform can compute the causal relations between variables for variously conditioned datasets and can construct a causal graph. As described herein, we visualize and analyze a causal graph of power consumption so that a person can intuitively discover and evaluate important elements of power consumption systems.

13.4.1 Visualization of Causality

In a causal graph of power consumption, variables represented as nodes and causal relations among variables are depicted as weighted and directed edges. The causal graph includes 45 nodes, corresponding to the electricity consumption of 43 building blocks, alongside temperature and time. The adjacency matrix W of the causal graph is an $N \times N$ matrix ($N = 45$), where the matrix elements w_{ij} represent the direct causal effect of variable x_i on variable x_j. Given the diverse scale of variables, we normalize the edge weights as $\tilde{w}_{ij} = w_{ij}\sigma_i/\sigma_j$, where σ_i is the standard deviation of variable x_i.

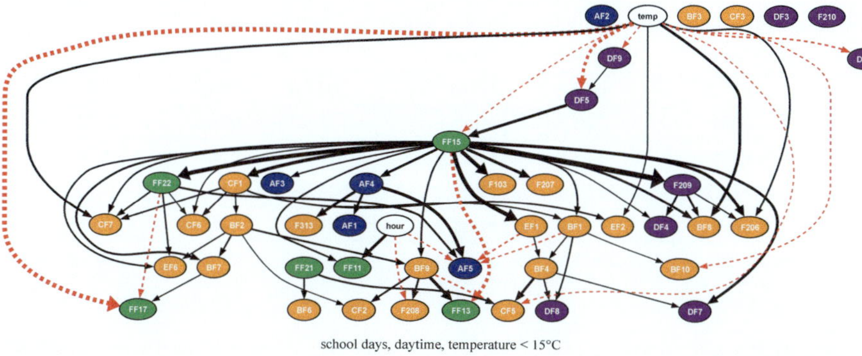

Fig. 13.4 Causal graph with conditions including school days, daytime, and temperature <15 °C

Figure 13.4 presents a causal graph computed from the dataset under conditions including school days, daytime, and temperature <15 °C. Edge thickness is proportional to the normalized weight, displaying only those edges having weight exceeding the threshold of 0.2. Positive and negative causal relations are shown respectively by solid black lines and dotted red lines, with arrows pointing from the cause to result. Node colors represent different campus sections.

Insights from the overall view of the causal graph are presented below. There are large negative causal relations from temperature (shown as temp in the figure) to FF17 and DF5. These findings indicate that electricity consumption in these building blocks increases concomitantly with decreasing temperature. Building block FF17 includes student dormitories. The Faculty of Science mainly uses DF5. It is not difficult to imagine that negative causality originates from the heating system. Most causal relations among electricity consumption of building blocks are positive, except for several negative causal relations, especially one observed from FF15 to FF13. The negative causality from FF15 to FF13 can be interpreted as the movement of people occurring between the two adjacent building blocks. Discussions must be conducted with actual users to elucidate the reasons underlying causal relations and to develop appropriate energy demand reduction strategies.

13.4.2 Network Centrality Measures

Applying network centrality measures to the causal graph enables us to evaluate the importance of each node [10]. First, we calculate the out-strength $s_i^{\text{out}} = \sum_j w_{ij}$. This measure represents the extent to which the power consumption of the result variables is altered because of the direct causal effect of variable x_i. The product $s_i^{\text{out}}\sigma_i$ [kWh] quantifies power consumption changes attributable to variable x_i. Next, the summation of the total causal effect, as propagated to other variables through downstream causal relations of the variable x_i, is captured by Katz centrality [11] in the

Table 13.3 Five nodes with the highest absolute values of centrality measures

Rank	Label	$s^{\text{out}} \times \sigma$	Label	$C^{\text{Katz}} \times \sigma$	Label	$\tilde{s}^{\text{out}}$
1	FF15	607.8	FF15	2041.4	FF15	9.82
2	AF4	225.7	temp	−1121.1	temp	6.31
3	BF4	174.8	DF5	1103.1	hour	3.26
4	FF22	151.1	DF9	811.2	FF22	2.77
5	hour	−150.8	hour	−698.8	CF1	2.52

absence of any attenuation as $C_i^{\text{Katz}} = \sum_n^{\infty} \sum_j (W^n)_{ij}$. Similarly to the out-strength, the product of this measure and the standard deviation $C_i^{\text{Katz}} \sigma_i$ [kWh] quantifies the effects of chains of causality initiated by building block i on the overall electricity consumption of the system. Furthermore, normalization of the edge weights introduces the concept of the modified out-strength, $\tilde{s}_i^{\text{out}} = \sum_j |\tilde{w}_{ij}|$. This measure shifts emphasis from evaluating the effects on power consumption to assessing the strength of relations which variable x_i maintains with other nodes in the causal graph.

Table 13.3 presents the top five nodes according to these centrality measures. It is noteworthy that FF15 outperforms the others to a considerable degree, suggesting that it can be expected to play an important role in reflecting fluctuations in electricity consumption. The ranking of the second place and below depends on the type of centrality measure. The difference between out-strength and Katz centrality reflects the difference in influence attributable to direct and indirect causal effects. By considering indirect causality, Katz centrality evaluates the importance of upstream variables in Fig. 13.4 more highly, such as temp, DF5, and DF9. By contrast, a node such as temp and hour with higher $\tilde{s}^{\text{out}}$ than another centrality measure has a large amount of non-negligible causal effects on each node in the causal graph. However, the total change to power consumption is small. Reasons for a node receiving such an evaluation include multiple causal relations, where positive and negative causal effects cancel each other out, or the low power consumption of the result nodes.

13.5 Summary and Discussion

Our summary of the current work is the following. First, based on a causal inference approach, looking into causality relations embedded in big and complex datasets can become possible. The causality-finding methodology was applied to the power grid serving Tohoku University. Furthermore, we analyze the causal graph using the measure of complex networks and interpreted causality relations within the power grid.

From the dataset, we extracted several perspectives of the causal discovery in two approaches: a focused analysis of causal relations between specific variables, and a comprehensive overview of causal networks. The first approach is an attempt to examine and clarify causal mechanisms among variables, as exemplified by how temperature changes affect electricity consumption. We demonstrated that negative causal effects were observed at lower temperatures, whereas positive effects were noteworthy at higher temperatures, indicating the strong influences of heating and cooling on electricity consumption. In addition, causality from the past electricity consumption (1, 2, and 3 h prior) suggests that electricity consumption reflects human movement patterns on campus. The second approach aims to provide a comprehensive overview of causal relations and to identify key variables and relations that contribute significantly to power consumption across the system. By capturing the relations affecting power consumption as a causal graph, we identified strong causal relations and influential variables.

We are just at the point of embarkation on this research. More extensive causality searches and effective visualization techniques must be applied to grasp all relevant aspects of the whole causality network. Centrality analysis is expected to contribute to identification of which node in what situation should be controlled from multi-objective optimization and game theory viewpoints. These approaches are attributable to the fact that electricity consumption on a university campus depends on campus activities.

As directions of future work, considering human mobility data will be useful to strengthen our model. Moreover, we are looking forward to extending our application areas to assessment of widely diverse problems, including crystal structures, electronic structures of functional materials, and sports sciences. Strengthening basic concepts and techniques such as combinatorics, mathematical logic, cognitive science, and philosophical approaches is unavoidable for answering the question of "what is understanding".

Acknowledgements The work is supported by the Fujitsu x Tohoku University Discovery Intelligence Laboratory, funded by Fujitsu Limited. This work was partially supported by the Council for Science, Technology and Innovation (CSTI), Cross-ministerial Strategic Innovation Promotion Program (SIP), the third period of SIP "Smart energy management system" Grant Number JPJ012207 (Funding agency: JST).

Competing Interests Takashi Yanase, Yuta Fujishige, Yuiko Ohta, and Hiroyuki Higuchi are employees of Fujitsu Limited.

References

1. Hamilton LD (1994) Time series analysis. Princeton University Press, Princeton, New Jersey
2. Murphy KP (2012) Machine learning—a probabilistic perspective. MIT Press, Boston, MA
3. Fujishige Y, Suzuki H, Koyanagi Y, Uemura K, Iwashita H, Chang S, Asai T, Maruhashi K, Kashima H (2023) Causal discovery platform in heterogeneous populations by subpopulation enumeration. In: Forum on information technology (FIT2023), F-051
4. Ikeda Y (2024) Data science of renewable energy integration. Springer Nature Singapore
5. Sun X, Wang X, Yan B, Wang B, Pan X, Guo J (2024) A systematic review of causal analysis in power systems: principles, application, and prospects. In: 2024 IEEE 7th international electrical and energy conference (CIEEC), pp 3944–3949

6. Iwashita H, Takagi T, Suzuki H, Goto K, Ohori K, Arimura H (2020) Efficient constrained pattern mining using dynamic item ordering for explainable classification. arXiv:2004.08015
7. Hello, Wide Learning! https://widelearning.labs.fujitsu.com/en/
8. Dong G, Li J (1999) Efficient mining of emerging patterns, discovering trends and differences. In: Proceedings of fifth international conference on knowledge discovery and data mining (KDD), pp 43–52
9. Shimizu S, Inazumi T, Sogawa Y, Hyvarinen H, Kawahara Y, Washio T, Hoyer PO, Bollen K (2011) DirectLiNGAM: a direct method for learning a linear non-Gaussian structural equation model. J Mach Learn Res 12:1225–1248
10. Newman MEJ (2010) Networks: an introduction. Oxford University Press, Oxford, UK
11. Katz L (1953) A new status index derived from sociometric analysis. Psychometrika 18:39–43

Part III
Reviews and Prospectives

Chapter 14
Generative AI-Driven Innovation Leads the Way to Coevolution Between Japan, India and the Global South Towards Decarbonization

Chihito Watanabe, Yuji Tou, Akira Nagamatsu, and Pekka Neittaanmaki

Abstract This article proposes to build a coevolution dynamism between Japan, India and the Global South by fusing Japan's manufacturing technology with India's DX (socio-economic transformation through digital technology) capabilities to (i) support the decarbonization of India, the world's third largest CO_2 emitter, (ii) accelerate Japan's DX improvement and decarbonization, and (iii) promote the global standardization of India's traditional innovation system, "Jugaad," and transfer it to the Global South. Focusing on the rapid growth trend of generative AI (artificial intelligence) and the highly compatible nature between Amazon Web Services' (AWS) AI/ML (machine learning)-based self-propagating high functionality that incorporates user learning and generative AI (GAI)-driven innovation, this article pursues the possibility of building a sustainable co-evolution relationship among Japan, India and the Global South through GAI-driven innovation by using AWS's co-evolutionary dynamism as a role model. Suggestions from pioneering successful cases of Japan-India collaboration are extracted thereon, and the direction of the renovation of Japan's manufacturing technology corresponding to the historical demand is suggested.

Keywords Decarbonization · India · Global South · GAI · AWS

C. Watanabe (✉)
Institute of Science Tokyo, Tokyo, Japan
e-mail: watanabe.c.pqr@gmail.com

International Institute for Applied Systems Analysis (IIASA), Laxenburg, Austria

Y. Tou
Tohoku Gakuin University, Sendai, Japan

A. Nagamatsu
Tohoku University, Sendai, Japan

P. Neittaanmaki
University of Jyvaskyla, Jyvaskyla, Finland

Lappeenranta University of Technology, Lappeenranta, Finland

T. Tuovinen et al. (eds.), *Challenges in Design Methods, Numerical Tools and Technologies for Sustainable Aviation, Transport and Industry*, Computational Methods in Applied Sciences 17, https://doi.org/10.1007/978-3-031-98675-8_14

14.1 Introduction

While Japan once took pride in its manufacturing technologies, they are now facing the turning point as satisfying (i) the development and embodiment of DX, (ii) the reconstruction of a resilient global supply chain (SC), (iii) harmony between growth and ESG (environmental, social, and governance) and (iv) contribution to the world requirements [28].

Under these circumstances, India, which has become the world's most populous country, is seeking to take the lead in reducing CO_2 emissions, the world's third-largest CO_2 emitter, by leveraging its world-class DX to become a leader in the Global South (GS) [7]. However, it is facing difficulties in backing up its manufacturing technology, and is at a crossroads where cooperation with Japan in such a way as leveraging its excellent manufacturing technologies is required to achieve decarbonization [29].

Decarbonization is now a top priority regardless of the North and South, and considering that GS's CO_2 accounts for two-thirds of global CO_2 emissions, the outcome of India's decarbonization will determine the success or failure of global climate change countermeasures. In addition, contribution to the decarbonization and growth of "climate change-vulnerable countries" including Africa has become an urgent competitive issue geopolitically [30]. However, while India has historically had deep ties with these countries, its ties with Japan [1] are generally shallow, and in this regard too, there are high expectations for Japan-India cooperation [12].

In light of the above, and taking into account the trend of generative AI (GAI)-driven innovation that is making great strides based on innovations in learning functions, this article aims to (i) support India's decarbonization by fusing Japan's manufacturing strengths with India's DX capabilities, (ii) accelerate Japan's improvement of DX and decarbonization by incorporating India's excellent DX capabilities, and (iii) evolve India's unique traditional innovation system, "Jugaad," which seeks frugal, flexible, and inclusive solutions in adverse circumstances [2, 4, 9], into a global standard (global universalization) and transfer it to "climate change-vulnerable countries" in the Global South.

Through the above, the dynamism of a virtuous cycle of growth and decarbonization of the three parties in a co-evolutionary way can be illustrated as shown in Fig. 14.1. Realizing this co-evolution will enable us to see the direction for innovating Japan's manufacturing technologies to address the four challenges as highlighted in the heading of this section.

In conceptualizing this co-evolutionary dynamism, it is important to understand that (i) emerging and developing countries also benefit from digital innovations [11], and (ii) this will lead awakening of "sleeping resources" in these countries such as renewable, mineral and human resources through the advancement of AI/ML-driven innovation that leads to creating new brand value (new functions) [23].

With this understanding, paying attention to the rapid growth trend of GAI and the highly compatible nature between Amazon Web Services' (AWS) AI/ML-based self-propagating high functionality that incorporates user learning, and GAI-driven

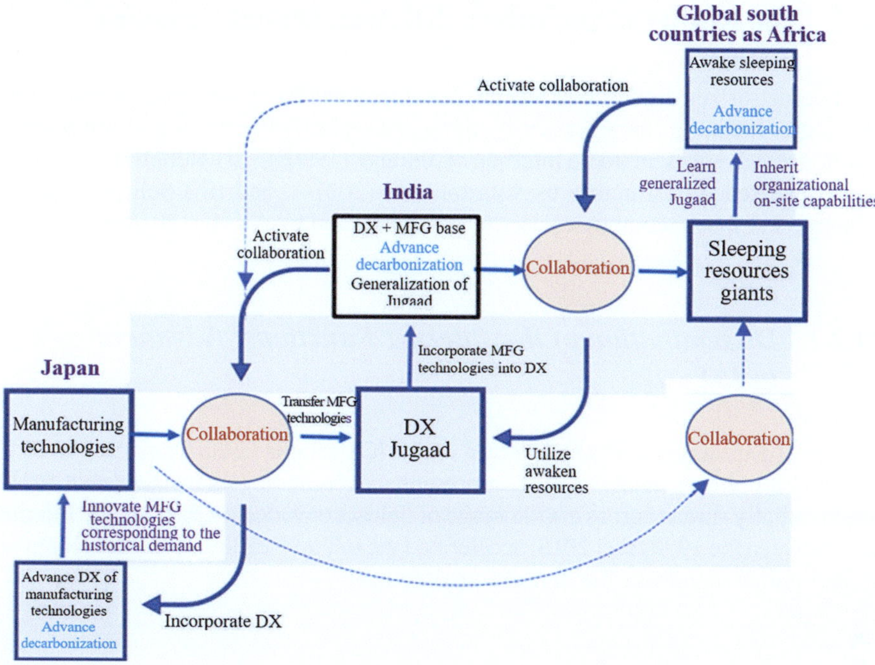

Fig. 14.1 Coevolution dynamism among Japan, India and the Global South

innovation [14, 32], the possibility of building a sustainable co-evolution relationship through GAI-driven innovation is pursued by using AWS's co-evolutionary dynamism as a role model.

By means of this approach, in line with a typical Japan-India decarbonization cooperation project, conceptualization of the co-evolutionary dynamism of the three parties of Japan, India, and the Global South shown in Fig. 14.1 can be expected thereby identify the direction of the renovation of Japan's manufacturing technology corresponding to the historical demand.

While there are many researches on the significance of decarbonization [27–29], no one has ever undertaken the possibility of GAI-driven co-evolution among three parties. In addition, no one has ever revealed the high compatibility between AWS and GAI in the development of the manufacturing industry.

Section 14.2 reviews the realities of the Internet-driven digital economy. Section 14.3 provides an overview of the actual state of leapfrog acceleration and the mechanism that induces it. Section 14.4 analyzes the dynamism of AI/ML-driven innovation creation led by Amazon, the darling of the Internet. Section 14.5 analyzes the dynamism of Japan-India co-evolution towards decarbonization based on this model. Section 14.6 provides a bird's-eye view of the cutting-edge challenges of the above trends. Section 14.7 summarizes the above, and shows the direction of Japan's manufacturing technology renovation over time based on the knowledge obtained.

14.2 The Reality of the Internet-Driven Digital Economy

The Internet-driven digital economy has developed rapidly and globally in both North and South. It can be characterized by three phases as (i) Penetration of the Internet functions: Advances in IoT (Internet of things) (1990s), (ii) Relief from dependency on fixed infrastructure by smartphones (2007–), and (iii) Self-propagating enhancement that encompasses user learning initiated by AI/ML (2010–).

14.2.1 Dissemination of the Internet Functions: Advances in IoT

The investigation on the R&D trends in major ICT-related fields from 1980 to 2015 through bibliometric analysis of 27 representative technologies revealed that R&D, which initially spread across a wide range of fields, converted on R&D in the Internet field between the 1990s and 2015, accelerated the IoT and leading to the development of the Internet that drove the digital economy [20].

14.2.2 Relief from Fixed Infrastructure by Smartphones

A careful tracing of the development trajectory of major ICT-related innovations since the 1970s, it is clearly demonstrated that, following the Internet in the 1990s, smartphones have led the IoT since the early 2000s, liberating people from inevitable dependency on fixed infrastructure, and have played a central role in laying the foundation for the development of AI/ML-driven digital innovation since the 2010s [26]. This liberalization has provided the basis for a rapid transition to a digital economy, especially in emerging countries [16].

14.2.3 Self-Propagating Enhancement that Encompasses User Learning by AI/ML

The hierarchical structure of learning recognition methods based on the core functions of AI and the corresponding development and utilization of AI clearly foreshadow the foundation of the broad development of ML that has become prominent since the 2010s [24].

As mentioned above, the commercialization of the Internet, that originated in the early 1990s, promoted the IoT throughout the 1990s, the spread of smartphones in 2007 led to relieving from fixed infrastructure, and since 2010, AI/ML can be seen that the breakthroughs in ML have led to self-propagating growth that incorporates not only developer innovation but also user learning outcomes [28].

ChatGPT, a conversational generative AI that was launched in November 2022, dramatically spurred "self-propagating growth that incorporates the learning outcomes of users," suddenly attracting attention to creator assets in emerging countries that had previously been overlooked as "sleeping resources."

The challenge of decarbonization is a "condensation of far-reaching system phenomena" and "a challenge for both the Global North and South," requiring the integration of diverse knowledge from stakeholders around the world [27, 28].

This clearly demonstrates the historical significance of the fusion of manufacturing capabilities with the explosive power of DX toward decarbonization [29].

14.3 Leapfrog Acceleration

14.3.1 Digital Transformation Leading to Leapfrog

Advances in the Internet and smartphones have brought the following socio-economic innovations to emerging countries, and have greatly contributed to the dramatic acceleration of the utilization of digital innovation [11]. AI/ML has promoted self-propagating high-performance by encompassing the learning of users particularly of emerging countries.

(i) Relieve from fixed infrastructure dependency,
(ii) Improve access to a wide range of fields such as information, finance, and education,
(iii) Drive innovation, reinvent business models, and accelerate growth,
(iv) Promote internal and external communication and collaboration, pioneering and expanding into new fields, and
(v) Enjoy a wide variety of remote-learning opportunities.

14.3.2 The Reality of Leapfrog

Correlation between growth (*V*), Internet use (*ID*), and smartphone dependence (*SP*) in 39 countries in the North and the South as tabulated in Table 14.1 is illustrated in Fig. 14.2.

Table 14.2 attempted the regression analyses of (i) *V* and *ID*, (ii) *ID* and *SP*, and (iii) *SP* and *V*.

All demonstrate high statistical significance. While *V* induced *ID* with slightly higher elasticity in the emerging countries, *SP* induced *V* with significantly higher elasticity in the developed countries. *ID* induced *SP* with similar elasticity in both economies.

Table 14.1 39 countries examined

Area	Developed countries	Emerging/developing countries	Number of countries
North America	USA, CAN		2
Europe	DEU, GBR, FRA, ITA, ESP	POL	6
Russia, Ukraine		RUS, UKR	2
Middle East	ISR	TUR, LBN, JOR	4
Asia, Pacific	AUS, KOR, JPN	MYS, CHN, IDN, VNM, PHL, IND, THA, PAK	10
Latin America		CHL, MEX, ARG, BRA, PER, VEN	6
Africa		ZAF, GHA, KEN, ETH, SEN, TZA, UGA, NGA, BFA	9
Total			39

These analyses demonstrate that digitalization has driven leapfrogging and formed a virtuous cycle with accelerating growth as demonstrated in Fig. 14.3.

Comparing the elasticities corresponding to the transitions (i) to (iii) above (representing a shift from emerging-country dominance to the same level to developed-country dominance), innovation inducement dynamism in response to the transition from the informal economy to the formal economy can be suggested [1, 3, 7, 12]. Noteworthy is the high elasticity of smartphone usage to GDP per capita across both emerging and developed countries as demonstrated in Fig. 14.4.

The virtuous cycle of accelerating growth described above can be attributed to a virtuous cycle (co-evolution) of rapid diffusion of advanced digital products in emerging countries comparable to that of developed countries, as demonstrated in Fig. 14.5 [5].

This virtuous cycle suggests that the breakthrough of AI/ML that drive post-smartphone digitalization may lead to expanded use of AI/ML in emerging countries [27]. This expansion translates into higher growth elasticity of smartphone ownership in developed countries, as clearly demonstrated in Fig. 14.4. This dynamism suggests the possibility of contributing to the expansion and sophistication of innovation in the creation of new functions [8].

14.3.3 The Realities of the Leapfrog Economy

These analyses support the postulate that the development of the leapfrog economy is evident in the following representative examples of the Global South as demonstrated in Table 14.3.

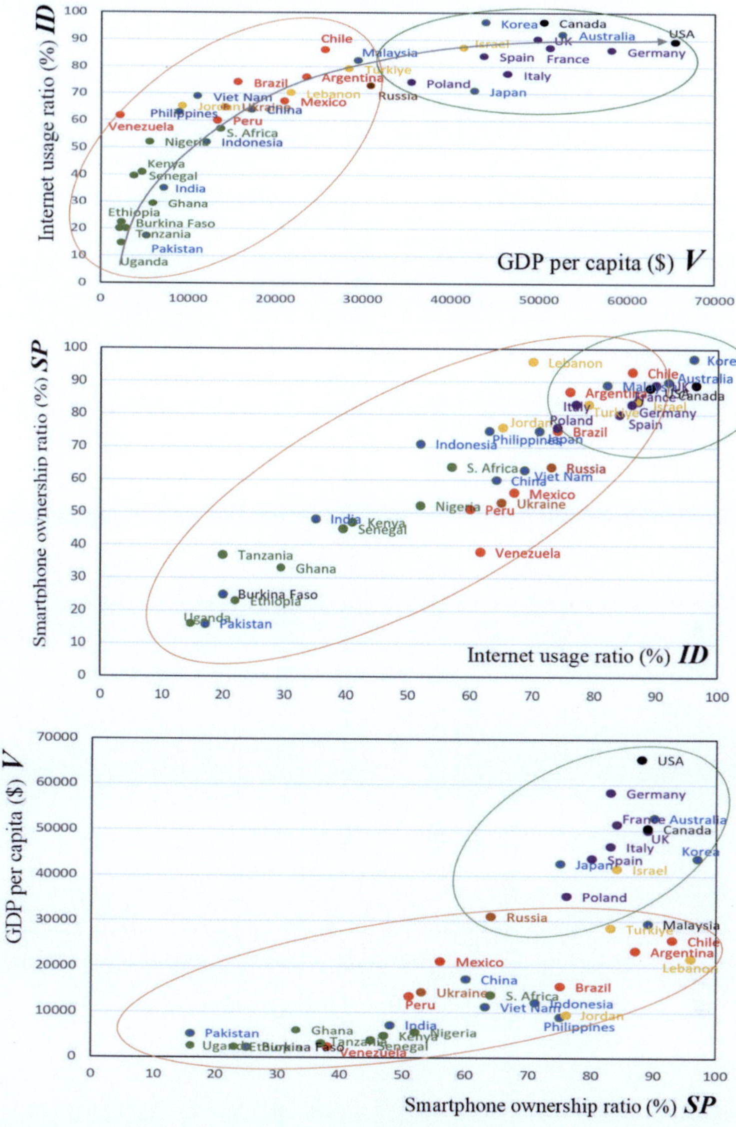

Fig. 14.2 Correlation between digitalization and growth (39 countries in both North and South in 2019)

Table 14.2 Correlation between *V*, *ID* and *SP* in 39 Countries in 2019

Equation		Relationship	Equation (ln form)	t-statistics	adj. R^2
1	Digitalization drives growth which accelerates th Internet dependence	V → ID	lnID = –1.46 + 0.58D1 lnV + 0.55D2lnV + 1.09D	(–3.12) (11.35) (12.47) (4.85)	0.836
2	Internet dependence induces smart phones	ID → SP	lnSP = 0.48 + 0.90D1lnID + 0.89D2lnID – 0.55D	(2.27) (18.30) (16.52) (–3.67)	0.907
3	Smart phones accelerate growth	SP → V	lnV = 3.94 + 1.32D1lnSP + 1.54D2lnSP – 1.01D	(6.48) (8.62) (11.03) (–2.52)	0.913

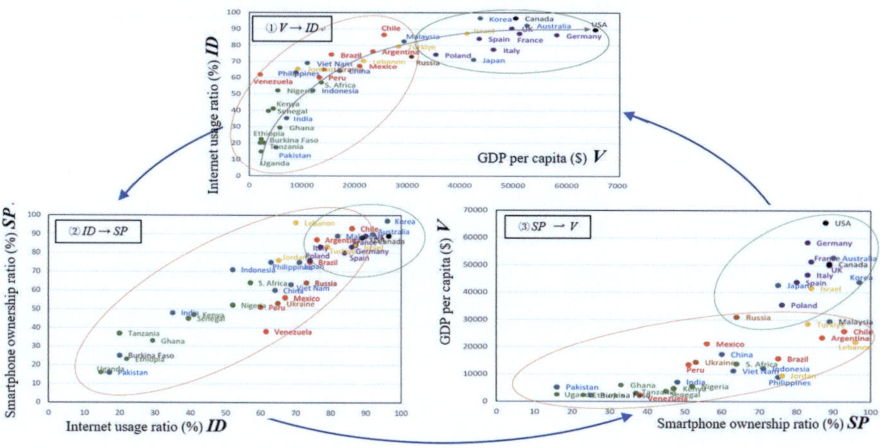

Fig. 14.3 Virtuous cycle between digitalization and growth (39 countries in both North and South in 2019)

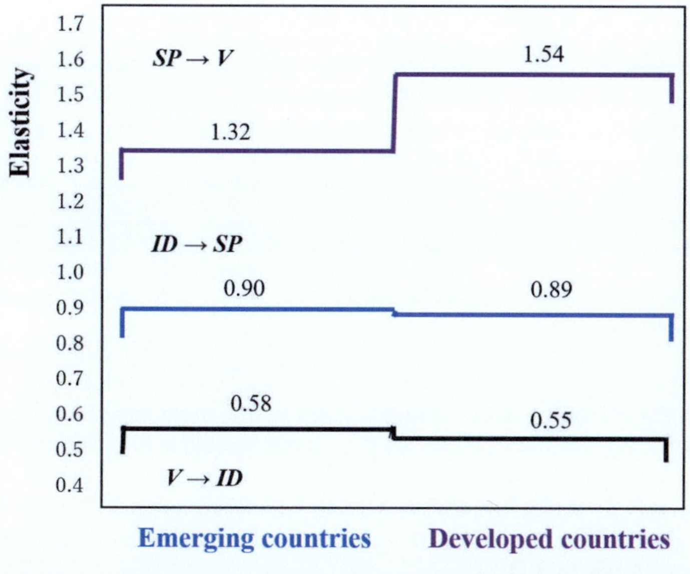

Fig. 14.4 Comparison of elasticity between emerging countries and developed countries by digitalization stage. *V*: GDP per capita, *ID*: Internet usage ratio, *SP*: Smartphone ownership ratio

14.4 The Dynamism of AI/ML-Driven Innovation

14.4.1 Coevolutionary Dynamism Initiated by Amazon

Amazon, the darling of the Internet, has developed a cloud service AWS that contributes to widespread use based on the infrastructure and apps built to develop its business in 2006 with R&D at the core of its management [17, 18, 22, 23, 26].

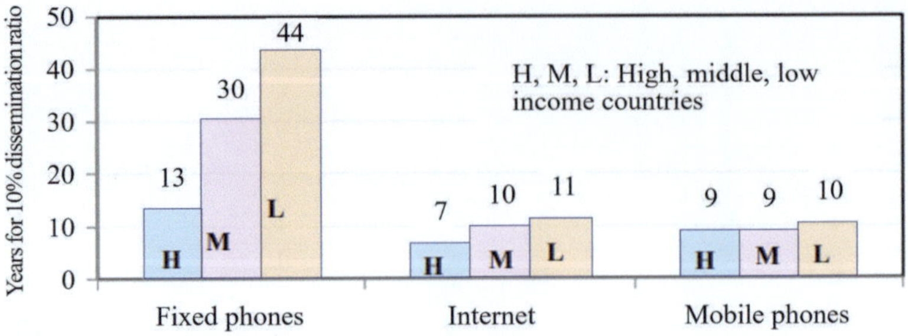

Fig. 14.5 Years for 10% dissemination ratio for fixed phones, internet and mobile phones in high, middle and low income countries. *Source* White Paper on Japan's Information and Communication (2011)

Table 14.3 Advancement of the digital economy and corresponding development of the leapfrog economy

Phase of digital economy	Years	Notable innovation	Typical examples
Dissemination of the Internet; Advances in IOT	1990s	Access to information and global markets; Create innovative economic opportunities	Internet in India and Brazil; IT business in India and the Philippines
Relief from fixed infrastructure by smartphones	After 2007	Mobile revolution (Relief from fixed lines); Enabling the emergence of new innovations	Mobile banking in Kenya; Ride-hailing services in Southeast Asia
Self-propagating enhancement encompassing user learning by AI&ML	After 2010	Improving service offering; Innovation in education and skill development	AI-driven agricultural management in India; Digital education in India

AWS is the crystal of R&D, which creates brand value and encompasses a co-evolution structure that also induces R&D as demonstrated in Fig. 14.6 and Table 14.4 [25].

Through this co-evolution, AWS's ML has been accumulated [25] and new functions (brand value) have been created in a self-propagating way as demonstrated in Fig. 14.7 and Table 14.5 [19].

Since successive investment of R&D constructs technology stock (T) while successive sales of AWS constructs knowledge stock of AWS (KS), both stocks are computed by the cumulative stock of R&D and sales of AWS, taking into account their respective lead times from investment to commercialization and obsolescence rates. Here, planned obsolescence management carefully manages the obsolescence rate (the reverse of lifetime) to maximize the return on the overall investment [21].

KS(T): Knowledge stock of AWS corresponding to T; T: Technology knowledge stock; N and Nk: carrying capacity; a,b,ak, bk; coefficients.

Fig. 14.6 Coevolution between R&D, AWS and Brand Value initiated by Amazon

Table 14.4 Coevolution between R&D, AWS and Brand Value initiated by Amazon (2008–2020: Quarterly)

X to Y	a	b	c	adj. R^2	DW	Dummy
R&D to AWS	–4.64	1.44		0.999	2.53	
	(–56.78)	(159.46)				
AWS to BV	–1.20	0.60		0.983	1.53	
	(–6.12)	(26.00)				
BV to R&D	–2.28	1.16	–0.27	0.993	2.18	2008, 2019, 2020 = 1; others = 0
	(–21.96)	(41.10)	(–3.35)			

lnY = a + blnX + cD
where X, Y: R&D, AWS, and BW; D: Dummy variable
Figures in parentheses are t-statistics: all are significantly at the 1% level

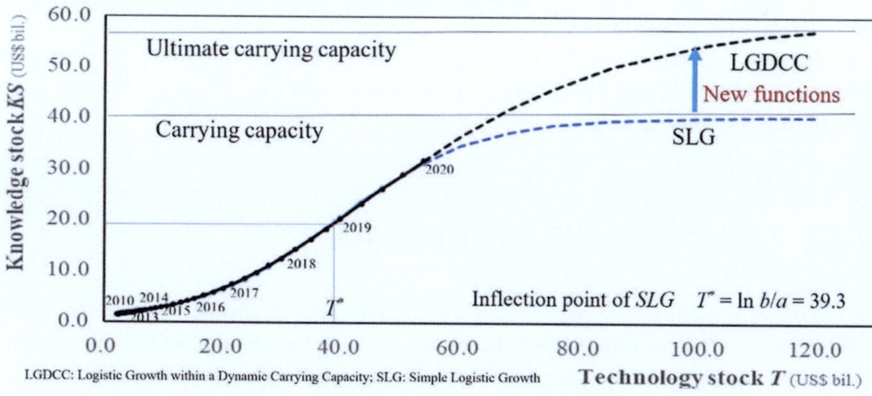

Fig. 14.7 Amazon's R&D-driven AWS stock accumulation trajectory (2010–2020)

Table 14.5 Amazon's R&D-driven AWS stock accumulation trajectory (2010–2020: Quarterly)

Model	N	A	B	Ak	Bk	Adj. R^2
SLG	40,467	8.65×10^{-5}	28.85			0.998
	(65.52)	(83.25)	(55.09)			
LGDCC	59,557	1.18×10^{-4}	40.42	4.58 × 10^–5	5.41	0.999
	(7.31)	(17.87)	(4.69)	(4.91)	(6.58)	

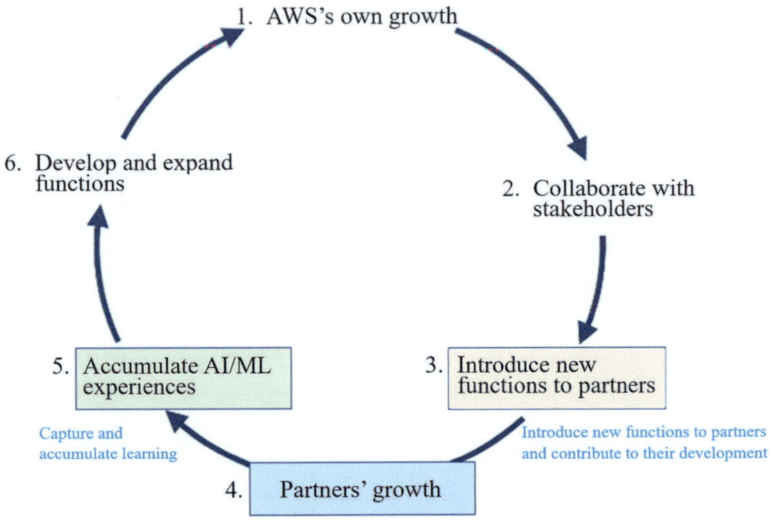

Fig. 14.8 Amazon's coevolution structure assimilating counterpart firm's growth

The new features penetrate the counterpart firms and contribute to their development. By assimilating this learning, AWS will further develop and expand as illustrated in Fig. 14.8 [25, 28].

In this way, Amazon rapidly expanded its R&D, as illustrated in Fig. 14.9, and grew into the world's top R&D firm as demonstrated in Fig. 14.10 by learning and assimilating the growth of its counterpart firms [28, 29].

14.4.2 Global Coevolutionary Dynamism

For the development of the manufacturing industry, AWS and generative AI (GAI) both pursue (i) improved operational efficiency, (ii) enhanced decision-making, (iii) accelerated innovation, (iv) supply chain optimization, and (v) data architecture construction, so in this context there are fundamental commonalities between AWS and GAI [14, 32].

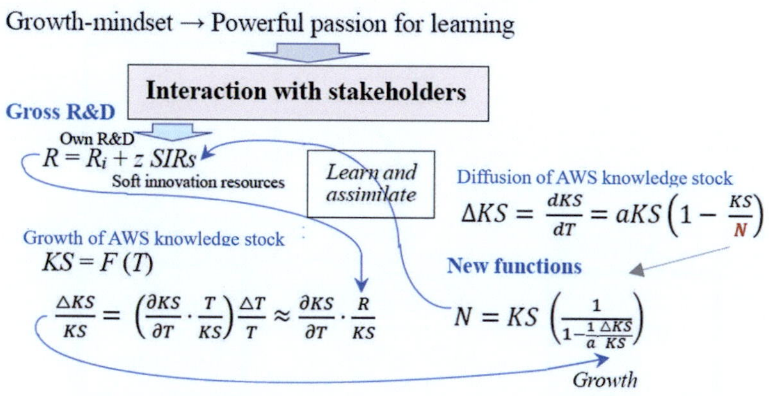

Fig. 14.9 Dynamism of Mutual Enlightenment Embodied in R&D

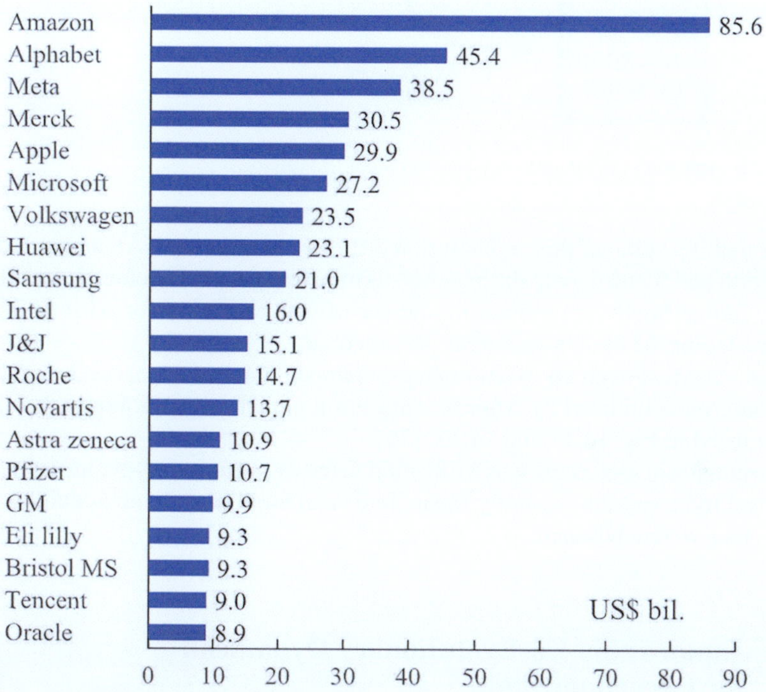

Fig. 14.10 R&D Top 20 Firms (2023)

At the same time, both are in a co-evolution relationship as evolution of cloud services such as AWS stimulates the evolution of GAI, which in turn induces further evolution of cloud services.

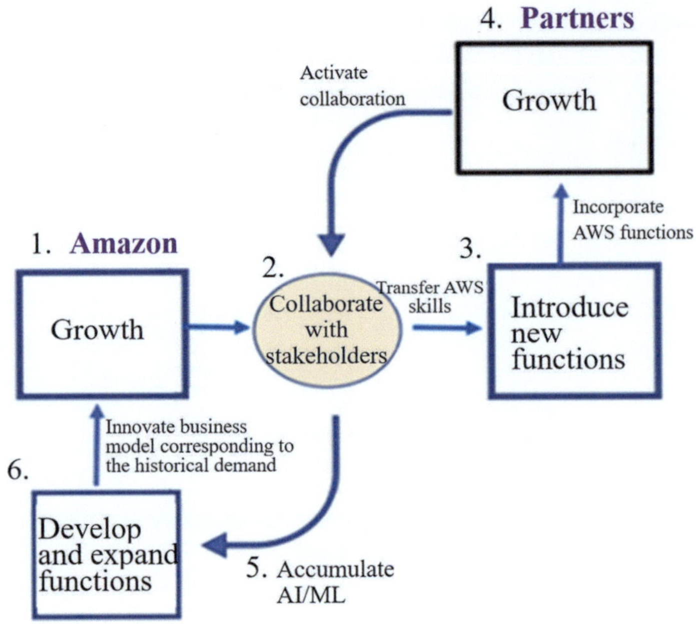

Fig. 14.11 Amazon's AI/ML-driven business model

The manifestation of this co-evolution is the key to utilizing GAI, and in order for both North and South to reap the benefits, (i) improving access to cloud infrastructure for the Global South, (ii) promoting the use of data necessary for AI learning, and (iii) developing AI human resources are essential.

Thus, AI/ML-driven co-evolutionary dynamism that encompasses the growth of the counterparts initiated by Amazon suggests a global co-evolutionary dynamism, as contrasted in Fig. 14.11 and 14.12. [26].

Learning from this contrast as a role model, the dynamics of co-evolution between Japan and India and subsequently Japan, India and the Global South as demonstrated in Fig. 14.1 were envisaged.

14.5 Japan-India Co-Evolutionary Dynamism for Decarbonization

India, which has the world's largest population of 1.4 billion people and is striving to promote manufacturing and digitalization under the banner of *Make in India* (2014) and *Digital India* (2015) policies under the Modi administration [14], is the center of attention of Japanese firms. As shown in Fig. 14.13, we have seen the vigorous expansion of Japanese firms, but with the progress of AI/ML-led innovation, the

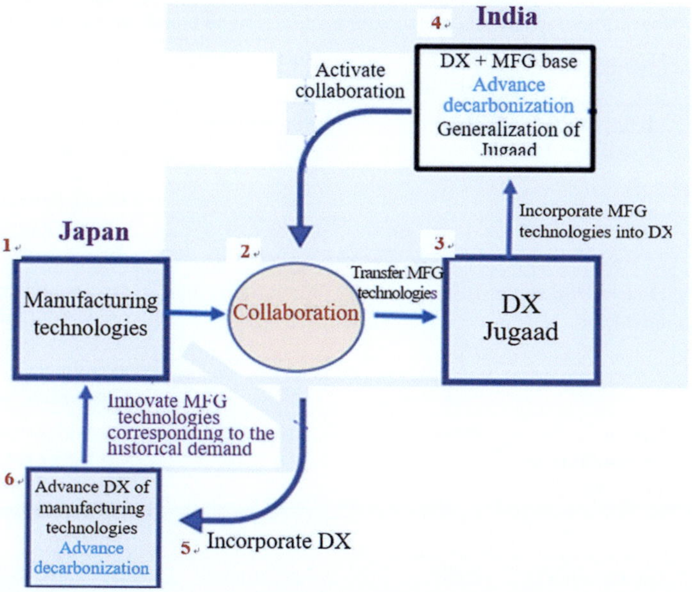

Fig. 14.12 Japan-India coevolution model for decarbonization

conventional pursuit of market and labor has shifted to brand value creation-oriented collaborations [6] such as those shown in Table 14.6.

Decarbonization has spurred this, and the importance of fusing Japan's manufacturing technology and India's DX capabilities has come into sharp focus (Table 14.7).

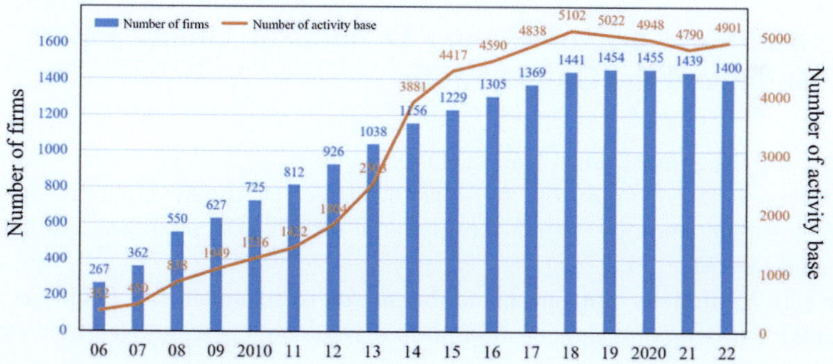

Fig. 14.13 Trend in Japanese Firms entering into Indian market (2006–2022). *Sources* Japanese Embassy in India, and JETRO (2024)

Table 14.6 Noteworthy Japan-India collaboration addressing to brand value creation

Japanese firm	Names in India	Year of establishment	Business focus	Brand values
Suzuki	Maruti Suzuki	1981	Automobile	Improving production efficiency; Predictive maintenance; Strengthening quality control; Customized service
Hitachi	Hitachi Digital Payments	1997	Financial system	Data analytics and prediction; Enhanced security; Automation and efficiency; Customized service
Daikin	Daikin Airconditioning India Pvt	2000	Air-conditioning	Predictive maintenance; Customer solutions; Smart control; Optimizing energy efficiency
Kubota	Escorts Kubota	2008	Agricultural machinery	Predictive maintenance; Smart agriculture; Customized solutions; Improving production efficiency
Nippon steel	AM/NS India	2019	Iron and steel	Improving production efficiency; Predictive maintenance; Strengthening quality control; Supply chain optimization

14.6 Building Co-Evolutionary Dynamism Among Japan, India and the Global South

Recognizing the above trends, Daikin (Daikin Industries, Ltd.) moved its core overseas expansion base from Europe to India and strived to fuse the strengths of its own manufacturing technology with the strengths of India's DX capabilities, as shown in Table 14.8 [13, 31].

Through continuous innovation and the pursuit of sustainability, Daikin is poised to meet the evolving demands of the Indian market and contribute to environmental protection.

This dynamism has been successful in both the absorption of India's DX excellence by Japan's firms and the learning of Japan's manufacturing excellence technology by Indian firms.

As a result, the on-site capabilities of Japan's manufacturing industry (high quality control, precise technical capabilities, efficient production systems, etc.) [10] have

Table 14.7 Fusion between Japan's Manufacturing Technology and India's DX in the Noteworthy Japan-India Collaboration

Company	Details
Maruti Suzuki	Subtle integration of Japan's advanced manufacturing technology and India's digital transformation capabilities to enhance operational efficiency, product quality, and market responsiveness. Leveraging digital twin technology, IoT, predictive analytics, and advanced automation to optimize production processes, improve supply chain management, and deliver innovative, high-quality vehicles tailored to Indian market. This strategic integration enabled maintaining leadership position in the Indian automotive industry and archieving sustainable growth
Daikin Air-conditioning India	Effectively leverage the synergy between Japan's manufacturing excellence and India's digital transformation strengths. Integrate digital twin technology, IoT, predictive analytics, and advanced automation to optimize manufacturing processes, increase supply chain efficiency, and revolutionize customer engagement. These strategic moves not only promoted operational excellence but also established a market leader position in the Indian air conditioning industry
Escorts Kubota	By strategically fusing Japan's manufacturing excellence with India's digital transformation capabilities, significant improvement of operations and market expansion were enabled in India. Adopt advanced manufacturing technology, digital twin technology, IoT, and data analytics to increase production efficiency, optimize supply chain operations, and improve customer engagement. This synergy not only established a leader position in the Indian market, but also established a benchmark for fusing traditional manufacturing strengths with cutting-edge digital innovation

Table 14.8 Fusion of Japan's manufacturing technology and India's DX initiated by Daikin

Contribution	Learning/Absorption
Active participation in Make in India	Collaboration with technology partnerships
Thorough localization	Adoption of digital tools and technology
Partnership with Indian Institute of Technology	Building a smart factory
Cooperation with local suppliers	Digitalization of customer experience
Job creation/technical education	Learning internal innovation culture

been transferred to a wide range of production sites in India as demonstrated in Fig. 14.14.

Consequently, co-evolution dynamism among Japan, India, and the Global South has constructed as illustrated in Fig. 14.1 [30]. The above process is based on Amazon's co-evolutionary dynamics, which learns and assimilates the growth of its counterparts, as reviewed earlier. It was modeled using hybrid AI analysis with ChatGPT and NotebookLM as illustrated in Fig. 14.15, and was then verified and evaluated before being systematized.

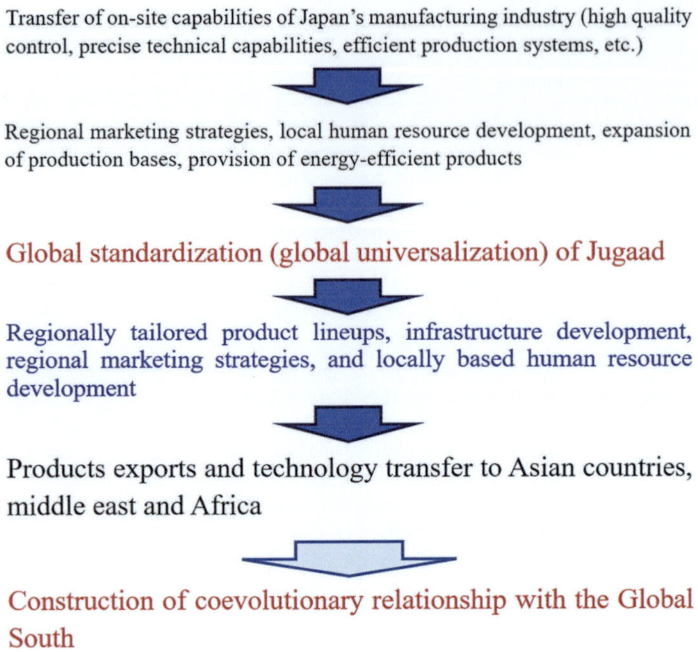

Fig. 14.14 Path to developing to the coevolution among Japan, India and the Global South

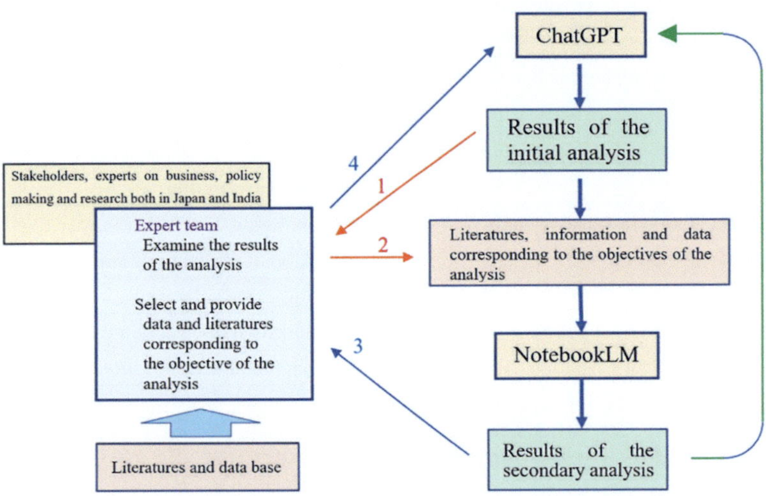

Fig. 14.15 Scheme of the Hybrid AI Analysis

14.7 Conclusion

Daikin, which has taken the lead in building a three-way co-evolution dynamism that encompasses the Global South based on the fusion of Japan and India, actively participated in *Make in India* and contributed to the enhancement of local production capacity. In partnership with the Indian Institute of Technology, it teaches Japan's technological capabilities and manufacturing know-how. Disseminate the on-site capabilities of the manufacturing industry at the grassroots to local suppliers. Strengthen on-site capabilities through technical education for local workers.

At the same time, Daikin learned about Tata Consultancy Services' IoT platform and data analytics capabilities, as well as Wipro's cloud solutions and cybersecurity measures. It applied SIRC (Osaka City University Startup) explosion-proof IoT technology to the Indian market to absorb India's DX power and put it into practical use. It acquired digital technology through the use of India's Production Linked Incentive (PLI) system while contributing to the advancement of this system by realizing the existing impediments associated with current scheme.

At the same time, it committed to digitizing the customer experience and fostering a culture of innovation within the company.

The accumulation of efforts to integrate grassroots manufacturing technology and DX capabilities has led to the global standardization (global universalization) of Jugaad, a traditional innovation measure that transfers the on-site capabilities of Japan's proud manufacturing industry and "seeks frugal, flexible, and inclusive solutions in adversity" unique to India. Based on this, it is also striving to promote co-evolution with African countries such as Nigeria, Algeria, Kenya, and Tanzania, as well as with neighboring countries in Asia and the Global South such as the Middle East, with a focus on co-evolutionary globalization and localization.

These initiatives suggest the direction of Japan's manufacturing technology towards the following renewal corresponding to the historical demand:

(i) **Progress and Embodiment of DX**

Taking advantage of the comparative advantages represented by high-quality control, precise technical capabilities, and efficient production systems, it is recommended to actively cooperate with DX-advanced countries and build a co-promotion structure oriented toward value co-creation that makes full use of AI/ML to embody DX and promote the digitalization of the entire manufacturing process.

(ii) **Rebuilding a Resilient Global Supply Chain**

In addition to decentralizing the supply chain that is heavily focused on China from the perspective of fusing Japan's manufacturing technology and India's DX capabilities, it is recommended to build a supply chain with the coevolution of globalization and localization in mind from a perspective on the entire Global South.

(iii) **Co-Evolution of Growth and ESG**

Based on the above process, it is recommended to build a co-evolution structure between Japan, India, and the Global South to awake the "sleeping resources" associated with decarbonization.

(iv) **Contribution to the World**

By using the above AI/ML-driven innovations it is recommended to provide innovative solutions in the fields of healthcare, education, and agriculture, and create a new global standard through deployment in the Global South.

Further deepening and expanding the analysis of empirical case studies is a focal issue for the future.

References

1. Charmes J, Gault F, Wunsch-Vincent S (2018) Measuring innovation in the informal economy—formulating an agenda for Africa. J Intellect Cap 19(3):536–549
2. Chatterjee S, Chaudhuri R, Vrontis D (2022) Managing knowledge in Indian organizations: An empirical investigation to examine the moderating role of Jugaad. J Bus Res 141:26–39
3. Elgin C, Kose MA, Ohnsorge F, Yu S (2021) Growing apart or moving together? Synchronization of informal and formal economy cycles. Munich Personal RePEc Archive (MPRA) Paper No. 109492
4. Frömmter J (2018) Use Jugaad and Localize: 7 Tips for global startups to expand into India Successfully. https://www.linkedin.com/pulse/use-jugaad-localise-7-tips-global-startups-expand-india-fr%C3%B6mmter (retrieved 0708 2024)
5. Gupta S, Kumar V, Karam E (2020) New-age technologies-driven social innovation: what, how, where, and why? Ind Mark Manage 89:499–516
6. Gupta S (2023) Japanese and Korean Firms in India. IIMB-WP No. 678/2023
7. Lakum AC, Namrata, Kumar H (2024) Diffusion dynamics of the informal sector sustainable innovations: Exploring cases of grassroots innovations in India. Environ Innov Soc Transit 52:100886
8. Okolo CT (2023) AI in the Global South: Opportunities and challenges towards more inclusive governance. Brookings Institution, Washington D.C. https://www.brookings.edu/articles/ai-in-the-global-south-opportunities-and-challenges-towards-more-inclusive-governance/ (retrieved 0909 2024)
9. Rangaswamy N, Densmore M (2016) Understanding Jugaad: ICTD and the tensions of appropriation, innovation and utility. https://www.microsoft.com/en-us/research/wp-content/uploads/2016/02/ICTD2013-Rangaswamy-Jugaad.pdf (retrieved 2024 0708)
10. Sawaguchi M (2016) How does Japanese "Kaizen Activities" collaborate with "Jugaad Innovation"? In: 2016 proceedings of PICMET 16
11. Sharma G, Dahlstrand AL (2023) Innovations, informality, and the global South: A thematic analysis of past research and future directions. Technol Soc 75:102359
12. Sharmelly R, Klarin A (2021) Customer value creation for the emerging market middle class: Perspectives from case studies in India. J Risk Financ Manag 14(10):455
13. Shibata T (2021) Evolution of global product development strategy Daikin's growth trajectory. Discussion Paper Series AGISSS Gakusyuin University, No. 2021–1, Tokyo
14. Srinivasa Rao L (2024) Harnessing the power of secure and scalable generative AI: A deep drive into AWS and SAP's cutting-edge collaboration. Int J Sci Res Comput Sci, Eng Inf Technol 10(5):221–232

15. Thite M (2016) Internationalisation of India Inc.: Tracing the global footprints of Indian Multinationals. In: Thite M, Wilkinson A, Budhwar P (eds) Emerging Indian multinationals: Strategic players in a multipolar world. Oxford University Press, New Delhi
16. Tou Y, Watanabe C, Moriya K, Neittaanmäki P (2019) Harnessing soft innovation resources leads to neo open innovation. Technol Soc 58:101114
17. Tou Y, Watanabe C, Moriya K, Naveed N, Vurpillat V, Neittaanmäki P (2019) The transformation of R&D into neo open innovation: A new concept of R&D endeavor triggered by amazon. Technol Soc 58:101141
18. Tou Y, Watanabe C, Neittaanmäki P (2020) Fusion of technology management and financing management: Amazon's transformative endeavor by orchestrating techno-financing systems. Technol Soc 60:101219
19. Watanabe C, Kondo R, Ouchi N, Wei H, Griffy-Brown C (2004) Institutional elasticity as a significant driver of it functionality development. Technol Forecast Soc Chang 71(7):723–750
20. Watanabe C, Tou Y, Neittaanmäki P (2018) A new paradox of the digital economy: Structural sources of the limitation of GDP statistics. Technol Soc 55:9–33
21. Watanabe C, Naveed N, Neittaanmaki P (2018) Digitalized bioeconomy: Planned obsolescence-driven circular economy enabled by co-evolutionary coupling. Technol Soc 56:8–30
22. Watanabe C, Tou Y (2020) Transformative direction of R&D: Lessons from amazon's endeavor. Technovation 88:102081
23. Watanabe C, Tou Y, Neittaanmäki P (2020) Institutional Systems Inducing R&D in Amazon: The role of an investor surplus toward stakeholder capitalization. Technol Soc 63:101290
24. Watanabe C, Akhtar W, Tou Y, Neittaanmäki P (2021) Amazon's initiative transforming a non-contact society: digital disruption leads the way to stakeholder capitalization. Technol Soc 65:101596
25. Watanabe C, Akhtar W, Tou Y, Neittaanmäki P (2021) Amazon's New Supra-omnichannel: Realizing growing seamless switching for apparel during COVID-19. Technol Soc 66:101645
26. Watanabe C, Tou Y, Neittaanmäki P (2021) Transforming the socio economy with digital innovation. Elsevier, Amsterdam
27. Watanabe C, Tou Y (2021) Digital awakening of the Japanese model towards carbon neutrality: ESG-oriented stakeholder capitalism following the example of amazon. In: Proceedings, annual general conference, Japan society for research policy and innovation management, Tokyo, pp 368–373
28. Watanabe C, Tou Y (2022) Renewing Japanese-style innovation for decarbonization: Fusing diverse knowledge following the example of amazon's climate pledge. In: Proceedings, annual general conference, Japan society for research policy and innovation management, Tokyo, pp 696–701
29. Watanabe C, Tou Y (2023) Next-generation technology management: Fusing manufacturing capabilities for decarbonization with digital transformation capabilities of the global South. In: Proceedings, annual general conference, Japan society for research policy and innovation management, Tokyo, pp 269–274
30. Watanabe C, Tou Y (2024) A revolution in manufacturing technology: fusing manufacturing capabilities for decarbonization with the explosive digital transformation potential of the Global South—Japan-India decarbonization cooperation to advance jugaad global standardization. In: Proceedings, annual general conference, japan society for research policy and innovation management, Tokyo, pp 255–260
31. Xu F (2018) A "Win-win" Model between Daikin's and Gree. Int J Manag Appl Sci 4(6):1–6
32. Yadav D, Zhang X, Jin BT, Krishnan P, Clarke D (2024) Generative AI based virtual assistant for reconciliation research. Amazon Science 77(28):1–4

Chapter 15
An Interpretation of China's Actions for Global Climate-Neutral Aviation

Zhang Shukun, Ji Yuhan, and Bai Wen

Abstract One major application area of the CM3—Transport 2023 conference is climate-neutral aviation. Recent actions of China government and aviation industry targeting carbon dioxide peaking before 2030 and carbon neutrality before 2060, are interpreted with a bird's eye view focusing on aviation industry, hopefully for fitting better the science & technology research requirements at large, and particularly in the field of Computational Multi-Physics, Multi-Scales and Multi Big Data.

Keywords Climate-Neutral Aviation · Government policy · MDAO · MBSE

15.1 Introduction

Climate change is a challenge for all of humanity. The People's Republic of China attaches great importance to its response to climate change. Tightly following the government's responding to climate change, national functional ministries and commissions, industry enterprises including aviation industry, formulate action plans for carbon peaking and neutrality, and take actions.

Following an interpretation of China's actions for global climate-neutral aviation, key S&T challenges are analyzed. It is concluded that the Multidisciplinary Design

---

The last author is appreciated for being invited to join the Scientific and Technical Committee of the CM3 Transport 2023 Conference and would like to dedicate the paper to the 80th anniversary of Mr. Jacques Periaux.

Z. Shukun
Quality and Safety Division, Chinese Aeronautical Establishment, Beijing, P. R. China

J. Yuhan
Aviation Industry Development Research Center of China, Beijing, P. R. China

B. Wen (✉)
New Concept Air Vehicle Technology Research Institute, Chinese Aeronautical Establishment, Beijing, P. R. China
e-mail: baiwen@cae.ac.cn

T. Tuovinen et al. (eds.), *Challenges in Design Methods, Numerical Tools and Technologies for Sustainable Aviation, Transport and Industry*, Computational Methods in Applied Sciences 17, https://doi.org/10.1007/978-3-031-98675-8_15

Analysis Optimization (MDAO) technology is crucial towards climate-neutral aviation, integrating MDAO into the aircraft lifetime through Model Based Systems Engineering (MBSE) has been well implemented and practiced, in which various applications of high-fidelity aeronautics numerical simulation technology are inherently multi-physics, multi-scales, and multi big data.

15.2 Government Actions

The State Council Information Office (SCIO) of China had regularly released a series of white papers titled "Responding to Climate Change: China's Policies and Actions" since 2008. The main chapters of two most recent versions are sited below for interpretation. The main chapters of 2021 version [1] published by SCIO include: China's New Responses to Climate Change, Implementing a National Strategy of Actively Responding to Climate Change, Significant Changes in China's Response to Climate Change, and Building a Fair and Rational Global Climate Governance System for Win–Win Results. It is concluded: "Challenges posed by climate change are real, severe and lasting. The response requires the joint effort of all the international community, if we are to leave a clean and beautiful world to future generations. China will honor its promises and continue to support multilateralism, however the global situation changes. It will work with other parties to achieve the full, balanced, effective and sustained implementation of the United Nations Framework Convention on Climate Change and the Paris Agreement, to fulfill its NDC goals, to control greenhouse gas emissions, and to increase its ability to adapt to climate change. It will redouble its efforts to promote a global community of shared future, and make a greater contribution to a better home planet for all humanity."

In October 2021, "Action Plan for Carbon Dioxide Peaking Before 2030" was formulated and published by SCIO [2]. The actions related to aviation sector include: to optimize the industrial structure, accelerate efforts to eliminate outdated production capacity, vigorously develop strategic emerging industries, and speed up green and low-carbon transformation in traditional industries; to spur industrial energy consumption to go low-carbon, promote clean and efficient use of fossil fuels, increase the proportion of renewable energy used, strengthen electricity demand-side management, and raise the level of industrial electrification; to thoroughly implement the green manufacturing project, vigorously promote green design, refine the green manufacturing system, and build green factories and industrial parks; to drive integrated development of digital, smart, and green technology in the industrial domain, and bolster technological transformation in key industries and sectors.

The actions to promote green and low-carbon transportation include: promoting low-carbon transformation of transportation vehicles and equipment, making airport operations based increasingly on electric power and smart technology, and developing new-energy aircraft, by 2030, the share of incremental vehicles fueled by new and clean energy will reach around 40%, carbon emission intensity of commercial vehicles measured on the basis of converted turnover will be cut by about 9.5%

compared with 2020; developing green, high efficiency transportation systems, developing smart transportation and promote rational division of functions and effective linkages between different means of transportation, so as to reduce empty-loading ratio and unreasonable turnover in passenger and freight transportation, boosting application of advanced and applicable technologies so as to make operation and management of civil aviation more efficient, guiding businesses in the aviation sector in making their operations smarter and saving energy and reducing carbon emissions in a systematic manner; accelerating construction of green transport infrastructure, implementing green and low-carbon thought throughout the whole process of planning, building, operating, and maintaining transport infrastructure, reducing lifecycle energy consumption and carbon emissions, carrying out green and low-carbon transformation of transport infrastructure, and make utilization of infrastructure more efficient by coordinating the use of route, land, and airspace resources in comprehensive transportation corridors, by 2030, all ground vehicles and equipment at civil airports will strive to be powered by electricity.

Main chapters of 2022 anniversary report [3] released by Ministry of Ecology and Environment (MEE) of China include: New Arrangements on Combating Climate Change, Actively Mitigating Climate Change, Actively Adapting to Climate Change, Improving Policy Systems and Support Ensuring, Actively Getting Involved in Global Governance in Response to Climate Change, Basic Positions and Propositions on the 27th Conference of the Parties to the United Nations Framework Convention on Climate Change. In the preface of the report, it is pointed out: "Since 2021, China has actively implemented the Paris Agreement on climate change, updating its Nationally Determined Contributions (NDCs) goals, earnestly carrying out effective and well-conceived actions and making significant progress in meeting the targets of carbon dioxide peaking and carbon neutrality. Based on preliminary calculations, in 2021, the carbon intensity of China dropped by 3.8 percent and 50.8 percent from the level respectively in 2020 and 2005, the share of non-fossil fuels in primary energy consumption rose to 16.6 percent, the total installed capacity of wind and solar power generation combine increased to 635 million kilowatts, the coal consumption per unit of GDP dropped significantly, and the forest coverage and stock have both risen the past 30 consecutive years. By the first anniversary of the national carbon trading system, the cumulative volume of Carbon Emission Allowances (CEA) was 194 million tonnes, and the cumulative turnover CNY 8.492 billion."

It is also pointed out that China has put in place "1 + N" policy framework for carbon dioxide peaking and carbon neutrality, developed a mid-term and long-term strategy for controlling greenhouse gas emissions, accelerating the development of a national carbon market, and formulated and implemented the National Strategy for Climate Change Adaptation. The "1 + N" policy framework comprises of two top-level design documents, i.e., the Working Guidance for Carbon Dioxide Peaking and Carbon Neutrality in Full and Faithful Implementation of the New Development Philosophy, the Action Plan for Carbon Dioxide Peaking Before 2030, and the action plans to achieve carbon emissions peak in key areas and sectors such as energy, industry, transport, urban and rural construction, iron & steel, non-ferrous metals, and cement, plus the support measures in technology, finance, standards, and talent development. In accordance with the Action Plan for Carbon Dioxide Peaking Before 2030 [2], China will take 10 actions: (1) for green and low-carbon energy transition; (2) for energy saving, carbon emissions reduction, and efficiency improvement; (3) for carbon emissions peaking in industry; (4) for carbon emissions peaking in urban and rural development; (5) for promoting green and low-carbon transport; (6) for facilitating carbon emissions reduction through the circular economy, (7) for encouraging innovation in green and low-carbon technology; (8) for consolidating and improving carbon sink capacity; (9) for advocating a green and low-carbon lifestyle nationwide, and (10) for reaching carbon emissions peak by different regions in a structured and orderly manner.

It is just our personal initial realization to site also the following highlights, which are more or less relevant to aviation industry: Adjusting the Industrial Structure, Strengthening Climate Change Monitoring, Early Warning and Risk Management, Accelerating the Development of Greenhouse Gas Statistical Accounting and Monitoring System, Strengthening the Support of Scientific and Technological Innovation, Deepening High-Level Exchanges on Climate Change, Promoting Multilateral and Bilateral Climate Change Negotiations, and Strengthening Pragmatic Cooperation on Climate Change.

SCIO published recently "China's Green Development in the New Era" [4]. The white paper elaborates on the underlying philosophy of China's green development in the new era, systematically summarizes China's actions and achievements in green development in the new era, and offers with a large number of facts on China joining hands with the rest of the world to build the Earth into a beautiful home. Only the titles of main chapters are cited here and full text could be easily accessed by the referenced link: Staying Firmly Committed to Green Development, A Basic Green Territorial Configuration Is in Place, Adjusting and Improving the Industrial Structure, Extensive Application of Green Production Methods, Eco-Friendly Living Becomes the Prevailing Ethos, Improving the Institutions and Mechanisms for Green Development, and Building the Earth into a Beautiful Home.

Ministry of Finance (MOF) released in May 2022 "Working Guidance for Financial Support of Carbon Dioxide Peaking and Carbon Neutrality" [5]. The main objectives are: by 2025, fiscal policy instruments have been enriched, a fiscal and tax policy framework conducive to green and low-carbon development has been preliminarily established, strongly supporting all regions and all sectors to speed up green and low-carbon transformation; by 2030, a fiscal and tax policy system conducive to green and low-carbon development has preliminarily taken shape, a long-term mechanism for promoting green and low-carbon development has been gradually established, pushing the smooth and successful realization of the carbon peak target; by 2060, the policy system of fiscal support for green and low-carbon development is mature and sound, pushing the smooth and successful achievement of the goal of carbon neutrality.

It is proposed to support key directions and areas, including: building a clean, low-carbon, safe and efficient energy system, green and low-carbon transformation in key industries and areas, green and low-carbon S&T innovation and basic capacity building, green low-carbon life and economical use of resources, consolidating and improving carbon sink capacity, perfectly improving the green and low-carbon market system. (This is an unofficial translation. In case of any divergence, the official text in the Chinese language shall prevail).

Ministry of Science and Technology (MOST), National Development and Reform Commission (MDRC) and (NDRC) and MEE released jointly "Implementation Plan for S&T Supporting Carbon Peaking and Carbon Neutrality" [6] in June 2022. Main actions include: (1) energy, green and low-carbon transformation supported by science and technology, (2) low-carbon and zero-carbon industrial process reengineering technology breakthrough action, (3) low-carbon and zero-carbon technologies for urban and rural development and transportation, (4) technology capacity enhancement initiatives for negative carbon and non-carbon dioxide greenhouse gas emission reduction, (5) innovation initiatives for cutting-edge disruptive low-carbon technology, (6) low carbon and zero carbon technology demonstration action, (7) support actions for carbon peak carbon neutral management decisions, (8) carbon peak carbon neutral innovation projects, bases, talents synergistic action, (9) cultivation and service action for green and low-carbon S&T enterprise, (10) international cooperation action for carbon peak carbon neutral S&T innovation. (This is an unofficial translation. In case of any divergence, the official text in the Chinese language shall prevail).

Ministry of Industry and Information Technology (MIIT), MDRC and MEE released jointly "Carbon Peaking Implementation Plan for Industrial Sector" [7] in July 2022. Key tasks and major actions related to aviation sector include: creating and building advanced manufacturing clusters with obvious effects of low-carbon transformation, creating and building digitalized and coordinated green supply chains; focusing on reengineering of high-value key components such as aeroengines, creating and building remanufacturing innovation carrier, accelerating innovation and industrial application of key remanufacturing technologies including additive manufacturing, flexible molding, special materials and nondestructive testing; responding to the green and low-carbon requirements of industry sector including aviation industry, focusing on key processes, strengthening the integration of advanced casting, forging, welding, heat treatment and other basic manufacturing processes with new technologies, implement intelligent and digital transformation; promoting green development of the next generation of domestically produced civil aircraft, actively developing electric aircraft and other new-energy aircraft. (This is an unofficial translation. In case of any divergence, the official text in the Chinese language shall prevail).

China has continuously improved its standard system of climate change response to the alignment and integration of new and existing standards and norms. The first national carbon measurement center has been approved for preparation in May 2022. State Administration for Market Regulation (SAMR), NDRC, and MIIT etc., released jointly "Implementation Plan for Establishing Sound Standard and Measurement System of Carbon Peaking and Carbon Neutrality" [8] in October 2022. The planned key project and actions include: implementing S&T innovation projects for carbon metering, implementing projects to elevate basic capacity of carbon metering, implementation of carbon metering benchmarking leading projects, carrying out projects for providing precise carbon metering services, implementing international exchange and cooperation projects on carbon metering, carrying out actions to strengthen the base of double carbon standards, launching a hundred initiatives to improve energy conservation and carbon reduction standards, leading the standards for cutting-edge low-carbon technologies, carrying out international cooperation actions on green and low-carbon standards. (This is an unofficial translation. In case of any divergence, the official text in the Chinese language shall prevail).

15.3 Aviation Transportation Industry Actions

Civil Aviation Administration of China (CAAC) released "2022 China Civil Aviation Green Development Policy and Action" [9] on December 30, 2022. In accordance with the China's policy and roadmaps to achieve the grand national goal of the carbon peaking and carbon neutrality, CAAC has issued and implemented a serious planning and policies to enhance the construction of a green, low-carbon and circular civil aviation industry, which have formed the planning and policy system for the green development of civil aviation in China and chart the course for the green, low-carbon and circular development of the industry: The 14th Five-Year Plan for Civil Aviation Development (December 2021); The 14th Five-Year Specialized Plan for Civil Aviation Green Development (December 2021); Action Plan for New Generation of Smart Civil Aviation (June 2021); Action Plan for the Construction of Safe, Green, Smart and Humanitarian Airports (2020- 2035) (December 2020); Guidance on Further Promoting Green Development of Civil Aviation; Guidance on Promoting New Infrastructure Construction for High-Quality Development of Civil Aviation (December 2020); Action Plans on the Construction of Safe, Efficient, Smart and Coordinated Air Transport Management; The Roadmaps to Construct a Smart Civil Aviation (January 2022).

In 2021, the total transportation turnover, passenger traffic and cargo and mail traffic of China's civil aviation reached 85.67 billion tons kilometers, 440 million passengers and 7.31 million tons respectively; with 4,054 aircrafts with an average age of 8.6 years. 60 cities in 41 countries and regions have been accessible, with 279 international routes. The number of civil airports reached 248; the annual passenger and cargo and mail throughput of transport airport reached 907 million passengers and 17.83 million tons respectively.

In 2019, China's civil aviation fuel consumption per ton kilometer was 0.285 kg, and the energy consumption and CO2 emissions per passenger at airports decreased by about 15.8% and 43.2%, respectively, compared with the average level of 2013–2015. In 2021, China's civil aviation fuel consumption per ton kilometer was 0.309 kg, with accumulative reduction of 400million tons of CO2 emissions compared with 2000; airport energy consumption per passenger increased by about

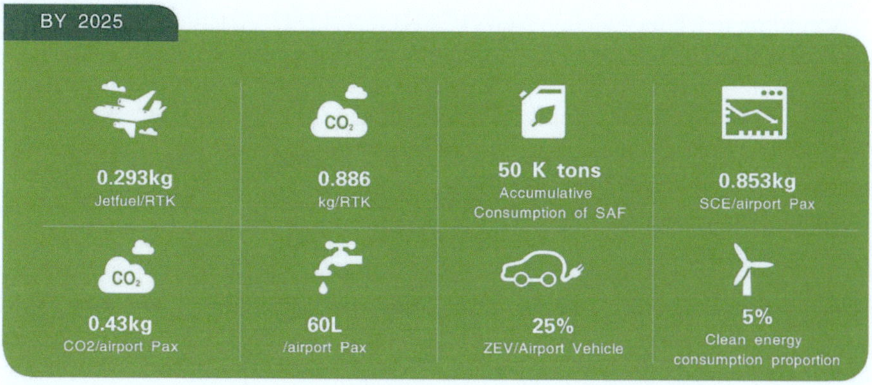

Fig. 15.1 CAAC goals for reducing pollution and carbon emissions by 2025

Fig. 15.2 CAAC goals for reducing pollution and carbon emissions by 2035

2.3% compared with the average level of 2013–2015, and CO2 emissions per passenger decreased by 22.1%.

CAAC goals for reducing pollution and carbon emissions by 2025 and by 2035 are illustrated in Figs. 15.1 and 15.2.

Notice of CAAC on guidelines for Construction and Development of Civil Aviation Big Data had been announced on October 9, 2022 [10]. In which, overall consideration and development goals, main tasks, safeguard measures, were described.

15.4 Aviation Manufacturing Industry Actions

China's aviation industry has been actively exploring technologies to achieve climate-neutral aviation, mainly focused on developing green civil aircraft technology, exploring new energy aircraft technology, strengthening green and low-carbon basic technology, and establishing green orientation for quality development.

According to China Aviation News (CAN) report [11], Aviation Industry Corporation of China (AVIC) deployed to comprehensively promote carbon peaking and carbon neutrality work. AVIC Work Plan for Carbon Peaking and Carbon Neutrality had been deliberated and approved. The so-called “Double Carbon” objectives of AVIC are: by 2025, green aviation constructing system has been preliminary formed, energy efficiency has been substantially improved; by 2030, green manufacturing system has been basically established, energy efficiency has been further improved; by 2060, an industry system of green zero-carbon aviation has been comprehensively established, energy efficiency has achieved international advanced level, the core competing advantage of green and low carbon emissions formed, fruitful progress has been made in ecological civilization, carbon–neutral objective comprehensively achieved. Focusing on the above goals, the plan proposed specific measures in such aspects as establishing and improving the carbon emission management mechanism, strengthening the carbon reduction action of green products, deploying the carbon reduction action of green industrial chain, promoting the carbon reduction action of enterprise sites, improving the management ability of carbon assets and carbon trading, accelerating the construction of green support platform, and improving the regulatory mechanism of green development.

AVIC implements its carbon peaking action plan through multiple measures to advance energy conservation and carbon reduction in aircraft manufacturing processes. Action No.1: strengthening carbon emission data management by promoting a unified system for carbon emission monitoring, inventory verification, and reporting. The corporation has issued the Guidelines for Greenhouse Gas Accounting and Reporting, requiring key energy-consuming entities to conduct carbon emission accounting, with verification of their calculations. Action No.2: releasing “Action Plan for Energy Efficiency Improvement in Key Energy-Consuming Systems (2024–2026)”, outlining four major tasks: conducting comprehensive energy audits, enhancing energy efficiency in power systems, improving process efficiency, and strengthening foundational energy-saving practices. Action No.3: enforcing legally compliant management of energy conservation reviews, environmental impact assessments, and energy/environmental acceptance checks for fixed-asset investment projects to ensure risk control at both source and process levels. Action No. 4: encouraging to expand green electricity adoption across scenarios and scales to reduce carbon emissions. Meanwhile, AVIC drives decarbonization in the aviation sector by advancing new-energy aircraft development, researching green and low-carbon critical technologies, and fostering international collaboration on green and low-carbon initiatives.

The Chinese Aeronautical Establishment (CAE) is a state-owned, government-sponsored overarching aeronautical research organization in China. CAE has also formulated an action plan for achieving carbon peaking and put forward the general idea of practicing the “double-carbon” strategy through scientific and technological innovation. The main objective by 2030 is to form landmark application achievements transformed through green aviation science and technology, to promote the institute’s own low-carbon and high-quality development through energy-saving and efficiency improvement, new energy utilization and other measures, and to improve measurement, standards, certification consulting and other basic services. In terms of new energy aircraft technology, research on key technologies and systems of electric vertical takeoff and landing flight platforms is being carried out, including low noise and efficient rotor design, rotor tilt and transition state control, high performance variable speed motor and other technologies. Also, on-going research on key technologies of electric commuter aircraft with distributed propellers. In terms of green and low-carbon fundamental technologies, research on common technologies related to green aviation, such as standards, metrology, airworthiness and intellectual property rights, keeps on-going. And the green orientation for quality development is establishing for green life style.

CAE applied the Life Cycle Assessment (LCA) methodology to develop a carbon footprint accounting model for typical aircraft components. Using this model, the institute analyzed CO_2 emission patterns from the "cradle-to-gate" phase (raw material extraction to factory exit) of representative aero-engine components. The study revealed that the raw material acquisition stage accounts for a significantly higher proportion of carbon emissions compared to the manufacturing phase in typical aviation engine components. This highlights the critical importance of establishing green supply chains to control emissions at the raw material stage for the aviation manufacturing sector. However, the most crucial task for the aviation industry remains reducing operational-phase emissions, necessitating accelerated development of new-energy aircraft.

Further findings indicate that aviation products involve complex material compositions and manufacturing processes, with existing databases proving inadequate for precise calculations. Additionally, there is a lack of standardized calculation rules and methodologies. To address these challenges, the project proposes establishing a civil aircraft carbon footprint management system. This includes advancing research on carbon accounting from individual components to complete aircraft, promoting unified carbon footprint calculation standards for civil aviation, and progressively building a foundational carbon database for the aviation industry.

Commercial Aircraft Corporation of China, Ltd. (COMAC) brings up building "Green COMAC" philosophy, through research and development of "Green Aircraft", establishing "Green Homeland", constructing "Green Industrial Chain", promoting sustainable development of green recycling, firmly promote the implementation of the Company's "14th Five-Year" Green Development Plan.

In a prospect of new energy aviation development under carbon peaking and carbon neutrality goals by JI Yuhan (Aviation Industry Development Research Center of China) and SUN Xiasheng (CAE) et al. [12], the successful stories of two SMEs in China had been marked. The two- seats electric aircraft, ER1E, produced by Rhyxeon General Aircraft Co., Ltd. (RGAC), obtained the production license in February 2015 and approved to put into operation in early 2017. It is also China's first new energy light-sport aircraft with completely independent intellectual property right. The range-extended version (RX1E-A) made its first successful flight in November 2017, while RX4E is the first four-seats electric aircraft in the world to apply for airworthiness certification of CCAR-23 aircraft. The second marked SME

"EHang" is an autonomous aerial vehicle technology platform company headquartered in Guangzhou city of China, honorably developed and produced "EHang 216", an electric vertical takeoff and landing aircraft with 8 cantilevers and 16 rotors, both two-seats manned and logistics version. The aircraft production has completed manned test flights in China, Austria, Qatar, Japan and other countries. The team leaded by North-West Technology University had achieved successful flight tests of NPU-BWB-300, an innovative Blended Wing Body concept model, on January 2023.

In the field of so-called "Low Altitude Economy", lots of low altitude air vehicles, which are supposed to be environmentally friendly, are under development and airworthiness certificate. Recently, in December 2024, AutoFlight's unmanned cargo eVTOL "CarryAll" becomes the world's first 2-Ton eVTOL aircraft to receive production license. EHang's EH216-S eVTOL operators obtained Air Operator Certificates (OC) for civil human-carrying pilotless aerial vehicles in March 2025 granted by CAAC. Before that, the vehicle had obtained type certificate (TC), standard airworthiness certificate (AC), and production certificate (PC). Several innovative concepts of high speed eVTOL for urban air mobility, cargo transportation, and emergency rescue, are at different development phases at CAE.

15.5 S&T Requirements of Global Climate-Neutral Aviation

A nice classification of S&T thematic areas that contribute to climate-neutral aviation from Horizon 2020 projects implemented by the European Union Innovation and Networks Executive Agency (INEA) has been presented by EC project/programme officers [13]: 1. Disruptive Airframe Configurations; 2. Sustainable Aviation Fuels; 3. Hybrid-Electric Propulsion; 4. Hydrogen Propulsion; 5. Lightweight Structures and Advanced Manufacturing Processes; 6. Operations Impact on Climate Change; 7. Air Quality and Noise.

An excellent site to find the clues of technology roadmaps leading to zero-carbon emission commercial aviation by 2030 is the project FlyZero led by the Aerospace Technology Institute and backed by the UK government. FlyZero has concluded that green liquid hydrogen is the most viable zero-carbon emission fuel with the potential to scale to larger aircraft of operational ranges up to 5,250 nautical miles (= 9723 km) utilizing fuel cell, gas turbine and hybrid systems, this range could enable all worldwide flight paths with one stop [14]. FlyZero has identified two propulsion architectures which could deliver large scale zero-carbon flight: A liquid hydrogen fuel cell electric aircraft, capable of scaling to regional aircraft, and A liquid hydrogen gas turbine aircraft, capable of scaling to midsize aircraft.

The FlyZero technology roadmaps outline developments which could be aligned to one or both architectures. Collectively they detail a route to transforming the aviation industry to utilize hydrogen as its primary fuel and potentially completely decarbonize aviation. The ATI FlyZero project team has identified 13 technology 'bricks' that are required to enable hydrogen flight. Six hydrogen aircraft bricks are revolutionary aerospace technologies fundamental to realizing liquid hydrogen fueled aircraft: Aerodynamic Structures, Thermal Management, Hydrogen Gas Turbines & Thrust Generation, Cryogenic Hydrogen Fuel System and Storage, Electrical Propulsion System, and Fuel Cells. Seven cross-cutting bricks are critical to ensuring that hydrogen aircraft are commercially and operationally enabled and deliver tangible sustainability improvements: Aircraft Systems; Lifecycle Impact; Hydrogen Infrastructure and Operations—Airports, Airlines, Airspace; Advanced Materials; Advanced Manufactoring; Compressed Design and Validation – Culture and Digital Tools; and Sustainable Cabin Design. A technology roadmap has been generated for each technology brick to enable technology readiness level (TRL) 6 of the first generation of hydrogen technology by 2025. Major findings for each brick as well as integration steps are outlined in the report [14], and the technology roadmaps of each brick are publicly available on the FlyZero website, https://www.ati.org.uk/flyzero-reports/. Take a browse through the EC-funded projects and the roadmaps proposed by FlyZero, it can be inferred quite clearly that multidisciplinary design optimization technology is crucial towards climate-neutral aviation, which should take into account of aerodynamics, structures, propulsions, flight control, and the whole lifetime operation in view of systems engineering.

The EU funded H2020 projects AGILE [15, 16] and AGILE 4.0 [17, 18] targets the digital transformation of main pillars of the aeronautical supply-chain: design, production, certification and manufacturing. The composition of the AGILE 4.0 consortium and capabilities available enable to address realistic development scenarios integrating multiple stakeholders and covering all the aspects of the development of complex aeronautical systems. The application of AGILE 4.0 technologies had been well demonstrated by different test cases, including multidisciplinary design of a more electric regional aircraft including certification constraints [19], where different on-board systems architectures with increasing electrification levels are considered, and the aircraft design includes aerodynamic, structural, systems and propulsion domains. The results show a reduction of 3% of MTOM and 3% of fuel mass depending on the systems architecture selected. From the certification side, the design has proven to be certifiable and the margins with the certification constraint can be controlled to improve the overall design. The follow-up project COLOSSUS aims at developing a system-of-systems design methodology to enable the combined optimization of aircraft, operations and business models, with specific solutions for sustainable intermodal transport and aerial wildfire-fighting [20].

A coherent, holistic and collaborative framework and toolbox for aviation impact assessment had been developed by the EU funded Horizon Europe project Impact Monitor consortium, to support EC in making science-based informed decisions for technology and policy assessment of the environmental, economic, and societal impact of European aviation R&I. Specifically, the assessments focus on greenhouse-gas emissions, local air-quality, and noise [21].

It is quite worth to cite another remarkable publication entitled "Together for Sustainable Aviation", which was created within the EU funded Horizon Europe project PULSAR framework [22]. The brochure presents a glimpse into research conducted on environmental aviation through EU collaborative projects that go beyond well-known high-TRL-focused EU Joint Undertakings, featuring 22 EU-founded projects and 5 national initiatives working towards a greener aviation industry with low environmental impacts through research and innovation. The effort illustrates the European Commission's dedication to achieving key aviation and environmental objectives.

In the field of aeronautical numerical simulations, Multi-physics and Multi-scales are no doubt the inherent nature. What is beyond this paper comes digital twins, artificial intelligence and intelligent deep learning, big data, uncertainty quantification.

References

1. SCIO White Paper (2021) Responding to climate change: China's policies and actions. Accessed 24 Feb 2023
2. SCIO White Paper (2021) Action plan for carbon dioxide peaking before 2030. Accessed 24 Feb 2023
3. MEE Report (2022) China's policies and actions for addressing climate change. Accessed 24 Feb 2023
4. SCIO White Paper (2023) China's green development in the new era. Accessed 24 Feb 2023
5. Ministry of Finance (2022) Working guidance for financial support of carbon dioxide peaking and carbon neutrality. Accessed 24 Feb 2023
6. MOST/NDRC/MIIT etc (2022) Implementation plan for S&T supporting carbon peaking and carbon neutrality. Accessed 24 Feb 2023
7. MOST/NDRC/MEE (2022) Carbon peaking implementation plan of industrial sector. Accessed 24 Feb 2023
8. SAMR/NDRC/MIIT etc (2022) Implementation plan for establishing sound standard and measurement system of carbon peaking and carbon neutrality. Accessed 24 Feb 2023
9. CAAC (2022) China civil aviation green development policy and action. Accessed 24 Feb 2023
10. CAAC (2022) Notice of the civil aviation administration of china on releasing guidelines on the construction and development of civil aviation big data. Accessed 24 Feb 2023
11. China Aviation News (CAN) (2022) AVIC deployed to comprehensively promote carbon peaking and carbon neutrality works. Accessed 24 Feb 2023
12. Ji Y, Sun X, Yu X et al (2022) Development prospect of new energy aviation under carbon peaking and carbon neutrality goals. Aeronaut Sci & Technol 33(12):1–11
13. Kousoulidou M, Violato D (2020) Towards climate-neutral aviation: Contributions from horizon 2020 projects implemented by INEA. ISBN 978–92–9208–100–3. https://doi.org/10.2840/047167, European Union. Accessed 19 Sep 2022
14. Hadnum L, Pacey M, Milne K (2022) FlyZero Technology roadmaps—technology pathways to enable zero-carbon emission flight. FZO-IST-MAP-0012. Accessed 26 Sep 2022
15. Ciampa PD, Nagel B (2020) AGILE Paradigm: the next generation collaborative MDO for the development of aeronautical systems. Prog Aerosp Sci 119(100643)
16. AGILE Project Consortium. AGILE Aircraft 3rd generation MDO for innovative collaboration of heterogeneous teams of experts. http://www.agile-project.eu. Accessed 24 Feb 2023
17. AGILE 4.0 Project Consortium. AGILE 4.0—Towards cyber-physical collaborative aircraft development. https://www.agile4.eu/. Accessed 24 Feb 2023
18. Boggero L et al (2022) The AGILE4.0 MBSE-MDAO development framework: overview and assessment. In: 33rd Congress of the International Council of the Aeronautical Sciences (ICAS), Stockholm. https://doi.org/10.5281/zenodo.7070940

19. Fioriti M et al (2022) Multidisciplinary design of a more electric regional aircraft including certification constraints. AIAA Aviation 2022, Chicago. https://doi.org/10.5281/zenodo.6801154
20. Prakasha PS et al (2024) COLOSSUS EU Project—collaborative system of systems exploration of aviation products, services and business models: overview and approach. ICAS, Florence
21. Alder M et al (2025) Impact monitor framework: development and implementation of a collaborative framework for aviation impact assessment. Eng Proc 90:61. https://doi.org/10.3390/engproc2025090061
22. PULSAR Project Consortium. Together for Sustainable Aviation—A glimpse into European collaborative research efforts on aviation emissions and noise. https://www.pulsar-project.eu/. Accessed 10 Sep 2024

Chapter 16
Prevalence of AI Misapplications and Their Implications for Our Future

Yoshiyasu Takefuji

Abstract The emergence of generative AI has brought forth significant concerns regarding AI misapplications on a global scale. This paper presents an explainable AI framework, leveraging SHAP (SHapley Additive exPlanations), to illuminate potential societal impacts and risks associated with these technologies. We emphasize the critical importance of understanding the tools and methodologies that researchers worldwide have implemented and distributed. Furthermore, we propose a paradigm shift in research practices, advocating for responsible and ethical AI applications that benefit society at large. Despite the urgency of addressing AI misapplications, prominent scientific publications, including Nature, Science, and Cell, have not adequately addressed these concerns, suggesting a potential gap in editorial awareness regarding the broader implications of AI technologies. To bridge this gap in academic discourse, we recommend robust statistical approaches for evaluating AI systems, including nonlinear and nonparametric pairwise methods such as Spearman's correlation with p-values, Kendalls tau with p-values, Goodman–Kruskal gamma with p-values, Somers' D with p-values, and Hoeffding's D with p-values, while Mutual Information (MI) analysis examines complex multivariate interactions. This comprehensive approach aims to establish more rigorous standards for AI research and promote responsible innovation in the field.

Keywords Feature importance · Random forests · SHAP values · Biases · Statistical methods · Machine learning

16.1 Introduction

Many editors and editorial teams often underestimate the significant impact of biases introduced by AI algorithms, leading to flawed conclusions when fundamental AI principles are misunderstood. These biases are frequently introduced inadvertently

Y. Takefuji (✉)
Faculty of Data Science, Musashino University, Tokyo, Japan
e-mail: takefuji@keio.jp

T. Tuovinen et al. (eds.), *Challenges in Design Methods, Numerical Tools and Technologies for Sustainable Aviation, Transport and Industry*, Computational Methods in Applied Sciences 17, https://doi.org/10.1007/978-3-031-98675-8_16

by users who do not fully grasp the assumptions underlying AI tools, and any violation of these assumptions can significantly distort outcomes. Although experts in their respective fields, researchers, editors, and reviewers often employ AI technologies—such as machine learning predictions, feature importance analyses, and feature reduction—without a full appreciation of their implications.

This paper seeks to clarify key AI concepts to enhance research integrity, focusing on the significance of feature importance and the critical role of cross-validation. It is crucial to recognize that target prediction accuracy and feature importance accuracy are distinct; high target prediction accuracy does not necessarily equate to reliable feature importances. In fact, while cross-validation is valuable for evaluating prediction performance, it does not inherently improve the accuracy of feature importance assessments.

Researchers must grasp the critical role of ground truth values in AI applications. In supervised machine learning, ground truth data serves to validate target prediction accuracy; however, feature importance metrics lack such definitive reference points. As a result, different models employ various methods for computing feature importances, leading to discrepancies among their outputs. This inherent variability introduces significant biases in machine learning models—even those with high target prediction accuracy—which must be carefully considered when interpreting results. Over 100 peer-reviewed articles documented non-negotiable biases in feature importances from machine learning models.

The authors made substantial contributions to the study across multiple dimensions: conceptualizing the research framework, designing the methodology, analyzing data, writing and editing the manuscript, overseeing the research process, securing funding, and ensuring the accuracy and integrity of the findings. However, in several high-profile studies published in journals such as Nature, Science, and Cell, some supervisors lack a clear understanding of the limitations inherent in explainable AI tools like SHapley Additive exPlanations (SHAP). Additionally, many researchers do not adequately distinguish among machine learning predictions, feature importances, and cross-validation techniques.

The command 'explain = SHAP(model)' indicates that SHAP not only reflects but can also amplify existing biases in feature importance calculations. This issue is further compounded by the segmented roles within research teams, which can lead to a disconnect in sharing fundamental AI principles. Such disconnects may result in misleading outcomes and evaluation errors that have significant, adverse societal impacts.

This paper highlights that over 80% of journals reject articles that either emphasize bias or present conclusions influenced by it, while fewer than 20% of submissions are accepted. This trend can be attributed to the fact that many editors may not recognize bias as a critical issue.

28 peer-reviewed articles were published in prestigious journals across diverse fields [3, 23, 25, 34, 48, 35, 49, 36, 50, 51, 37, 52, 38, 39, 53, 54, 40, 41, 42, 55, 43, 56, 57, 44, 45, 46, 58] to support this paper.

16.2 Fundamental Theoretical Principles of Machine Learning

This paper delineates essential principles of artificial intelligence crucial for robust research methodology. While machine learning algorithms excel at predicting target outcomes with high accuracy, they often generate potentially biased feature importances, selections, and reductions. This bias stems from the fundamental challenge that feature importance calculations lack ground truth validation, unlike target predictions whose accuracy can be directly verified. Consequently, high predictive accuracy does not necessarily translate to reliable feature importance measurements. Statistical analysis addresses this challenge through three critical components: understanding data distribution, evaluating statistical relationships between variables, and validating statistical significance through p-values. Traditional robust methods, including Spearman's correlation, Kendall's tau, Goodman–Kruskal gamma, Somers' D, and Hoeffding's D, incorporate these components and provide statistical validation through p-values. However, these methods are limited to pairwise analysis, making them insufficient for understanding complex multivariate interactions. To address this limitation, Mutual Information (MI) analysis emerges as a powerful tool capable of capturing and quantifying intricate relationships among multiple variables simultaneously, thereby offering a more comprehensive understanding of feature interactions in AI systems.

In addition, cross-validation is a powerful tool that helps verify the accuracy of predictions. Unfortunately, it does not extend to evaluating the reliability of feature importances derived from machine learning models. This limitation can lead to misleading conclusions if researchers assume that high prediction accuracy automatically translates to trustworthy feature importances.

The integrity of research outcomes can be severely compromised when methodological assumptions are violated in AI applications. Specifically, the misapplication of linear methods to nonlinear data or parametric approaches to nonparametric distributions can lead to substantial distortions in results. These methodological misalignments, combined with inherent algorithmic biases, underscore a crucial point: biases and errors in AI systems are fundamentally human-induced phenomena, stemming from decisions made during model design, implementation, and interpretation. The root cause of these issues often lies in researchers' insufficient understanding of fundamental machine learning principles, leading to the introduction of significant biases and errors that can propagate through the entire analytical process. This emphasizes the critical need for robust theoretical grounding and appropriate methodological choices in AI research to ensure reliable and valid scientific conclusions.

Based on these AI principles, this paper presents examples from esteemed journals to illustrate how such biases and methodological errors can impact research findings. As diligent researchers, it is essential to pursue unbiased algorithms and robust methodologies. Relying on potentially biased calculations can lead to incorrect conclusions, adversely affecting the validity of data analyses and, ultimately, the broader implications of our work. Be aware that employing linear or parametric methods for analyzing unknown or biological data should be avoided, as such data typically exhibit nonlinear and nonparametric characteristics.

16.3 Case Studies of Biased Calculations

These biases can lead to outcomes that deviate significantly from true results. Importantly, this issue is not confined to isolated studies [21], it is a widespread concern affecting 34 articles published in Nature and other esteemed journals such as Cell, Nature Methods and others. Each of these articles addresses the bias issues stemming from feature importances in machine learning, which can lead to incorrect conclusions. Therefore, it is crucial to recognize that feature importances are model-specific and do not necessarily reflect true associations between the target variable and individual features. This highlights the necessity for careful consideration and mitigation of biases to ensure the reliability and validity of machine learning results. If left unaddressed, more researchers may rely on flawed feature importances, drawing inaccurate conclusions.

Schade et al. developed a classifier to predict sensitivity in triple-negative breast cancer, presenting a promising therapeutic approach for this aggressive tumor type [27]. Their classifier utilizes feature selection through Random Forest (RF) and Support Vector Machine (SVM) algorithms. However, feature selection with RF and SVM can introduce inherent biases due to their model-specific characteristics. Random Forest may prioritize features with greater variability or those characterized by complex interactions, potentially overlooking relevant but less prominent features. In contrast, SVM focuses on maximizing the margin between classes, which can skew feature selection toward those features that most significantly influence the decision boundary, thereby missing other important associations. Such biases can result in the misidentification of key features, compromising the accuracy and reliability of the classifier. The term "model-specific" indicates that different machine learning models yield distinct sets of selected features. Schade et al. illustrate this model-specific nature by demonstrating how various algorithms prioritize different features, even in the presence of true associations.

In another study, Shenhav et al. trained a gradient-boosted decision tree model to differentiate between children diagnosed with asthma at age three and healthy controls using data from nasal and gut microbiomes [29]. They employed tenfold cross-validation for training and testing and combined microbial trajectories with components from human milk, evaluating prediction accuracy using area under the receiver operating characteristic (auROC). Feature importance was assessed by ranking microbial taxa and human milk components based on their contributions to model performance. This study raises a critical alarm regarding the use of feature importance in machine learning applications due to inherent biases that could lead to incorrect conclusions. It emphasizes that models such as gradient-boosted decision trees (GBDT), when used to compute feature importances, may not accurately reflect true associations between the target variable and features, as these importances are model-specific and can mislead interpretations.

Almet et al. examined the top upstream transcription factors ranked by their regulatory effects on outflowing Sonic Hedgehog (SHH), utilizing random forest feature (Gini) importance to measure these effects [2]. The feature importances were calculated over 10 runs, with data presented as the mean $\pm$ standard deviation. The methods employed in this study can lead to incorrect conclusions.

Interpretable machine learning (IML) is essential for extracting meaningful insights from complex biological data. Chen et al. [8] identified three key pitfalls in the application of IML within biological contexts and offered strategies to mitigate these challenges. First, relying solely on a single IML method can yield biased assessments of feature importance. While cross-validation is effective for evaluating prediction accuracy, using multiple IML methods for cross-validation does not necessarily clarify target-feature relationships and may lead to misleading conclusions. Secondly, many IML techniques inadequately capture genuine associations between targets and features, often resulting in an overemphasis on irrelevant variables.

Even within the realm of safety science, the critical issue of bias has been overlooked by the editor. Lian et al.'s Data-Knowledge Hybrid-Driven Ensemble Feature Selection (DKH-EFS) method, while promising, relies on feature importances derived from Random Forest (RF) and Support Vector Utility Vector (SUV) algorithms, both of which are inherently biased. The RF algorithm tends to favor high-cardinality and high-variance features, while the SUV algorithm is sensitive to feature scaling and multicollinearity [19]. These inherent biases can result in the selection of misleading features for health assessments, ultimately jeopardizing the accuracy and reliability of the DKH-EFS method.

Dong et al. developed a sophisticated model to identify inefficient public open spaces (POSs) [10]. They employed a random forest approach, which constructs multiple decision trees through random sampling and the selection of random features. This powerful technique enables the model to effectively fit training samples and evaluate the significance of input features, ultimately generating predictions based on a voting or averaging mechanism. To quantify the importance of various indicators, they utilized the SHAP method. In this framework, a higher absolute SHAP value signifies greater relevance of the feature in the model's predictive process.

However, it is imperative to recognize that many researchers, including Dong et al., may overlook the critical distinction that machine learning feature importances do not necessarily represent true associations between the target variable and the features, due to the model-specific nature of these computations. This model-specificity implies that different algorithms yield varying feature importance scores, suggesting that the computed importances are not genuine associations but rather distortions influenced by the models' characteristics. In other words, random forest consistently and inherently suffers from biased feature importances [7, 22, 24, 33, 59].

This paper argues for a clear distinction between using machine learning for accurate prediction and employing it for feature importance calculations. While the use of machine learning for predicting target variables is valuable, relying on it for determining feature importance consistently leads to inherent biases and erroneous conclusions.

Additionally, although SHAP is widely embraced by researchers, its reliance on machine learning models makes it susceptible to the same biases inherent in those models. Consequently, when models produce skewed feature importances, SHAP inadvertently inherits these biases, leading to potentially misleading interpretations [5, 32]. Therefore, it is crucial for researchers to exercise caution when interpreting feature importances derived from machine learning techniques and to differentiate clearly between predictive and explanatory uses of these models.

Random forest is widely recognized for producing feature importances that are inherently biased. In fact, over 100 peer-reviewed articles have explored this issue in detail [7, 22, 24, 33, 59]. To address the shortcomings associated with biased feature importances, researchers should consider employing robust statistical methods [1, 14, 18, 30, 60], such as Spearman's correlation with p-values and Kendall's tau with p-values. These approaches provide a more reliable assessment of feature significance, independent of machine learning models and free from the biases that can distort interpretations. This paper advocates for employing robust statistical methods and discourages the use of feature importances from machine learning models. This paper addresses why random forest generates feature importance biases.

Understanding the biases associated with random forests and SHAP requires a closer look at their algorithmic structures and underlying mechanisms. Both methodologies have distinctive algorithms that contribute to the emergence of biased feature importances and interpretations.

Random forests are an ensemble learning method that constructs multiple decision trees during training. Each tree is trained on a random subset of the data, using the bootstrap aggregation method, which introduces randomness but also potential biases based on how samples are selected. Since each sample may favor particular instances, especially if the dataset has some inherent imbalance or noise, this randomness can be a source of bias in the feature importances reported by the model.

When building decision trees, random forests randomly select a subset of features to determine the best split at each node. This feature selection process can disproportionately favor certain features that yield better splits by chance. If a feature consistently appears in multiple trees due to this randomness, it may be assigned an inflated importance score, even if its contribution to prediction is not fundamentally strong. This phenomenon is compounded when features are correlated, as trees can erroneously attribute importance to one correlated feature over another based on random occurrences in the dataset.

To measure the "goodness" of a split within decisions, random forests we often use metrics like Gini impurity or entropy. These metrics evaluate how well a feature separates the classes, and if a feature provides significant splits in the majority class, it can lead to the model placing excessive importance on that feature. This situation creates a bias in the interpretation of feature relevance. Moreover, in imbalanced datasets, features that are related to the majority class can dominate the measure, while features relevant to the minority group may be overlooked altogether.

While it is true that random forests are generally less prone to overfitting compared to individual decision trees, they can still overfit to noise in the training data. Such overfitting skews feature importances, especially if the noise correlates with the target variable, thereby distorting the model's assessment of which features are truly important.

On the other hand, SHAP values are derived from cooperative game theory, treating features as players contributing to a prediction. The calculation of SHAP values is based on considering all possible combinations of the features, which can be computationally intensive. However, this approach does not inherently correct for biases that may exist in the underlying model. SHAP values are contingent on the predictions of the model they explain. If the model exhibits bias, such as inflated feature importances due to correlation or selection bias, the resulting SHAP values will reflect and potentially amplify these biases.

Moreover, SHAP provides a local approximation of feature contributions based on the area around a specific prediction. The method employs a linear model to approximate how each feature influences the prediction for that instance. However, if the underlying model is biased or exhibits complex interactions not captured in the model training, the SHAP values may misrepresent the true importance of those features in broader contexts, leading to skewed interpretations.

Finally, SHAP assumes that features contribute independently to the prediction outcome. In reality, interactions between features can complicate this assumption, which can lead to biases in the assessment of feature contributions. If a feature interacts with others in ways not taken into account during model training, the corresponding SHAP values can misrepresent its importance, further perpetuating any underlying biases.

In summary, from an algorithmic perspective, the biases in random forests arise from their ensemble approach, feature selection randomness, and the impact of noisy data on decision-making. In contrast, SHAP amplifies these biases through its dependence on model-specific outputs and assumptions regarding feature contributions. Recognizing these algorithmic intricacies is crucial for researchers aiming to accurately interpret feature importances and make informed decisions based on model outputs. By understanding the limitations of both random forests and SHAP, practitioners can navigate potential pitfalls in their analyses and consider complementing these methods with more robust statistical techniques to obtain clearer insights.

Three articles published in 2023 and 2024 in the International Journal of Information Management [9, 15, 16] highlight pitfalls associated with feature importances. These studies should reevaluate the associations between the target and features using Spearman's correlation with p-values, rather than relying on feature importances in machine learning. This week, two articles in the Nature Portfolio have been identified as having these pitfalls [61, 28].

Instead of investigating individual articles [9, 15, 16, 61, 28], this paper addresses the broader issue of why feature importances in machine learning are biased and do not represent true associations between the target and features. Feature importances are byproducts of machine learning models, where the primary goal is to improve prediction accuracy between the target and features, rather than to calculate associations between the target and individual features.

In other words, the problem lies in the fact that feature importances derived from machine learning models do not necessarily represent true associations between the target and features. These feature importances are model-specific and can induce biases. The following summarized reasons substantiate this claim:

1. Model-Specificity: Different algorithms may assign varying importance scores to the same features, leading to discrepancies in the perceived importance of variables [6, 31].
2. Correlation versus Causation: Feature importance measures the correlation between a feature and the target variable, but it doesn't establish a causal relationship [11, 26].
3. Feature Interactions: Machine learning models can capture complex interactions between features, which might not be fully accounted for in feature importance methods [8, 13].
4. Overfitting: Overfitting can inflate feature importance scores, leading to features being deemed more important than they actually are [4, 17].
5. Data Distribution: The distribution of data can influence feature importance, with features that are more informative for distinguishing between classes in a particular dataset having higher importance scores [12, 20].

Given the current prevalence of explainable AI tools like SHAP, it is prudent to temporarily rely on robust statistical methods, such as Spearman's correlation and Chi-squared tests, to assess associations between targets and features. These traditional statistical approaches come with well-established metrics, including p-values, which help in determining the significance of relationships without the inherent biases that can arise from certain machine learning algorithms.

In addition to these methods, it is essential to seek out algorithms that minimize bias, ensuring that our analyses are grounded in reliable and objective data interpretations. By prioritizing approaches that are free from bias, we can cultivate a more holistic and accurate understanding of complex systems. This shift not only enhances the integrity of our findings but also builds greater trust among stakeholders who rely on our results for decision-making.

Ultimately, while explainable AI tools are invaluable for interpreting complex models, the foundational robustness provided by established statistical methods will safeguard against potential misinterpretations. As we navigate the evolving landscape of data analysis, it will be crucial to balance modern techniques with traditional methodologies to foster comprehensive and unbiased computing practices.

16.4 Discussion

The presence of biases in machine learning studies is a pressing concern that significantly impacts the reliability of research outcomes. These biases arise from a variety of sources, including the model-specific nature of feature importance calculations and the misunderstandings surrounding fundamental AI principles. Many editors and editorial teams often underestimate the significant impact of biases introduced by AI algorithms, leading to flawed conclusions when these principles are misunderstood. These biases are frequently introduced inadvertently by researchers who do not fully grasp the assumptions underlying AI tools, and any violation of these assumptions can distort outcomes significantly.

A critical factor contributing to these biases is the failure to distinguish between target prediction accuracy and feature importance accuracy. While machine learning models may achieve high accuracy in predicting outcomes, this does not guarantee that the feature importances derived from these models are valid. For instance, models like Random Forest and Support Vector Machines may over-prioritize features based on the algorithms' inherent characteristics, resulting in inflated importance scores for certain features. Additionally, the training data can introduce biases if it contains imbalances or noise, further distorting the feature importance outputs. Thus, researchers must understand that failures to maintain methodological assumptions—such as applying linear methods to nonlinear data—can exacerbate these issues.

Moreover, while cross-validation is a valuable technique for assessing prediction performance, it does not inherently enhance the accuracy of feature importance analyses. This misconception can lead researchers to draw erroneous conclusions, believing that an accurate model necessarily produces reliable feature importances. As a result, discrepancies among feature importance outputs from various models become significant, further complicating interpretations and increasing the likelihood of flawed conclusions.

The issue is further compounded by the segmented roles within research teams, which can create a disconnect in sharing fundamental AI principles. Supervisors and researchers alike, despite being experts in their respective fields, may lack a clear understanding of the limitations inherent in explainable AI tools like SHapley Additive exPlanations (SHAP). Such disconnects can lead to misleading outcomes and evaluation errors that may have significant adverse societal impacts.

This paper seeks to clarify key AI concepts to enhance research integrity by emphasizing the importance of ground truth values and the critical role of feature importance and cross-validation. In supervised machine learning, ground truth data validates target prediction accuracy; however, feature importance metrics lack definitive reference points, making their reliability questionable. Over 100 peer-reviewed articles have documented non-negotiable biases present in feature importances derived from machine learning models, highlighting the urgency of addressing these concerns.

It is notable that over 80% of journals reject articles that emphasize bias or present conclusions influenced by it, while fewer than 20% of submissions are accepted. This trend suggests that many editors may not recognize bias as a critical issue, which can perpetuate flawed analyses and misinterpretations across published research.

Furthermore, this paper outlines several fundamental principles of artificial intelligence essential for sound research practices. Relying on biased calculations can lead to incorrect conclusions, negatively affecting the validity of data analyses and the broader implications of research findings. For example, employing linear or parametric methods to analyze biological data—which typically exhibits nonlinear and nonparametric characteristics—can introduce significant distortion in results.

In conclusion, understanding and addressing the biases inherent in machine learning methodologies is vital for preserving research integrity. Through careful engagement with these principles, researchers can improve the accuracy of their findings and avoid common pitfalls related to feature importance and predictive modeling. By prioritizing robust methodologies and a comprehensive understanding of AI tools, the research community can work toward producing reliable, impactful outcomes that genuinely reflect the underlying data associations.

References

1. Aguilar-Elena R, Agún-González JJ (2024) Chi-square automatic interaction detection (CHAID) analysis of the use of safety goggles and face masks as personal protective equipment (PPE) to protect against occupational biohazards. J Biosaf Biosecurity 6(2):125–133. https://doi.org/10.1016/j.jobb.2024.05.001
2. Almet AA, Tsai YC, Watanabe M et al (2024) Inferring pattern-driving intercellular flows from single-cell and spatial transcriptomics. Nature Method. https://doi.org/10.1038/s41592-024-02380-w
3. Arif M, Takefuji Y (2025) Why AI image generators cannot afford to be blind to racial bias. AI & Soc. https://doi.org/10.1007/s00146-025-02258-1
4. Babyak MA (2004) What you see may not be what you get: A brief, nontechnical introduction to overfitting in regression-type models. Psychosom Med 66(3):411–421. https://doi.org/10.1097/01.psy.0000127692.23278.a9
5. Bilodeau B, Jaques N, Koh PW, Kim B (2024) Impossibility theorems for feature attribution. Proc Natl Acad Sci USA 121(2):e2304406120. https://doi.org/10.1073/pnas.2304406120
6. Breiman L (2001) Random forests. Mach Learn 45(1):5–32. https://doi.org/10.1023/A:1010933404324
7. Chen J, Ooi LQR, Tan TWK, Zhang S, Li J, Asplund CL, Eickhoff SB, Bzdok D, Holmes AJ, Yeo BTT (2023) Relationship between prediction accuracy and feature importance reliability: An empirical and theoretical study. Neuroimage 274:120115. https://doi.org/10.1016/j.neuroimage.2023.120115
8. Chen V, Yang M, Cui W et al (2024) Applying interpretable machine learning in computational biology: Pitfalls, recommendations, and opportunities for new developments. Nat Methods 21(12):1454–1461. https://doi.org/10.1038/s41592-024-02359-7
9. Căpățînă A, Patel NJ, Mitrov K, Cristea DS, Micu A, Micu A-E (2024) Elevating students' lives through immersive learning experiences in a safe metaverse. Int J Information Management, 75:102723.https://doi.org/10.1016/j.ijinfomgt.2023.102723

10. Dong X, Zhang X, Jing Y, Zhou Q, Bai L, Du S (2024) Does every public open space (POS) contribute to sustainable city development? An assessment of inefficient POS in Beijing. Sustain Cities Soc: 105980. https://doi.org/10.1016/j.scs.2024.105980
11. Friedman JH (2001) Greedy function approximation: A gradient boosting machine. Ann Stat 29(5):1189–1232. https://doi.org/10.1214/aos/1013203451
12. Guyon I, Elisseeff A (2003) An introduction to variable and feature selection. J Mach Learn Res 3:1157–1182. https://doi.org/10.1162/153244303322753616
13. Hawkins DM (2004) The problem of overfitting. J Chem Inf Comput Sci 44(1):1–12. https://doi.org/10.1021/ci0342472
14. Jiang J, Zhang X, Yuan Z (2024) Feature selection for classification with Spearman's rank correlation coefficient-based self-information in divergence-based fuzzy rough sets. Expert Syst Appl, 249(B):123633. https://doi.org/10.1016/j.eswa.2024.123633
15. Joung J, Kim H (2023) Interpretable machine learning-based approach for customer segmentation for newproduct development from online product reviews. Int J Information Management, 70:102641.https://doi.org/10.1016/j.ijinfomgt.2023.102641
16. King KK, Wang B (2023) Diffusion of real versus misinformation during a crisis event: a big data-driven approach. Int J Information Management, 71:102390.https://doi.org/10.1016/j.ijinfomgt.2021.102390
17. Kuhn M, Johnson K (2019) Feature engineering and selection: A practical approach for predictive models. CRC Press. https://doi.org/10.1201/9781315108230
18. Li X, Ma Y, Zhou Q, Zhang X (2024) Sparse large-scale high-order fuzzy cognitive maps guided by Spearman correlation coefficient. Appl Soft Comput 167(A):112253. https://doi.org/10.1016/j.asoc.2024.112253
19. Lian Z, Zhou ZJ, Hu CH, Wang J, Zhang CC, Zhang CL (2024) A health assessment method with attribute importance modeling for complex systems using belief rule base. Reliab Eng Syst Saf 251:110387. https://doi.org/10.1016/j.ress.2024.110387
20. Lundberg SM, Lee SI (2017) A unified approach to interpreting model predictions. Adv Neural Inf Process Syst 30:4765–4774. https://doi.org/10.5555/3295222.3295230
21. Lynch CJ, Elbau IG, Ng T et al (2024) Frontostriatal salience network expansion in individuals in depression. Nature. https://doi.org/10.1038/s41586-024-07805-2
22. Nguyen TT, Huang JZ, Nguyen TT (2015) Unbiased feature selection in learning random forests for high-dimensional data. Sci World J 2015:471371. https://doi.org/10.1155/2015/471371
23. Nguyen MH, Takefuji Y (2025) Reassessing predictive modeling for emergency department return in COVID-19 patients. Am J Emerg Med 43. https://doi.org/10.1016/j.ajem.2025.01.009
24. Oh S (2022) Predictive case-based feature importance and interaction. Inf Sci 593:155–176. https://doi.org/10.1016/j.ins.2022.02.003
25. Pan H, Takefuji Y (2025) Enhancing feature importance analysis with Spearman's correlation with p-values: Recommendations for improving PHLF prediction. Eur J Surg Oncol 51(7):109687. https://doi.org/10.1016/j.ejso.2025.109687
26. Pearl J (2009) Causality: Models, reasoning, and inference. Cambridge University Press. https://doi.org/10.1017/CBO9780511803161
27. Schade AE, Perurena N, Yang Y et al (2024) AKT and EZH2 inhibitors kill TNBCs by hijacking mechanisms of involution. Nature. https://doi.org/10.1038/s41586-024-08031-6
28. Shen M, Chen M, Chen Y, et al (2024) Mitophagy related diagnostic biomarkers for coronary in-stent restenosis identified using machine learning and bioinformatics. Scientific Reports, 14:24137. https://doi.org/10.1038/s41598-024-74862-y
29. Shenhav L, Fehr K, Reyna ME et al (2024) Microbial colonization programs are structured by breastfeeding and guide healthy respiratory development. Cell 187(19):5431–5452.e20. https://doi.org/10.1016/j.cell.2024.07.022
30. Shi X, Xu M, Du J (2023) Max-sum test based on Spearman's footrule for high-dimensional independence tests. Comput Stat Data Anal 185:107768. https://doi.org/10.1016/j.csda.2023.107768
31. Shmueli G (2010) To explain or to predict? Stat Sci 25(3):289–310. https://doi.org/10.1214/10-STS330

32. Singhal A, Neveditsin N, Tanveer H, Mago V (2024) Toward fairness, accountability, transparency, and ethics in AI for social media and health care: Scoping review. JMIR Med Inform 12:e50048. https://doi.org/10.2196/50048
33. Strobl C, Boulesteix A-L, Zeileis A, Hothorn T (2007) Bias in random forest variable importance measures: Illustrations, sources and a solution. BMC Bioinformatics 8(1):25. https://doi.org/10.1186/1471-2105-8-25
34. Takefuji Y (2024a) Addressing bias in feature importances derived from XGBoost. Br J Anaesth 133(3):351–359. https://doi.org/10.1016/j.bja.2024.11.033
35. Takefuji Y (2024c) Evaluating feature importance biases in logistic regression: Recommendations for robust statistical methods. Eur J Intern Med. https://doi.org/10.1016/j.ejim.2024.11.022
36. Takefuji Y (2024e) Mitigating feature importance bias in regression models for clinical decision-making. Am J Obstet Gynecol. https://doi.org/10.1016/j.ajog.2024.12.010
37. Takefuji Y (2024h) Reply to the editor. Clin Nutr. https://doi.org/10.1016/j.clnu.2024.11.031
38. Takefuji Y (2024j) Unveiling hidden biases in machine learning feature importance. J Energy Chem 102:49–51. https://doi.org/10.1016/j.jechem.2024.10.032
39. Takefuji Y (2025a) Addressing bias in feature importance: A hybrid approach for risk prediction in prognostic survival models. JCO Precis Oncol. https://doi.org/10.1200/PO-24-00785
40. Takefuji Y (2025d) Critical evaluation of feature importance assessment in FFNN-based models for predicting Kamlet-Taft parameters. Green Chem Eng. https://doi.org/10.1016/j.gce.2025.01.003
41. Takefuji Y (2025e) Model-specific feature importances: Distinguishing true associations from target-feature relationships. J Affect Disord 369:390–391. https://doi.org/10.1016/j.jad.2024.10.019
42. Takefuji Y (2025f) Reevaluating feature importances in machine learning models for schizophrenia and bipolar disorder: The need for true associations. Brain Behav Immun 124:123–124. https://doi.org/10.1016/j.bbi.2024.11.036
43. Takefuji Y (2025h) Reevaluating feature importance in machine learning models for CO2 photoreduction: A statistical perspective. Appl Catal B 368:125145. https://doi.org/10.1016/j.apcatb.2025.125145
44. Takefuji Y (2025k) Critical evaluation of feature importance assessment in proteomic analysis using skin microdialysis. J Allergy Clin Immunol. https://doi.org/10.1016/j.jaci.2024.12.1096
45. Takefuji Y (2025l) Challenges in feature importance interpretation: Analyzing LSTM–NN predictions in battery material flotation. J Ind Inf Integr 45:100809. https://doi.org/10.1016/j.jii.2025.100809
46. Takefuji Y (2025m) Enhancing machine learning in gas–solid interaction analysis: Addressing feature selection and dimensionality challenges. Coord Chem Rev 534:216583. https://doi.org/10.1016/j.ccr.2025.216583
47. Takefuji Y (2025o) Reevaluating feature importance in machine learning: Concerns regarding SHAP interpretations in the context of the EU Artificial Intelligence Act. Water Res 280:123514. https://doi.org/10.1016/j.watres.2025.123514
48. Takefuji Y (2024b) Chi-Squared and P-values vs. machine learning feature selection. Ann Oncol. https://doi.org/10.1016/j.annonc.2024.10.013
49. Takefuji Y (2024d) Mitigating biases in feature selection and importance assessments in predictive models using LASSO regression. Oral Oncol 159, Article 107090. https://doi.org/10.1016/j.oraloncology.2024.107090
50. Takefuji Y (2024f) Reassessing feature importance biases in machine learning models for infection analysis. J Infect 89(6), Article 106357. https://doi.org/10.1016/j.jinf.2024.106357
51. Takefuji Y (2024g) Reevaluating statistical methods in metabolomic studies: A case for Spearman's correlation. Molecular Plant 17(1). https://doi.org/10.1016/j.molp.2024.12.014
52. Takefuji Y (2024i) Unveiling feature importance biases in linear regression: Implications for protein-centric cardiovascular research. Atherosclerosis, Article 119049. https://doi.org/10.1016/j.atherosclerosis.2024.119049

53. Takefuji Y (2025b) Addressing feature importance biases in machine learning models for early diagnosis of type 1 Gaucher disease. J Clin Epidemiol 178, Article 111619. https://doi.org/10.1016/j.jclinepi.2024.111619
54. Takefuji Y (2025c) Beyond XGBoost and SHAP: Unveiling true feature importance. J Hazard Mater 488, Article 137382. https://doi.org/10.1016/j.jhazmat.2025.137382
55. Takefuji Y (2025g) Reevaluating feature selection in phase field models for battery performance: A call for robust statistical approaches. Energy Storage Mater 75, Article 104060. https://doi.org/10.1016/j.ensm.2025.104060
56. Takefuji Y (2025i) Reevaluating feature importance in machine learning for food authentication: Addressing bias and enhancing methodological rigor. Trends Food Sci & Technol, Article 104853. https://doi.org/10.1016/j.tifs.2024.104853
57. Takefuji Y (2025j) Beyond principal component analysis: Enhancing feature reduction in electronic noses through robust statistical methods. Trends Food Sci & Technol, 104919. https://doi.org/10.1016/j.tifs.2025.104919
58. Takefuji Y (2025n) Reevaluating feature importance in gas–solid interaction predictions: A call for robust statistical methods. Coord Chem Rev 534, Article 216584. https://doi.org/10.1016/j.ccr.2025.216584
59. Thakur D, Biswas S (2024) Permutation importance based modified guided regularized random forest in human activity recognition with smartphone. Eng Appl Artif Intell 129:107681. https://doi.org/10.1016/j.engappai.2023.107681
60. Vierra A, Razzaq A, Andreadis A (2023) Chapter 28—Categorical variable analyses: Chi-square, fisher exact, and mantel-haenszel. In: Eltorai AEM, Bakal JA, Newell PC, Osband AJ (eds) Handbook for designing and conducting clinical and translational research, pp 171–175. Academic Press. https://doi.org/10.1016/B978-0-323-90300-4.00095-1
61. Wu Y, Shi Z, Zhou X, et al (2024) scHiCyclePred: a deep learning framework for predicting cell cycle phases from single-cell Hi-C data using multi-scale interaction information. Communications Biology, 7:923.https://doi.org/10.1038/s42003-024-06626-3

Index

T. Tuovinen et al. (eds.), *Challenges in Design Methods, Numerical Tools and Technologies for Sustainable Aviation, Transport and Industry*, Computational Methods in Applied Sciences 17, https://doi.org/10.1007/978-3-031-98675-8

MIX
Papier aus verantwortungsvollen Quellen
Paper from responsible sources
FSC® C105338

If you have any concerns about our products,
you can contact us on
ProductSafety@springernature.com

In case Publisher is established outside the EU,
the EU authorized representative is:
Springer Nature Customer Service Center GmbH
Europaplatz 3, 69115 Heidelberg, Germany

Printed by Libri Plureos GmbH
in Hamburg, Germany